MPI Series in
Biological Cybernetics

Vol. 57

MPI Series in Biological Cybernetics

Vol. 57

Editor:

Prof. Dr. Heinrich H. Bülthoff

Max Planck Institute for Biological Cybernetics
Max-Planck-Ring 8
D-72076 Tübingen
Germany

Diane Cleij

Measuring, modelling and minimizing perceived motion incongruence for vehicle motion simulation

Logos Verlag Berlin

Bibliographic information published by the Deutsche Nationalbibliothek

The Deutsche Nationalbibliothek lists this publication in the Deutsche Nationalbibliografie; detailed bibliographic data are available on the Internet at http://dnb.d-nb.de .

ISBN 978-3-8325-5044-8
ISSN 1618-3037

Logos Verlag Berlin GmbH
Comeniushof, Gubener Str. 47,
D-10243 Berlin, Germany

phone: +49 (0)30 / 42 85 10 90
fax: +49 (0)30 / 42 85 10 92
http://www.logos-verlag.com

Measuring, Modelling and Minimizing Perceived Motion Incongruence

FOR VEHICLE MOTION SIMULATION

Proefschrift

ter verkrijging van de graad van doctor
aan de Technische Universiteit Delft,
op gezag van de Rector Magnificus prof. dr. ir. T.H.J.J. van der Hagen,
voorzitter van het College voor Promoties,
in het openbaar te verdedigen op
dinsdag 4 februari 2020 om 15:00 uur

door

Diane CLEIJ

Ingenieur in de Luchtvaart en Ruimtevaart,
Technische Universiteit Delft, Nederland,
geboren te Rotterdam, Nederland

Dit proefschrift is goedgekeurd door de promotoren.

Samenstelling promotiecommissie bestaat uit:

Rector Magnificus,	voorzitter
Prof. dr. ir. M. Mulder,	Technische Universiteit Delft, promotor
Prof. dr. H.H. Bülthoff,	Max-Planck-Institut für biologische Kybernetik, promotor
Dr. ir. D.M. Pool,	Technische Universiteit Delft, copromotor

Onafhankelijke leden:

Prof. dr. ir. P. Breedveld	Technische Universiteit Delft
Prof. dr. J.E. Bos	Vrije Universiteit Amsterdam
Prof. dr. A. Kemeny	Arts et Métiers ParisTech, Frankrijk
Dr. ir. M. Wentink	Desdemona B.V.
Prof. dr. D.G. Simons	Technische Universiteit Delft, reservelid

Dr. ir. J. Venrooij en dr. P. Pretto hebben als begeleiders in belangrijke mate aan de totstandkoming van het proefschrift bijgedragen.

Dit onderzoek is gefinancierd door het Max-Planck-Institut für biologische Kybernetik, en is deels ondersteund door de Technische Universiteit Delft.

"The best material model for a cat is another, or preferably the same cat."

Norbert Wiener

Contents

Summary

Motion-based vehicle simulators are frequently used in research and development, e.g., for human factors evaluations and vehicle design, as well as for pilot/driver training, as such simulations provide a safe and cost-effective testing environment. Visual and physical motion cues are combined to provide occupants with a feeling of being in the real vehicle. While visual cues are generally not limited in amplitude, physical cues certainly are, due to the limited simulator motion space. A motion cueing algorithm (MCA) is used to map the vehicle motions onto the simulator motion space. This mapping inherently creates mismatches between the visual and physical motion cues.

Due to imperfections in the human perceptual system, not all visual/physical cueing mismatches are perceived. However, if a mismatch is perceived, it can impair the simulation realism and even cause simulator sickness. For MCA design, a good understanding of *when* mismatches are perceived, and ways to *prevent* these from occurring, are therefore essential. While most other research tries to predict perceived mismatches based on complex non-linear models of human perception, in this thesis a data-driven approach, using continuous subjective measures of Perceived Motion Incongruence (PMI), is adopted. PMI refers to the effect that perceived mismatches between visual and physical motion cues have on the resulting simulator realism. When a mismatch is perceived, but does not influence the simulation realism, the PMI is low, while a mismatch that is detrimental to simulator realism results in a high PMI. In this thesis we focus on car driving, but the proposed methods can also be applied to other vehicles.

One often-occurring type of mismatch between visual and physical motion cues is referred to as scaling errors. Such errors are caused by a pure (down) scaling of the vehicle physical motion, such that it fits in the simulator motion space. MCAs also make use of tilt-coordination, where the gravitational force and a non-zero rotation with a rotational rate below human perception threshold, are used to simulate sustained accelerations. This mechanism can cause shape differences between the visual and physical motion signals, resulting in other types of cueing errors, i.e., missing or false cues. It is well known that missing or false cues are more likely to be perceived than scaling errors with a similar amplitude and are often more detrimental to simulator realism. Thus, not only the magnitude of the mismatch, but also the *type of mismatch* is important information when designing and optimizing an MCA. While this is widely accepted knowledge and is implicitly used by experts to tune MCAs, currently this knowledge is not explicitly used in MCA optimization due to its qualitative, rather than quantitative, nature.

Another characteristic of simulator realism is that it is inherently *time-varying*. While a simulation might feel mostly realistic, momentary manoeuvres requiring a large motion space, such as driving a roundabout, can cause a sudden decrease in realism. Currently, experts often apply worst-case MCA tuning, resulting in suboptimal physical motions for those parts of the simulation that do not require this large motion space. Certain MCAs, such as those based on model predictive control, on the other hand, can optimize the simulator realism at each simulation time step. This thesis aims to connect the benefits of expert knowledge on motion cue mismatches with the advantage of optimization algorithms in dealing with the time-varying aspect of simulation realism. It aims to develop an MCA-independent, offline prediction method for time-varying PMI during vehicle motion simulation, with the purpose of improving motion cueing quality. To this end, the thesis is divided in three parts, dedicated to *measuring*, *modelling* and *minimizing* PMI, respectively.

Part I focuses on the development of a novel method to *measure* time-varying PMI using a continuous subjective rating. Two human-in-the-loop experiments were performed, where participants were asked to rate the PMI continuously throughout several repetitions of a passive driving simulation. The first experiment, Experiment 1, assessed the reliability and validity of the method itself. Comparing the ratings of several repetitions of the same simulation showed consistency in participants' ratings, verifying the reliability of the method. The validity of the method was assessed by comparing the continuous ratings to a more established time-independent rating method and to expert knowledge from literature on the different cueing error types. The continuous ratings correlated well with a time-independent rating method for each segment of the simulation and was also consistent with expert knowledge on the relative PMI between several scaled, missing and false cues.
In a second experiment, Experiment 2, the continuous rating method was applied to compare the performance of two motion cueing algorithms in a highly realistic vehicle motion simulation. Again, participants were able to provide consistent continuous ratings across several repetitions of the same simulation and their time-independent ratings for each tested MCA setting compared well to their average continuous rating. This confirmed the reliability and validity of the continuous rating method, also for more realistic vehicle simulations.

In **Part II**, the data obtained with the two experiments described in Part I were used to develop PMI *models*, to predict the time-varying PMI within and between experiments. A general model structure was designed to map visual and physical motion cues onto a Motion Incongruence Rating (MIR), which represents the time-varying rating of PMI obtained with the continuous rating method. First, the model translates the visual and physical cues into different types of cueing errors that are combined into one measure of PMI and then filtered to obtain the modelled Motion Incongruence Rating (MIR). A wavelet-based Cueing Error Detection Algorithm (CEDA) was developed to differentiate between scaled, missing and false cues, and its parameters were tuned using data from Experiment 1. Applying the algorithm showed that the CEDA could distinguish between scaled, missing and false cues as hypothesized.
To determine within-experiment prediction capabilities, three models of different complexity were derived from the general model template. These models were fitted to the first half of the data from Experiment 1, after which their prediction power was assessed using the second half of this dataset. The prediction results showed that all models could predict important PMI features and that the prediction improved with increasing model complexity. An interesting observation was also that false cues were modelled as being two times more detrimental than scaled cues for the same cueing error magnitudes. Overall it was shown that the models can indeed link different type of cueing errors to decreases in cueing quality and predict such decreases for data within one experiment.
To compare datasets from different experiments, first a method for estimating a Model

Transfer Parameter (MTP) was developed, with which ratings from one experiment can be mapped onto the ratings of a second experiment. The MTP needed to align the ratings from Experiment 1 and 2 was estimated with this method and used to assess the between-experiment prediction capabilities of the three derived models. Good prediction capabilities were obtained only when a rich enough dataset was used for model fitting. The hypothesis that better models can be obtained when increasing the richness of the estimation dataset was supported by the fact that models fitted to aggregated data from both experiments were more accurately matched to the measured ratings then those fitted to either dataset.

Part III focuses on *minimizing* PMI. The capabilities of the PMI models in predicting decreased cueing quality opens up opportunities to improve this quality. The predictions can, for example, be used to tune MCAs such that the most critical drops in cueing quality are avoided. Additionally, developing these PMI models, by correlating the time-varying PMI to different cueing errors, can help in gaining a better understanding of what exactly causes decreased cueing quality. In this thesis, a PMI model was used in an optimization-based MCA. The weights for linear acceleration and rotational velocity visual-physical cues differences in the cost function of this MCA were estimated using a static version of the least complex PMI model from Part II and data from both Experiments 1 and 2.
In a third human-in-the-loop experiment, Experiment 3, the cueing quality of the MCA with the PMI model weights was compared to the cueing quality of the MCA with its original weights, which accounted solely for the differences in unit between linear acceleration and rotational velocity. The results showed that only a small group of participants, all with prior simulator experience, preferred the MCA with the PMI model weights. The preference of the remaining, larger group seemed to mainly be based on a preference for "lower than unity gains" between vehicle and simulator motions, which is consistent with earlier literature, but was not yet accounted for in the PMI models. Overall, the results indicate that for MCA optimization a PMI model needs to be fitted to a much richer dataset in terms of, among others, number and variety of participants, cueing errors and simulators.

In this thesis a novel approach to improve perceived cueing quality of motion cueing algorithms was introduced. A complete roadmap, describing how to *measure* and *model* PMI and how to apply such models to predict and *minimize* PMI in motion simulations was presented. The results presented in this thesis show the potential of this novel approach. For future research it is recommended to adapt the developed PMI measurement method for use in active driving simulations and improve the PMI models by designing algorithms to detect additional cueing error types. It is also recommended to gather more and richer PMI rating data via human-in-the-loop experiments to improve the parameter estimation of these models. Finally, a systematic investigation on how and under which circumstances these models can be used to improve cueing quality should also be performed. With these advances, the approach outlined in this thesis can enable major improvements in simulator cueing and realism.

Samenvatting

Bewegingssimulatoren worden vaak gebruikt in onderzoek en ontwikkeling, voor bijvoorbeeld de evaluatie van menselijke factoren en het ontwerp van voertuigen, alsook voor de opleiding van piloten/bestuurders, omdat dergelijke simulatoren een veilige en kosteneffectieve testomgeving bieden. Visuele en fysieke bewegingsstimuli worden gecombineerd om de inzittenden het gevoel te geven dat ze zich in het echte voertuig bevinden. Alhoewel visuele bewegingsstimuli over het algemeen niet beperkt zijn in amplitude, zijn fysieke bewegingsstimuli dat zeker wel, vanwege de beperkte bewegingsruimte van de simulator. Een bewegingsalgoritme, een zogenaamd 'Motion Cueing Algorithm (MCA)', wordt gebruikt om de bewegingen van het voertuig te projecteren op de bewegingsruimte van de simulator. Deze projectie creëert van nature discrepanties tussen de visuele en inertiële bewegingsstimuli.

Door onvolkomenheden in het menselijke perceptuele systeem worden niet alle visuele/inertiële bewegingsstimuli discrepanties waargenomen. Als een discrepantie echter wél wordt waargenomen, kan dit het ervaren realisme van de simulatie aantasten en zelfs simulatieziekte veroorzaken. Voor het ontwerpen van MCAs is een goed begrip van *wanneer* discrepanties worden waargenomen en hoe deze kunnen worden *voorkomen*, daarom essentieel. Terwijl de meeste andere onderzoeken proberen om waarneembare discrepanties te voorspellen op basis van uitgebreide niet-lineaire modellen van menselijke perceptie, wordt in dit proefschrift een datagestuurde benadering toegepast, gebruikmakend van continue subjectieve metingen van de waargenomen bewegingsincongruentie (PMI). PMI verwijst naar het effect dat waarneembare discrepanties tussen visuele en inertiële bewegingsstimuli hebben op het resulterende realisme van de simulator. Wanneer een discrepantie wordt waargenomen, maar niet als erg storend ervaren wordt in het simulatierealisme, is de PMI laag, terwijl een discrepantie die schadelijk is voor het simulatorrealisme resulteert in een hoge PMI. In dit proefschrift richten we ons op het autorijden, maar de voorgestelde methoden kunnen ook worden toegepast op simulaties van andere voertuigen.

Een vaak voorkomende vorm van discrepantie tussen visuele en inertiële bewegingsstimuli zijn schalingsfouten. Dergelijke fouten worden veroorzaakt door een pure (terug) schaling van de fysieke beweging van het voertuig, zodat deze in de bewegingsruimte van de simulator past. Om aanhoudende versnellingen te simuleren, maken MCAs gebruik van "tilt-coordination", i.e., het langzaam kantelen van de simulator. Hierbij kantelt de simulator met een rotatiesnelheid onder de menselijke waarnemingsdrempel, zodat een component van de zwaartekracht leidt tot een ervaren versnelling van het lichaam. Dit mechanisme kan vormverschillen veroorzaken tussen de visuele en inertiële bewegingssignalen, wat resulteert in andere soorten fouten, zoals ontbrekende of foutieve signalen. Het is bekend dat ontbrekende of foutieve signalen eerder worden waargenomen dan schalingsfouten met een vergelijkbare amplitude en dat ze vaak schadelijker zijn voor het ervaren realisme van de simulator. Daarom geeft dus niet alleen de grootte van de discrepantie, maar ook het type van de discrepantie belangrijke informatie voor het ontwerpen en optimaliseren van een MCA. Hoewel dit algemeen aanvaarde kennis is en impliciet door experts wordt gebruikt om MCAs af te stemmen, wordt deze kennis momenteel niet expliciet gebruikt in MCA-optimalisatie

vanwege het veelal kwalitatieve, in plaats van voor optimalisatie vereiste kwantitatieve, karakter ervan.
Een ander kenmerk van simulatorrealisme is dat het van nature varieert over tijd. Ook als een simulatie voor het merendeel van de tijd realistisch aanvoelt, kunnen kortstondige manoeuvres die een grote bewegingsruimte vereisen, zoals het rijden over een rotonde, een plotselinge daling van het realisme veroorzaken. Op dit moment stemmen experts de MCAs vaak af zodat de projectie van de grootste bewegingsamplitudes in de bewegingsruimte past, wat resulteert in suboptimale inertiële bewegingen voor die delen van de simulatie die deze grote bewegingsruimte helemaal niet nodig hebben. Bepaalde MCAs, zoals die op basis van "Model Predictive Control", kunnen daarentegen het realisme van de simulator bij elke stap in de simulatie optimaliseren.
Dit proefschrift streeft ernaar de voordelen van deskundige kennis over de discrepanties tussen visuele en fysieke beweginsstimuli te combineren met het voordeel van deze moderne optimalisatiealgoritmes in het omgaan met het tijdsveranderende aspect van het simulatierealisme. Het doel is om een MCA-onafhankelijke, offline methode te ontwikkelen om de tijdsafhankelijke PMI tijdens de simulatie van voertuigbewegingen te voorspellen, met de intentie de kwaliteit van de bewegingssimulatie van het voertuig vervolgens te verbeteren. Hiertoe is het proefschrift opgedeeld in drie delen, respectievelijk gewijd aan het *meten*, *modelleren* en *minimaliseren* van PMI.

Deel I richt zich op de ontwikkeling van een nieuwe methode voor het *meten* van de tijdveranderende PMI, die gebruik maakt van een continue subjectieve waardering. Er werden twee mens-in-de-loop experimenten uitgevoerd, waarbij de deelnemers werd gevraagd de PMI continu te waarderen gedurende verschillende herhalingen van een passieve rijsimulatie.
Het eerste experiment, Experiment 1, beoordeelde de betrouwbaarheid en validiteit van de methode zelf. Het vergelijken van de waarderingen van verschillende herhalingen van dezelfde simulatie toonde consistentie in de waarderingen van de deelnemers en verifieerde de betrouwbaarheid van de methode. De validiteit van de methode werd geanalyseerd door de continue beoordelingen te vergelijken met een meer gevestigde tijdsonafhankelijke beoordelingsmethode en met de kennis van deskundigen uit de literatuur over de verschillende typen bewegingsstimuli fouten. De continue waarderingen correleerden goed met een tijdonafhankelijke waarderingsmethode voor elk segment van de simulatie en waren ook consistent met de kennis van de relatieve PMI tussen verschillende geschaalde, ontbrekende en foutieve bewegingsstimuli.
In een tweede experiment, Experiment 2, werd de continue waarderingsmethode toegepast om de prestaties van twee MCAs in een zeer realistische bewegingssimulatie van het voertuig te vergelijken. Opnieuw waren de deelnemers in staat om consistente continue waarderingen over meerdere herhalingen van dezelfde simulatie te geven en waren hun tijdonafhankelijke waarderingen voor elk geteste MCA goed te vergelijken met hun gemiddelde continue waarderingen. Dit bevestigde de betrouwbaarheid en validiteit van de continue waarderingsmethode, ook voor meer realistische voertuigsimulaties.

In **Deel II** werden de gegevens verkregen met de twee experimenten beschreven in Deel I, gebruikt voor de ontwikkeling van PMI *modellen*, om de tijd variërende PMI in en tussen experimenten te voorspellen. Een algemene modelstructuur werd ontworpen om visuele en inertiële bewegingsstimuli te vertalen naar een bewegingsincongruentie

waardering, de Motion Incongruence Rating (MIR). De MIR vertegenwoordigt de tijd-variërende waardering van PMI, verkregen met de continue waarderingsmethode. Eerst vertaalt het model de visuele en inertiële bewegingsstimuli in verschillende typen discrepanties die gecombineerd worden in één maat van PMI en vervolgens gefilterd worden om de gemodelleerde MIR te verkrijgen. Een op wavelet-gebaseerde bewegingsstimuli discrepantie detectie algoritme (CEDA) werd ontwikkeld om onderscheid te maken tussen geschaalde, ontbrekende en foutieve bewegingsstimuli, waarbij de parameters werden geschat met behulp van gegevens van Experiment 1. Het toepassen van het algoritme toonde aan dat de CEDA onderscheid kon maken tussen geschaalde, ontbrekende en foutieve bewegingsstimuli zoals verondersteld.
Om de intra-experiment voorspellingscapaciteiten te bepalen, werden drie modellen van verschillende complexiteit afgeleid van de algemene modelstruktuur. De eerste helft van de gegevens van Experiment 1 is gebruikt voor de parameter schatting van de drie modellen, waarna hun voorspellend vermogen is geanalyseerd met behulp van de tweede helft van deze dataset. De voorspellingsresultaten toonden aan dat alle modellen belangrijke PMI-kenmerken konden voorspellen. De voorspelling verbeterde met de toenemende complexiteit van het model. Een interessante observatie was ook dat foutieve bewegingsstimuli werden gemodelleerd als twee keer schadelijker dan geschaalde bewegingsstimuli voor dezelfde bewegingsstimuli discrepantie magnitudes. In het algemeen werd aangetoond dat de modellen inderdaad verschillende soorten bewegingsstimuli fouten kunnen koppelen aan de gemeten dalingen in bewegingsstimuli kwaliteit en dat dergelijke dalingen te voorspellen zijn voor data binnen één experiment.
Om datasets uit verschillende experimenten te vergelijken, werd eerst een methode ontwikkeld voor het schatten van een Model Transfer Parameter (MTP), waarmee de waarderingen van het ene experiment kunnen worden geprojecteerd op de waarderingen van een tweede experiment. De MTP die nodig was om de waarderingen van Experimenten 1 en 2 te vergelijken werd geschat met deze methode en gebruikt om het voorspellende inter-experiment vermogen van de drie afgeleide modellen te analyseren. Goede voorspellingsmogelijkheden werden alleen verkregen wanneer er een voldoende rijke dataset werd gebruikt voor de parameter schatting. De hypothese dat betere modellen kunnen worden verkregen naarmate de schattingsdataset rijker is, werd ondersteund door het feit dat de modellen die op geaggregeerde gegevens van beide experimenten werden geschat, de gemeten waarderingen beter volgden dan de modellen die op een van beide datasets werden geschat.

Deel III, richt zich op het *minimaliseren* van de PMI. Het vermogen van de PMI-modellen om een verminderde bewegingsstimuli kwaliteit te kunnen voorspellen biedt mogelijkheden om deze kwaliteit te verbeteren. De voorspellingen kunnen bijvoorbeeld worden gebruikt om de kortstondige bewegingsstimuli kwaliteit zo af te stemmen dat de meest kritische kwaliteitsdalingen worden vermeden. Bovendien kan de ontwikkeling van deze PMI modellen, door het correleren van de tijdveranderende PMI met verschillende bewegingsstimuli fouten, helpen om een beter begrip te krijgen van wat precies de verminderde bewegingsstimuli kwaliteit veroorzaakt.
In dit proefschrift werd een PMI model gebruikt in een op optimalisatie gebaseerd MCA. De gewichten voor de verschillen tussen de visuele en inertiële bewegingsstimuli voor lineaire versnelling en rotatiesnelheid in de kostenfunctie van dit MCA werden geschat met behulp van een statische versie van het minst complexe PMI model van deel II en gegevens van zowel Experimenten 1 en 2.

In een derde mens-in-de-loop experiment, Experiment 3, werd de bewegingsstimuli kwaliteit van het MCA met de op het PMI model gebaseerde gewichten vergeleken met de kwaliteit van het MCA met zijn oorspronkelijke gewichten, die alleen rekening hielden met de verschillen in eenheid tussen lineaire versnelling en rotatiesnelheid. De resultaten toonden aan dat slechts een kleine groep deelnemers, allen met ervaring in de simulator, de voorkeur gaf aan het MCA met de op het PMI model gebaseerde gewichten. De voorkeur van de resterende, grotere groep, leek vooral gebaseerd te zijn op een voorkeur voor 'lager-dan-eenheid-schaling' tussen voertuig- en simulatorbewegingen, wat consistent is met eerdere literatuur, maar nog niet in de PMI modellen is verwerkt. In het algemeen, wijzen de resultaten erop dat voor MCA optimalisering een PMI model op een dataset moet worden geschat die veel rijker is in termen van, onder andere, aantal en verscheidenheid van deelnemers, bewegingsstimuli fouten en simulatoren.

In dit proefschrift werd een nieuwe aanpak geïntroduceerd om de waargenomen bewegingsstimuli kwaliteit van MCAs te verbeteren. Een compleet stappenplan werd gepresenteerd, waarin wordt beschreven hoe PMI te *meten* en te *modelleren* en hoe dergelijke modellen toe te passen om PMI in bewegingssimulaties te voorspellen en te *minimaliseren*. De resultaten die in dit proefschrift worden beschreven, tonen het potentieel van deze nieuwe aanpak.
Voor toekomstig onderzoek wordt aanbevolen om de ontwikkelde PMI meetmethode aan te passen voor gebruik in actieve rijsimulaties en de PMI modellen te verbeteren door algoritmes te ontwerpen om meer types bewegingsstimuli fouten te detecteren. Het wordt verder aanbevolen om meer en rijkere PMI waarderingsdata te verzamelen via mens-in-de-loop experimenten om de parameterschatting van deze modellen te verbeteren. Ten slotte moet ook een systematisch onderzoek worden uitgevoerd naar hoe en onder welke omstandigheden deze modellen kunnen worden gebruikt om de kwaliteit van de bewegingsstimuli te verbeteren. Met deze vooruitgang kan de in dit proefschrift geschetste aanpak belangrijke verbeteringen in simulator bewegingsstimuli realisme mogelijk maken.

1

Introduction

Humans always wanted to go faster and higher than their own legs could carry them. This led them to invent numerous types of vehicles to move fast over land, water and air. As training how to handle such vehicles and testing new developments can be dangerous and costly, vehicle motion simulators were invented. In 1910 the first vehicle motion simulator, the Antoinette trainer (Figure 1.1(a)), was developed to safely train pilots how to control an aircraft while staying on the ground. Since then, motion simulator technology has evolved tremendously and many different types of motion simulators have been developed for a range of vehicle types (Figure 1.1(b)).
In the aerospace industry, motion simulators have increased flight safety by providing a safe and cost effective way for pilot training, while also reducing the environmental impact as less airborne training is required [1]. Simulators used for pilot training often consist of a Stewart platform [2], or hexapod platform (see Figure 1.1(b) TU Delft simulator), to provide the physical motions cues and a cabin with display to host the pilot and provide visual motion cues. Aircraft manufacturers such as Airbus and Boeing, but also large airlines such as Air France - KLM and Lufthansa, operate several training centers with dozens of full motion flight simulators especially for training purposes. Apart from pilot training, flight simulators are used for aerospace research and development, such as display design [3], handling quality assessment [4, 5] and even accident investigation [6].
In the automotive industry, the focus of this thesis, also increasing use is made of motion simulators. For race car driving physical motion cues during high translational vehicle acceleration are important for proper driver training [7, 8]. Simulator-based eco-driving training for truck and bus drivers can help to decrease fuel consumption [9, 10], while simulator-based investigations into driving behaviour under dangerous conditions can help to improve driver safety [11, 12]. Also during the car design process motion simulators are used for, for example, chassis testing [13], evaluation of steering feel [14] or development of driver assistance systems [15, 16]. Due to the importance of linear motion during car manoeuvres, the motion simulators of big car manufacturers such as Daimler [17], Renault [18], Toyota [19] and soon also BMW [20], consist of a hexapod platform on top of a linear track or X-Y table, to expand the horizontal and lateral motion limits. The cabin is usually large enough to house a real size car and often contains a 360 degrees display.

(a) (b)

Figure 1.1: (a) Antoinette Trainer. (b) current vehicle motion simulators (top left: SIMONA Research Simulator, TU Delft, top right: Daimler Simulator, bottom: CyberMotion Simulator, MPI for Biological Cybernetics).

In the aerospace industry specialized motion simulators have also been developed to simulate specific parts of space flight. The vertical motion simulator at NASA Ames [21], for example, was designed to simulate the vertical take-off and landing of air- and spacecraft. The Desdemona, operated by Desdemona B.V. and AMST, having a centrifuge design motion platform and gimbaled cabin, was initially designed for disorientation training [22].
Other novel simulator designs such as the Cybermotion [23] and CableRobot [24] simulators at the Max Planck Institute for Biological Cybernetics are, for example, used for motion perception research. In the DriverLab at Toronto Rehabilitation center [25] the influence of physical and mental health on driving performance is investigated. This simulator has additional realistic features such as a weather and glare simulator to simulate rain and oncoming headlights.
While new technologies and simulator designs have greatly improved the realism of vehicle motion simulators, generating realistic physical motion cues while staying within the simulator workspace remains one of the grand challenges of vehicle motion simulation. Realistic motion cues are needed for many aspects of transfer of training in aircraft [26, 27]. Transfer of training related to fuel consumption reduction was also greater when providing eco-driving training in motion-base simulators compared to a fixed-base simulator [28]. Motion cueing has also been shown to significantly affect driving behaviour during, for example, braking [29] and curve driving [30, 31]. Especially for driving behaviour research, and vehicle and human support system development which rely on simulating realistic driving behaviour, realistic motion cueing is extremely important. However, while the addition of physical motion cues increases simulation realism [32, 33], poor cueing can actually cause a significant *reduction* in realism and even lead to simulator sickness [34]. In cases of very poor motion cueing no motion is often preferred to motion [27, 35, 36]. Much research is therefore done in improving the realism of physical motion cues, such as taking into account human perception models [37–39], implicitly [40] or explicitly [41, 42] accounting for simulator constraints or accounting for future simulator motions [43, 44] when generating phys-

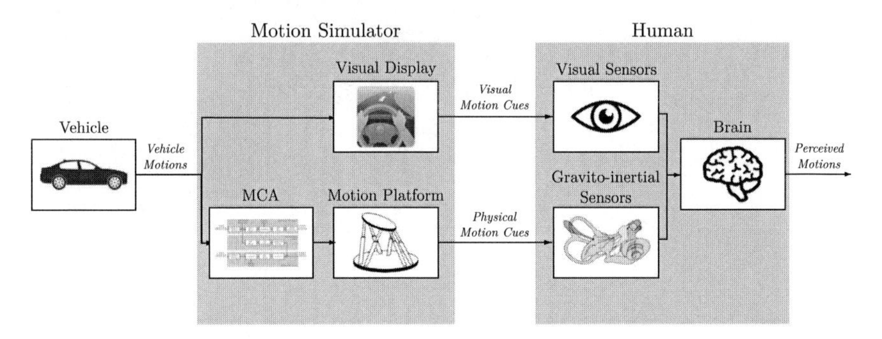

Figure 1.2: General scheme of motion cueing in a vehicle motion simulator.

ical motion cues, or specializing the generation of physical motion cues for specific simulation scenarios [45, 46].

1.1. MOTION CUEING ALGORITHMS

In Figure 1.2 a general scheme for motion cueing in a vehicle simulator is presented. The simulator motions, usually taken as the linear acceleration and rotational velocity, are used unrestricted to generate a visual scene that is displayed inside the motion simulator. From these visuals, visual motion cues are derived and sensed by the human visual system. A parallel path is shown for the physical motion cues, which are presented via the motion platform and sensed by the human gravito-inertial sensors such as the vestibular and somatosensory systems. The motion platform workspace, however, is restricted. Therefore, the vehicle motions are run through a motion cueing algorithm (MCA), which maps the vehicle motions onto the limited simulator workspace, *before* being send to the motion platform. Finally, the sensed visual and physical motion cues are then combined in the human brain and a percept of self motion is obtained.
Most simulators use an MCA that is based on the Classical Washout Filter (CWF) [47]. This MCA uses high pass filters to extract the high-frequency content of the linear accelerations and rotational velocities and sends only those to the simulator motion system. Low-pass filters are used to extract the low-frequency content of the lateral and longitudinal accelerations for what is called "tilt-coordination" [48], i.e., rotating instead of translating the cabin to simulate prolonged accelerations. If the rotations occur at a rate below the human perceptual thresholds, the physical rotation angle and the visual cues for linear acceleration combined are perceived as sustained acceleration rather than rotation [49]. To avoid the simulator hitting its limits, worst-case tuning of the MCA parameters is usually applied [50]. This type of tuning involves scaling down all motions, i.e., global scaling, such that those parts of the simulation that are not 'worst-case' are suboptimal. For the trade-off between hitting limits and simulator realism, experts are needed to tune the MCA parameters. This expert tuning often involves using human-in-the-loop experiments to obtain a 'feeling' of what is optimal [50, 51].
In attempts to avoid this inherently subjective manual tuning, adjustments to the CWF have been made. With the Optimal Washout Filter (OWF) [52], for example, filter orders and parameters are optimized off-line using an optimization algorithm that minimizes

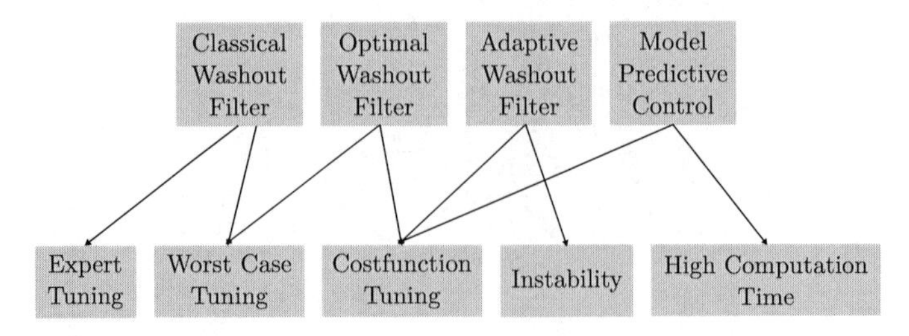

Figure 1.3: Challenges per MCA type.

a specific cost function. The cost function is often based on calculating the difference between vehicle and simulator linear acceleration and rotational velocities, but can also include models of, for example, the human vestibular system to account for thresholds in the human perceptual system [53]. With this type of MCA, however, the algorithm is again always tuned for the expected worst-case and the optimization needs to be repeated when different manoeuvres or vehicles are being simulated.
Adaptive Washout Filters (AWF) [54] were designed in another attempt to avoid global scaling. This MCA implicitly accounts for simulator limits by adjusting the filter gains in real time, based on minimizing a cost function that penalizes the difference between simulator and vehicle motions, the motion magnitude and the gain parameter change. It is a non-linear and much more complex procedure than the Classical or Optimal washout filters, but does not lead to significant improvements. Additionally, the adaptive filters are prone to instability, which, all together, makes that they are not widely used.
Lately, with increasing availability of computation power, MCAs based on model predictive control (MPC) [42, 55–57] have been introduced. Here a model of the simulator is used to predict the simulator motions for a given set of simulator inputs over a specified prediction horizon. By minimizing a cost function, the optimal simulator inputs for a given reference motion are found. Each time step the first simulator input is sent to the simulator, after which the optimization is repeated. With these MCAs the simulator limits are explicitly accounted for such that worst-case tuning is no longer needed, while algorithm stability is obtained with the combination of a well-designed cost function and a sufficiently long prediction horizon. Additionally, the cost function can be designed for optimal perceived simulation realism, by taking into account human percept.
In Figure 1.3 an overview of the different types of MCAs and their challenges are shown. From this figure it is clear that the one challenge that all types of MCAs share is the tuning of its parameters, may it be filter parameters or cost function parameters. Generally this tuning is done by experts, often using human-in-the-loop experiments. As these experiments are expensive and time consuming, often a limited number of parameter sets is tested [46, 58–61]. To make a well-grounded choice for the best MCA, it is imperative that these experiments provide a maximum of information on the cueing quality. Improving the MCA outside of the limited sets that are tested also requires knowledge on what causes the differences in perceived cueing quality.

1.2. CUEING QUALITY

Simulator fidelity, simulation realism, MCA performance and cueing quality, are just a few of the terms that exist in literature that aim to capture how realistic a vehicle motion simulation is. Simulation realism can be influenced by many things, such as the quality of the outside visuals and sounds, the vehicle mock up and the motion cue quality. In this thesis the term cueing quality will only refer to the effect of the physical motion cues generated by a motion cueing algorithm on the perceived simulation realism. A high cueing quality thus results in a more realistic simulation than a low cueing quality, when all other simulator and experimental conditions are considered equal.

As here we consider *perceived* realism, high cueing quality does not necessarily mean that the vehicle motions need to be replicated one to one with the simulator motions. Using tilt-coordination to simulate sustained acceleration while applying below human perception threshold rotations, for example, can result in the same cueing quality as would a one to one sustained acceleration, because the human subject would not perceive the difference.

As the human sensory system is subject to noise, also differences between visual and physical motion cues in the same motion channel cannot always be distinguished. In fact, in [62] and [63] coherence zones, ranges which indicate how much a visual motion cue can differ in magnitude from a corresponding physical motion cue while still being perceived as coherent, were identified for different motion channels. In this thesis, the term **Perceived Motion Incongruence (PMI)** refers to visual-physical motion cue pairs that are outside of these coherence zones, i.e., that are not coherent or, more generally, not congruent. The level of PMI then refers to the magnitude of its effect on the simulator realism, i.e., motion cue pairs that are perceived as incongruent but not detrimental to simulator realism will have a lower PMI than motion cue pairs that are perceived as incongruent and very detrimental to simulator realism. In Figure 1.4 the mapping from vehicle motions to this PMI is shown schematically. The vehicle motions usually consist of lateral, longitudinal and vertical linear accelerations and rotational velocities in roll, pitch and yaw. The simulator presents these vehicle motions to the human via visual motion cues, which are similar to the vehicle motion cues, and via physical motion cues, which differ from the vehicle motion cues. The human senses these motions cues with its visual and gravito-inertial sensors. Given the instruction, the human can use its car driving experience and preferences to compare the visual-physical motion cueing pair to real vehicle motions and generate one percept of motion incongruence.

Much research has been done on when motion cue pairs, i.e., visual and physical motion cues, are perceived as different [64–68], as these differences can be the most detrimental to motion simulation, due to the fact that they can induce simulator sickness [69, 70]. But also if a motion cue pair has magnitudes that lie within a coherence zone, i.e., differences are small, it can still affect the cueing quality negatively. In this case, it is possible that the vehicle motion magnitude that is being simulated is perceived incorrectly. Most cue integration models, such as [71] and [72], show that the perceived magnitude of a motion is some weighted average of the motion cues perceived by the different sensory organs. If an physical motion cue is within the coherence zone of a visual motion cue, but lower, a weighted average would thus imply that the perceived motion is somewhat lower than the vehicle motion that is being simulated.

Another aspect of cueing quality is that it is *time-varying*. Generally, the magnitude of the differences between vehicle and simulator motions already vary over time, logically resulting in differences in cueing quality over time. But also when the magnitude of the difference between simulator and vehicle motions is similar, the cueing quality can

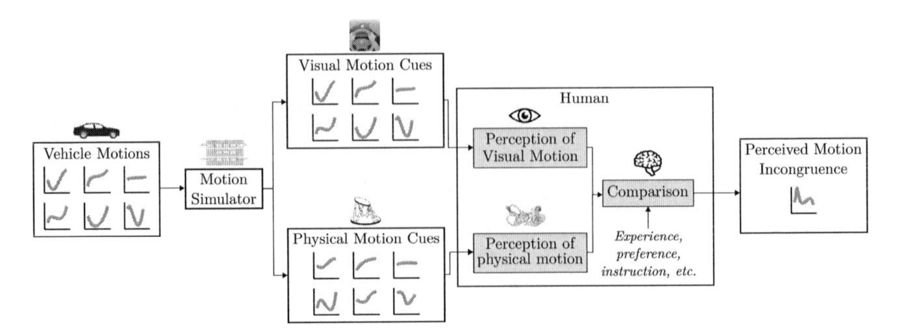

Figure 1.4: Scheme showing the mapping from vehicle motions to perceived motion incongruence in a motion simulator.

still differ. In [51] an overview is given of different cueing error types and their varying influences on the cueing quality. In Figure 1.5(a) an example of two of such cueing error types, a missing and a false cue, is shown. Both cueing errors have exactly the same objective quality, i.e., their euclidean distances are the same, as can be seen in Figure 1.5(b), but one is the result of *missing* motion, while the other is the result of *added* motion where no motion was expected. The false cue is known to be perceived as more detrimental to the cueing quality than the missing cue [51], as indicated in the fictional cueing quality detriment in Figure 1.5(b).
The quality of an MCA therefore strongly depends on how often the most detrimental cueing errors occur during a particular simulation. Information on the time variations in cueing quality is therefore essential when trying to understand why certain MCAs result in a low cueing quality, i.e., was there just one very detrimental cueing error or was the cueing quality constantly low? In the former case, the particular manoeuvre causing the large drop in quality could for example be removed, or if caused by hitting a limit, the gains can be scaled down. Knowing when a drop in cueing quality occurred can also help significantly in determining its exact cause related to the simulator motions.

1.2.1. Measuring Cueing Quality

Generally, a high quality MCA would cause the perception of being in a real moving vehicle, such that the participant behaves, i.e., controls the vehicle, in a similar way as in a real vehicle. For training purposes especially, it is important that the subject reacts to the perceived motions in exactly the same way as (s)he would in a real vehicle. One way of objectively measuring MCA quality is therefore to examine the control behaviour and compare it to control behaviour in a real vehicle, such as was done in [73] for flying and in [74, 75] for driving behaviour. This, however, is very time consuming and only a limited set of safe manoeuvres can be tested in this way.
Examining the difference in control behaviour with different MCA settings is also done, but it remains difficult to determine which control behaviour is desired without a real life example to compare it to. Another problem with examining control behaviour is that humans are very good at adapting [76], i.e., it is possible that the measured control behaviour in the simulator and real vehicle is similar, while the perceived motion differed. If this is the case during training, the participant would develop incorrect

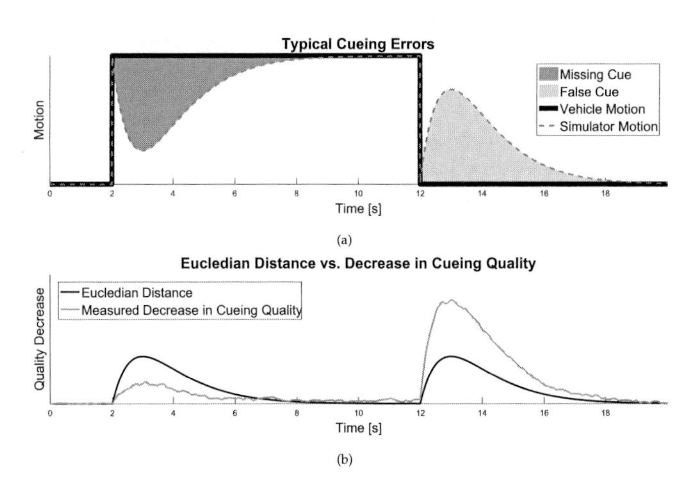

Figure 1.5: Simplified vehicle and simulator motions as the result of motion washout with corresponding cueing errors (a) and the fictional corresponding decrease in cueing quality and the actual euclidean distance between simulator and vehicle motions (b).

associations between certain perceived motions and their manoeuvres and, with that, fail to develop appropriate control behaviour.

Finally, a more practical issue with measuring control behaviour is that differences in control behaviour between participants will result in different motions between experiments. It is therefore difficult to perform a human-in-the-loop experiment where each participant is subjected to exactly the same motion cueing. Conclusions on differences in control behaviour are therefore generally made by analysing the behaviour over longer periods of time, and either fitting parameters of a control behaviour model to it [77] or averaging certain aspects [75, 78], such as control effort, over time. While this averaging over time can reduce the effects of small differences in motions between participants, it also removes all time information for the analysis. This time information, however, is particularly useful when analysing what caused the differences in MCA quality.

To avoid dealing with the adaptive nature of humans, one could try to measure the perceived motions or perceived cueing quality. As perception happens in the brain, however, it cannot be measured directly. Up till now the only objective measures related to cueing quality are physiological measures, such as measured in [79], showing simulator sickness. Simulator sickness, however, only occurs with very bad cueing quality and develops slowly over time, making it unsuitable to determine which part of the simulation really caused the sickness.

Instead, therefore, the perceived cueing quality is often measured subjectively. While subjective measures are generally disfavoured compared to objective measures due to their large variability, when obtaining a sufficient number of measurements, reliable results can be obtained. Many studies use subjective measurements such as questionnaires [36, 80], magnitude estimation [39] and paired comparison methods [81, 82] to determine cueing quality. While here a direct measure of the overall cueing quality is obtained, still important time information on when the cueing quality was high or low

is missing. While some have tried to include more time information via questionnaires [36], it remains difficult to directly relate such results to the provided visual and physical motion cues.
Finally, also off-line methods to determine MCA quality have been developed. Most of these methods, however, are related to the often used Classical Washout Filters. The Sinacori-Schroeder criterion [50], for example, determines acceptable gain and phase shift regions for the high-pass filter of a Classical Washout Filter. The Advani-Hosman criteria [83] instead provide such regions for the transfer function of the entire motion system. The more elaborate Objective Motion Cueing Test (OMCT) [84] uses similar criteria to determine the motion quality of the entire system, such that simulators can be compared. Both the Advani-Hosman criteria and the OMCT, however, assume a mostly linear system and analyse this system in the frequency domain. These methods are therefore suitable to compare systems using a Classical Washout Filter that keeps the motions within the simulator limits, but due to the lack of *time* information these methods are less suitable for highly non-linear MCAs such as the MPC-based MCA. Additionally, it is difficult to pinpoint exactly when or why an MCA has a lower quality.

1.2.2. IMPROVING CUEING QUALITY

While each type of MCA can be improved on different aspects, as shown in Figure 1.3, one challenge that all MCAs face is parameter tuning for optimal cueing quality. While human-in-the-loop experiments can be used to determine the best out of a limited group of MCA settings, such as was done in [46, 58, 60], it is very time-consuming to use such experiments to actively tune the parameters. Instead, off-line methods, such as using the Sinacori-Schroeder criterion, Advani-Hosman criterion or OMCT, are often used to perform an initial analysis and tune the parameters. As mentioned before, however, these methods were designed for mostly linear algorithms, and are not suitable for highly non-linear algorithms such as MPC-based MCAs.
Analysing and improving the cueing quality of such highly non-linear MCAs is often done with MCA independent methods such as visual analysis of the resulting simulator motions for a specific set of test manoeuvres [7, 85, 86] or comparing different parameter settings using a cost function that takes into account the difference between simulator and vehicle motions [41, 87, 88]. While the latter method is very time efficient, its effectiveness depends on the choice of cost function.
Cost functions used for MCA optimization, either used within the algorithm itself or used as an analysis tool of its results, come in many different forms. The simplest version is a weighted sum of the (squared) differences between simulator and vehicle motions [40, 89]. While such cost functions are easy to implement in a cueing algorithm, they lack important information related to the perception of motion. In an attempt to include the perceptual system in such cost functions, many have instead first ran the simulator and vehicle motions through simplified models of our vestibular system [41, 53, 56]. Such models, for example, account for the washout effect on the rotational velocity of our vestibular organs, i.e., we do not perceive sustained rotational velocity [90]. By running the simulator and vehicle motions through such models before comparing them, differences between simulator and vehicle motions from simulating sustained rotational velocity with a washout filter would, correctly, not be penalized.
While many claim that including such models improve the perceived cueing quality [37, 41, 55, 91], not many have actually tested this. In [37] the influence of different perception models in two types of MCAs was only investigated off-line, by analysing

the algorithm responses, i.e., the effect of using different perception models on the perceived cueing quality was not investigated using human-in-the-loop experiments. Moreover, not only the perception models changed between conditions, but also parameter retuning was applied, making it difficult to identify the influence of perception model changes alone. In [37] human-in-the-loop experiments were done to evaluate some of the different types of cueing algorithms described in [37], however, here the focus was on differences between an optimal and a non-linear cueing algorithm, rather than differences between perception models. In [41, 55, 91] the effect of using vestibular models was not evaluated at all. In [92], however, human-in-the-loop experiment results were analysed using such vestibular system models and they concluded that comparing motions after running them through such models does not explain the experiment results better than simply comparing the original motions signals.
Research has also been done on using more elaborate perception models that, for example, include models of the visual system [37, 93]. For now these models mainly contain low level sensory systems and not the higher level cognitive functions of the brain related to motion perception. While such models can be useful for gaining a better understanding of human motion perception, they only map actual simulator motions to perceived motions, and do not determine how this effects the perceived cueing quality. For example, they do not provide information on how to weigh and combine cueing errors between perceived vehicle and simulator motions from different motion channels into one measure of cueing quality. As also stated in [59], rather than the use of vestibular models, the cueing quality results of an optimization of any algorithm mainly depends on the choice of cost function and corresponding weighing constants. Moreover, current motion perception models cannot explain why different cueing error types, such as scaled and false cues, are not equally detrimental to the cueing quality. While the model driven bottom-up approaches are very useful in fully understanding how we perceive self-motion, they are not yet directly applicable to the problem of predicting cueing quality.

1.3. RESEARCH GOALS

The mapping of vehicle motions onto the simulator workspace, while maintaining a high simulation realism, remains one of the main challenges in vehicle motion simulation. To make an *MCA-independent* method to analyse *time-varying* motion cueing quality *off-line* is important when trying to improve cueing quality and the effects different types of cueing errors have on this quality, but is currently not yet available. The research goal of this thesis is therefore:

To develop an MCA-independent off-line prediction method for time-varying perceived motion incongruence during vehicle motion simulation, to improve motion cueing quality

The focus is put on Perceived Motion Incongruence (PMI) as the differences between visual and physical motion cues are assumed to be the most detrimental for simulator realism. No specific MCA type is assumed, such that any developed prediction method will be MCA-independent. The aim of the PMI prediction method is that it can be used for tuning of MCA parameters, such that the resulting motion cueing quality can be optimized.

1.4. APPROACH

The research goal is addressed in three steps: *measuring*, *modelling* and *minimizing* perceived motion incongruence. The thesis aims to provide a complete roadmap that describes how to *measure* and *model* PMI and how to apply such models to predict and with that *minimize* PMI in motion simulations.
Measuring time-varying PMI is essential for the development of PMI prediction models. While measurement methods such as paired comparison might be used to measure overall PMI, no method exists yet that can measure the time-varying aspect of PMI. As a first step in this thesis, a subjective method to measure time-varying PMI during a passive driving experiment was therefore developed and validated with two human-in-the-loop experiments. For subjective measurements the accuracy of the measurement strongly depends on the number of participants. As more participants with driving experience are available than, for example, with piloting experience, all experiments in this thesis are performed with car driving simulations.
The data obtained in the experiments were subsequently used for the development of a method to design data-driven PMI prediction *models*. A data-driven top-down modelling approach was chosen, as this would lead to the goal of PMI prediction more efficiently than a model-driven bottom-up approach. Via a data-driven approach the focus of the model automatically steers to those aspects of perception that influence PMI most. A model-driven approach, on the other hand, would require modelling all aspects of human self-motion perception and cognition, including those aspects that may not significantly affect simulator realism.
The developed models were subsequently analysed with respect to their explanatory and prediction power. For the analyses of the prediction power, first a prediction of new data within one experiment was made. Next, a method to compare PMI rating data between experiments was developed and used to analyse the prediction power of the PMI models between experiments.
While off-line PMI predictions can directly be used to *minimize* PMI via manual tuning of MCA parameters, a more efficient use of PMI prediction models would be to implement them in MCA optimization algorithms. Hence, in the last step of this thesis approach a simple PMI prediction model was implemented as the cost function of an optimization-based MCA, and its effectiveness was analysed in a human-in-the-loop experiment.

1.5. SCOPE

As simulator realism and the corresponding motion cueing quality have many aspects, a number of assumptions to limit the scope of this thesis were made.
First of all, it is assumed that humans can make a reasonable comparison between vehicle and simulator motions while experiencing a vehicle motion simulation. For this comparison it is assumed that the vehicle motions are perceived via some combination of visual cues and prior experience in car driving. For this reason, only participants that were in the possession of a valid driving license were allowed to participate in the experiments.
Assuming that vehicle motions can accurately be perceived from visual information and experience disregards important aspects of human self-motion perception. Visually perceived motion is strongly influenced by aspects such as field-of-view [94, 95] and visual scene content [95, 96]. Additionally, not all motions can be perceived with the same accuracy. As the measurement method developed in this thesis only involves

passive driving, i.e., the participant is not requested to provide any vehicle control inputs, also the influence of car driving style experience on the perceived motion is significant. Expected car motions that are difficult to derive from visual cues, such as longitudinal acceleration, might be influenced by the driving style of the participant which likely differs from the driving style of the 'automatic driver' used in the experiment. For those motions extra care should be taken when deriving conclusions from the corresponding PMI measurements. It is, however, expected that in general the visual motion cues are sufficient to derive a reasonable estimate of the vehicle motions.
A second assumption is that the time-variation of cueing quality is purely due to the time variation of the inputs, and that any PMI prediction model itself is therefore time-invariant. It is likely that some time variation is present in the human perception system, for example, due to the changing physical state of a participant related to fatigue or stress. However, by properly instructing the participant to, for example, take breaks before loosing focus, minimizing the experiment duration, and averaging over multiple measurements of PMI, such time variations are expected to be small when compared to the time variations caused by inputs.
A third assumption is that the PMI measurement is a good indication of the cueing quality of the vehicle motion simulation, and minimizing the PMI would thus improve cueing quality. This measure only includes perceived incongruences between vehicle and simulator motions, while congruent motion cue pairs that result in incorrectly perceived vehicle motions are not measured. While PMI is not the only aspect of cueing quality, any incongruences between vehicle and simulator motion negatively affect the cueing quality and can thus be taken as a measure of cueing quality.

1.6. OUTLINE

The first two chapters of this thesis are based on scientific publications, while other chapters are written to be included in this thesis first. All chapters can be read independently, although some back references occur. All nomenclature and references to literature are made uniform throughout the thesis.
The thesis is divided in three parts: measuring, modelling and minimizing perceived motion incongruence. In Figure 1.6 a schematic overview of the thesis is shown. It has a sequential structure, with each chapter providing data and/or developments for the next chapters. Part I yields a PMI measurement method and corresponding data sets. In Part II these data are used for the development of a PMI model. Part III uses both the developed measurement method, and a PMI model to optimize an MCA.
Chapter 2 introduces a newly developed subjective method based on continuous rating to measure time-varying PMI. The resulting motion incongruence rating is checked for reliability and validity in a human-in-the-loop experiment. Data collected in this experiment are also used in Chapters 4, 5 and 6.
Chapter 3 compares an optimization-based MCA, developed at the MPI for Biological Cybernetics, to the MCA used by the Daimler motion simulator which is based on classical washout filters. The two MCAs are evaluated using both the newly developed rating method, to determine its performance, and an off-line analysis describing the different strategies used by each MCA. The rating data from this experiment are also used in Chapter 6.
Chapter 4 addresses the varying influences which different cueing error types can have on cueing quality. A cueing error detection algorithm is developed using data from Chapter 2 and tested using data from an experiment performed outside this thesis, re-

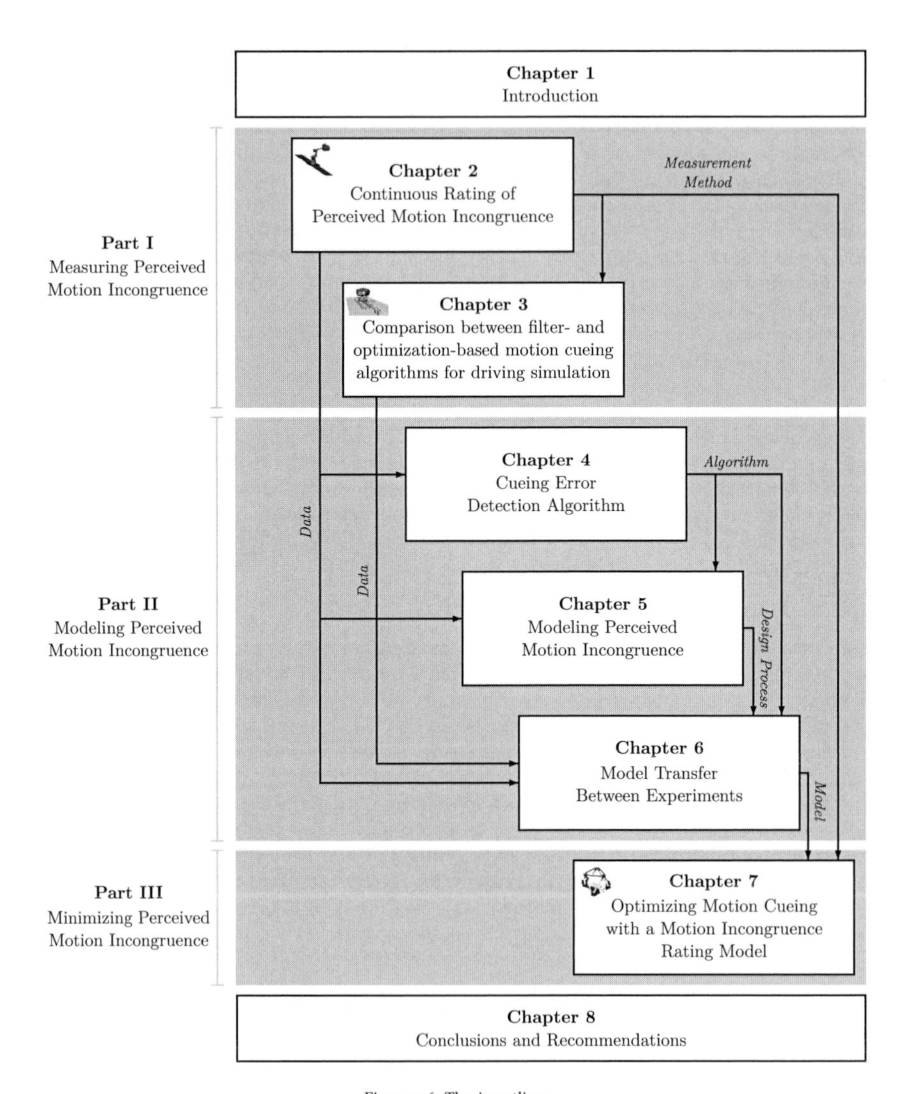

Figure 1.6: Thesis outline.

ported in [97].
Chapter 5 presents a system identification process for the development of PMI models. The algorithm developed in Chapter 4 is used in the non-linear part of such PMI models.
Chapter 6 shows how motion incongruence ratings from different experiments can be used to analyse the prediction capability of PMI models, or can be properly aggregated into a larger dataset. The introduced Model Transfer parameter is validated, and different PMI models are parametrized and analysed using data obtained in Chapters 2 and 3.
Chapter 7 describes the implementation of one of the PMI models from Chapter 6 as a cost function in an optimization-based MCA. The performance of this new cost function is compared to the original cost function using motion incongruence rating results of a human-the-loop experiment.
The thesis ends with conclusions and recommendations for future research.

I

Continuous Rating of Perceived Motion Incongruence

2

Continuous Subjective Rating of Perceived Motion Incongruence during Driving Simulation

In this chapter a method is presented to measure Perceived Motion Incongruence (PMI) continuously throughout a motion simulation. The method, which is based on continuous rating, was validated in an experiment. Subjects were requested to continuously provide a subjective rating of PMI during a vehicle simulation in the CyberMotion Simulator through constantly adjusting a rotary knob. Participants demonstrated that they could rate repetitions of the same simulation consistently. The resulting time-varying ratings were consistent with overall ratings of the same simulation and with literature on the typical cueing error types presented in this experiment. The time information contained in the rating data obtained with this method is essential for development of PMI prediction models as described in Chapter 5.

This chapter is based on the following publication:
Cleij, D., Venrooij, J., Pretto, P., Pool, D. M., Mulder, M., and Bülthoff, H. H. (2017). "Continuous Subjective Rating of Perceived Motion Incongruence during Driving Simulation." in *IEEE Transactions on Human-Machine Systems*, vol. 48, no. 1, pp. 17-29

2.1. Introduction

Motion-based vehicle simulators are used for a wide variety of applications. They are an increasingly important tool for training, research and vehicle system development in both the car [98] and aerospace industry [1]. However, one of the main challenges in motion-based simulation is to cope with the typically limited workspace of the simulator. To map the vehicle physical motions onto the simulator motion space, a Motion Cueing Algorithm (MCA) is used [85]. As the simulator motion space typically is much smaller than the vehicle motion space, this process inherently results in motion mismatches: differences between the unconstrained visual and the constrained physical motion cues. These mismatches result in a decrease of simulator motion fidelity and unrealistic simulations [51].

For motion simulation fidelity, a distinction is made between physical and perceptual motion fidelity [99]. Physical fidelity is defined as the match between objectively measured motion cues in the simulator and in the vehicle. Perceptual fidelity is defined as the match between simulator and vehicle motion cues as perceived by the human. The main reason for using a vehicle simulator is not to replicate the physical vehicle motions, but rather replicate the human perception of these motions [100]. Van der Steen [62] investigated the effect of physical incongruence between visual and physical motion on the perceived realism of the combined motion in a passive flight simulation. He introduced the term coherence zone for the range of physical motion amplitudes that were still perceived as coherent with a given visual motion amplitude. In [101] the effect of motion frequency on these coherence zones in passive flight simulation is investigated and in [102] the term phase coherence zone is introduced as the range of phase shifts for which physical and visual motion are still perceived as realistic. As in real vehicles, where all motion stimuli are congruent, motion simulators should provide physical motions that are within these coherence zones. If this is not possible, at least the perceived incongruence between different motion stimuli should be minimal. The current study therefore focuses on measuring any, linear or non-linear, incongruence between visual and physical motion that is perceived in a passive vehicle simulation. The degree to which this incongruence results in unrealistic motion is hereby called the Perceived Motion Incongruence (PMI).

To improve motion cueing we need to understand how this PMI is related to the physical motion mismatches presented in the simulator. Currently there are methods to directly or indirectly measure PMI, but they only provide time-invariant overall results. These discrete results can be used to quantify and compare the overall quality of an MCA, but cannot be correlated to the time-varying short-duration motion mismatches. It therefore remains unclear which motion mismatches are responsible for the overall PMI. A time-varying measure of PMI, that can be correlated to these mismatches, is therefore needed. Relevant motion mismatches can then be identified and, eventually, minimized. Besides being instrumental to improve motion cueing, such a measure can also be used to gain a better understanding of human motion perception.

Perceptual fidelity is measured using human-in-the-loop experiments. During these experiments participants are usually subjected to vehicle simulations using different MCA tunings. This fidelity can currently be measured directly via questionnaires or subjective ratings on the MCA quality. In [36] information on MCA quality during car motion simulation was obtained via questionnaires after each simulation run and overall MCA quality ratings at the end of the experiment. In [80] the Simulation Fidelity Rating scale together with an overall Motion Fidelity Rating were used to subjectively rate the motion fidelity of a helicopter motion simulation for different MCAs. In both

cases the only time varying information of MCA quality was obtained via questionnaires on specific parts of the simulation. In [39] an off-line rating method based on magnitude estimation with cross-modality matching was developed and used to detect differences between MCAs during car motion simulation. The MCA rating results obtained in these studies are time-invariant and can thus not easily be correlated with the time-varying motion mismatches. Direct objective and time-varying measures of PMI could possibly be done via physiological measures. Currently though, only physiological measures related to motion sickness, a possible effect of sustained or extreme PMI, have been found. Physiological measures such as heart rate and skin temperature were measured and compared with the continuously rated subjective estimate of discomfort during a car simulation in [103], while in [79] similar physiological measures during a driving and flying simulation were compared to off-line ratings of motion sickness. Instead of direct measurements, objective indirect measurements of PMI can be attained by observing the induced control behaviour for different MCAs. In [78] different MCAs were analysed based on objective measures such as overall control activity and tracking performance throughout an active driving simulation, while in [77] similar measurements during a flight simulation were used to determine the effect of heave washout filter settings on the parameters of a pilot model. To understand which differences in control behaviour indicate a 'better' MCA, in [104] and [61] control behaviour in a simulator for different MCAs is compared to real in-flight and in vehicle recordings respectively. In [105] the effect of time-varying filter gains on pilot control model parameters is investigated. However, the changes in control behaviour described by both time-invariant and time-variant models currently available, do not have the temporal resolution needed to identify the relevant short-duration motion mismatches.
This paper therefore presents a novel subjective measurement method which allows for measurement of the time-varying PMI continuously during a vehicle simulation (first described in [106]). The method is based on continuous subjective rating used in other research fields, such as 3D television [107]. The validation of the novel method is done by analysing the results of a human-in-the-loop experiment, where continuous rating of PMI was performed in a motion-based simulator during a passive driving simulation. First the measurements are tested for reliability and validity. Subsequently, the applicability of this method is analysed to determine if relevant short-duration motion mismatches can indeed be identified from the measured PMI. More information on the method and the validation process is given in Section 2.2. Section 2.3 describes the experiment set up, while the results with respect to reliability, validity and applicability of the method are presented in section 5.5. A discussion of these results and the corresponding conclusions are presented in Sections 5.6 and 5.7, respectively.

2.2. Continuous Rating Method

2.2.1. Background

Continuous rating (CR) refers to an on-line subjective rating, based on the method of magnitude estimation [108]. This method allows for the measurement of the perceived intensity of any physical stimulus. In the more traditional off-line rating method (OR), the observer provides a single rating (magnitude) to a certain property of the sensory stimulus via a dedicated rating interface. In the CR method, the observer is asked to provide this rating continuously throughout the sensory stimulus, resulting in a rating that varies over time.
In the field of 2D/3D television, CR methods are used to assess video quality by rating

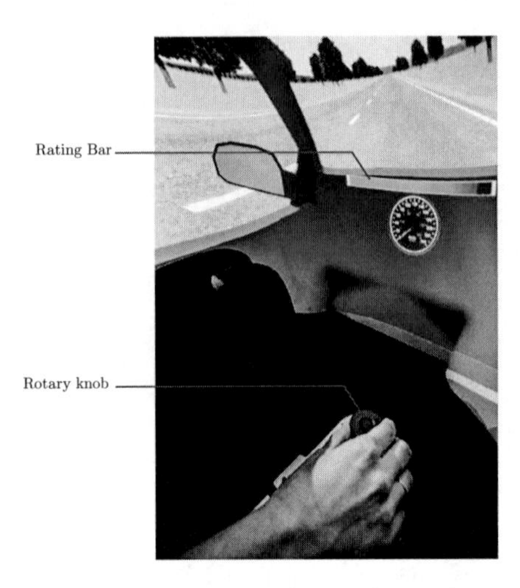

Figure 2.1: Rating interface consisting of a rating device and a rating bar.

the visual stimuli on visual comfort. In [109] this method is used to relate the measured visual comfort to disparity and motion, while in [110] the influence of 3D video properties, such as perceived depth, on the feeling of presence are rated. In the field of music analysis CR is also used. In [111] the method is used to measure the predictability of music over time, while in [112] a CR method is used to relate levels of emotion to specific aspects of music. Finally, in [113] and [114], a CR method is used to gauge strain and workload in, respectively, a motion-based and a fixed-based driving simulator, continuously.

2.2.2. PROCEDURE

The proposed rating method is based on the rating methods described above and used to measure perceived motion incongruence during a passive vehicle motion simulation in a motion-based simulator. The participants did thus not use the steering wheel or pedals, but were instead asked to continuously rate the perceived motion incongruence, i.e. judge the mismatch between the physical motions in the simulator and the motions you would expect in a real vehicle based on the simulator visuals. The resulting motion incongruence rating (MIR) is a measure for the perceived incongruence between visual and physical motion cues presented in the simulator.

The CR is performed using a dedicated rating interface, shown in Figure 2.1, consisting of a rotary knob to express the rating and a rating bar displayed on the screen, serving as visual feedback on the current rating. A maximum rating of one is given by turning the rotary knob fully to the right and will result in a fully coloured rating bar. A minimum rating of zero, given by turning the knob fully to the left, will result in a fully black rating bar. The method makes use of simulation trials: vehicle simulations of manoeuvres of interest, that each include the complete range of motion incongruence that will be presented during a specific experiment. In the experiment described in this paper, the simulation trials all consist of the same segments, combinations of manoeuvre and MCA, but ordered differently for each trial. To anchor both ends of the

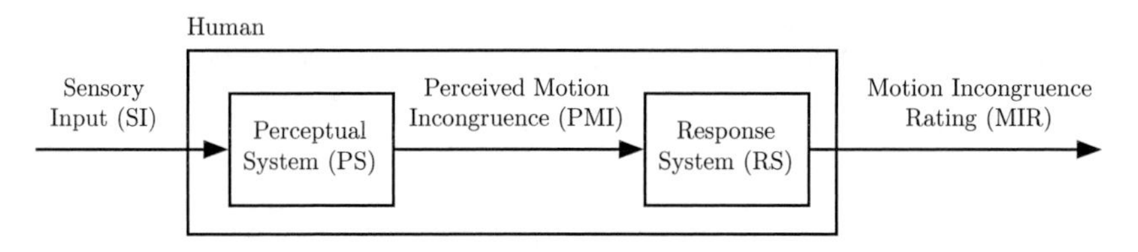

Figure 2.2: Simplified block diagram of the human subject during the continuous rating of perceived motion incongruence in a simulator.

rating scale, participants are instructed to provide the minimum rating of zero when no motion incongruence is perceived. When motion incongruence is perceived the rating should increase proportional to the incongruence intensity, with the maximum rating anchored at the highest incongruence perceived during the simulation trial.
Participants can only use such a rating scale properly, if the maximum incongruence during the experiment is known. The complete range of motion incongruence presented during an experiment should thus be observed at least once, before a proper rating can be performed. Therefore, participants first receive training that consists of two procedures: rating interface training and congruence range training, based on [115] and [109, 110], respectively. During the rating interface training, participants familiarize themselves with the rating interface via a simple control task, where they are asked to follow a second automatically adjusted rating bar. Subsequently, in the congruence range training, the participants familiarize themselves with the full range of motion incongruence that can occur during the experiment. They also familiarize themselves with the task of rating this incongruence continuously. To this end, participants are instructed to continuously rate the motion incongruence during a simulation trial. The training is repeated several times to check if the participant can provide a consistent rating. At the end of this training the participants should thus have learned to use the full range of the rating bar, i.e., when no motion incongruence is felt provide a rating of zero and when the maximum motion incongruence during the simulation trial is felt provide a rating of one.
In the measurement part of the experiment, participants are asked to continuously rate the motion incongruence in a simulation trial, using the rating interface. For verification of consistency of the rating, this procedure is repeated three times.
During the experiment described here, a second measurement, a retrospective off-line rating (OR), will be done. For this off-line rating, the simulation trial is split into several smaller segments. After observing a segment, participants are asked to provide one overall rating of the perceived motion incongruence during this segment using again the rating interface shown in Figure 2.1. This off-line rating method is commonly used to measure MCA quality [36, 61, 80] and is here assumed to be an accepted measure of perceived motion incongruence.

2.2.3. Measurements

To better understand what is measured with the CR method, a simplified block diagram of the human subject during the continuous rating task is shown in Figure 2.2. This diagram shows that the sensory input (SI), generated by the simulator motion and visualization systems during the simulation trial, is processed by the human perceptual system (PS) into, among other signals, the perceived motion incongruence (PMI). Here, subjects use their response system (RS) to translate the PMI into a continuous motion incongruence rating (MIR). The latter is the continuous rating data obtained during the experiment. When using the block diagram for the off-line rating (OR) task mentioned above, the response system will yield a time-invariant rating, the off-line MIR.

2.2.4. VALIDATION

The most important properties of a measurement method are reliability and validity of the measurements [116]. The main advantage of this measurement method in particular, is the possibility to correlate the various physical motion mismatches to the measured MIR. In the following paragraphs, the validation process and the applicability of the method in finding these correlations, are further explained.

RELIABILITY

To validate the novel measurement method, the results need to show that participants gave consistent ratings. The reliability analysis will determine within-subject consistency by comparing the three consecutive ratings of the same simulation trial by the same participant. The between-subject reliability analysis will be done by comparing all mean continuous ratings across participants. The reliability estimate Cronbach's Alpha [117] is calculated in both cases. This parameter measures internal consistency and serves as a metric for the expected correlation between the ratings. A value of 0.7 or higher is generally considered to reflect acceptable reliability [118].

VALIDITY

In addition to reliability, the CR method should also provide a valid measure of perceived motion incongruence. One way of analysing this validity is to compare the continuous MIR to a generally accepted measure of perceived motion incongruence. The continuous MIR will therefore be compared to the off-line MIR introduced in Section 2.2.2. To pass the validity test, the continuous MIR should show a significant correlation with the off-line MIR per segment. For this correlation calculation the continuous MIR, containing measurements for each time step, should be reduced to one value per segment. The reduction method will be chosen based on the measurement results and could, for example, include the mean or the maximum MIR per segment. The resulting correlation coefficient between the off-line and continuous MIR per segment will be tested for significance [119]. The t-test used to calculate the significance is shown in equation (2.1), where N is the amount of test items and r the correlation coefficient.

$$t = \frac{r\sqrt{N-2}}{\sqrt{1-r^2}} \tag{2.1}$$

In this paper it is assumed that participants can use the visual motion cues presented in the simulator, together with their real world driving experience, to derive the desired vehicle motion, while the simulator motion is represented by the perceived physical motion cues generated by the motion platform. This means that motion mismatches as defined previously, can be represented by the difference between desired vehicle motion and obtained simulator motion. In the experiment, multiple MCAs and manoeuvres are used to generate specific physical motion mismatches between vehicle and simulator in different motion channels. It is hypothesized that, if the continuous MIR is indeed a valid measure of perceived motion incongruence, these motion mismatches can be clearly identified from the continuous MIR. This second validity check will be done via

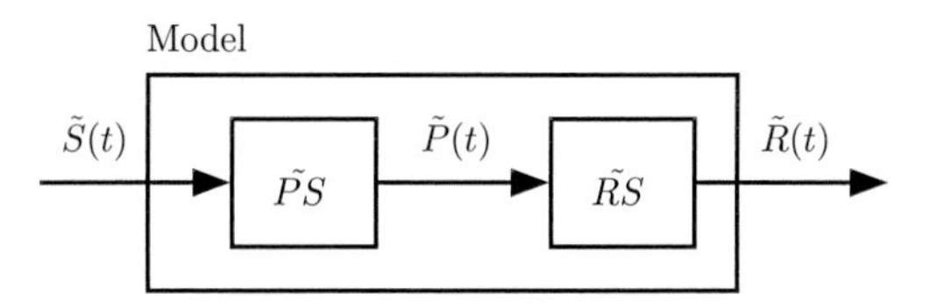

Figure 2.3: Block diagram of a model mimicking a human subject during the continuous rating of perceived motion incongruence in a simulator.

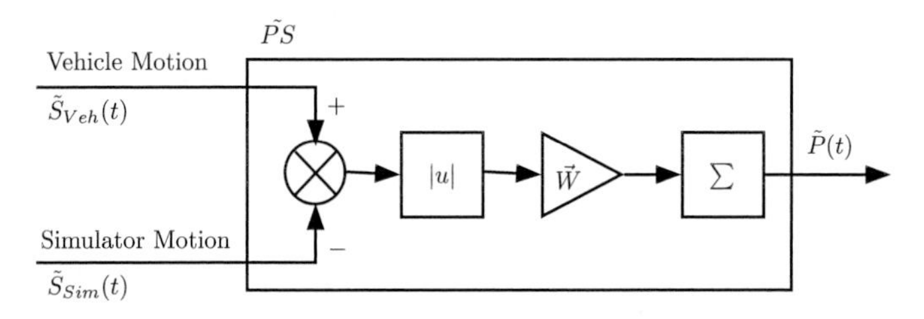

Figure 2.4: Block diagram of the perceptual system model, when rating perceived motion incongruence in a motion simulator.

visual comparison of the mean continuous MIR over all participants and the induced physical motion mismatches.

APPLICABILITY

As mentioned in Section 2.1, a major advantage of this measurement method is that its results can be used to obtain a deeper insight in the correlation between the sensory input generated by the simulator and the perceived motion incongruence. Subsequently, this correlation can be used to identify relevant short-duration motion mismatches and, eventually, minimize them. For this purpose, the block diagram of Figure 2.2 is transformed into the model shown in Figure 2.3. The measurement method provides a continuous MIR $R(t)$ which can be compared to the modelled continuous MIR $\tilde{R}(t)$ to provide insight in the correlation between SI and PMI. As the model presented here is merely a first example of the applicability of the measurement method, the models $\tilde{PS}$ and $\tilde{RS}$ will be kept simple and in accordance with previous literature.
In the field of motion simulation, sensory input $\tilde{S}(t)$ is often described as specific force and rotational velocity in longitudinal (x), lateral (y) and vertical (z) direction [51], which, for simplicity, will also be done here. The perceptual system $\tilde{PS}$ translates these sensory inputs into motion mismatches that together form the perceived motion incongruence $\tilde{P}(t)$. In literature, these motion mismatches are often described as the absolute difference between vehicle and simulator sensory inputs in individual degrees of freedom [40, 89], which will also be used here. The perceived motion incongruence is then calculated as the weighted sum of these motion mismatches, resulting in the perceptual system $\tilde{PS}$ shown in Figure 2.4. The modelled response system $\tilde{RS}$ should account for certain dynamics in the human rating process. In previous research where a CR method was used for perceived positive emotion [120] and melody predictability [111], the continuous rating was found to be a smoothed and delayed version of the expected signal. Hence, in this paper the continuous MIR is expected to be a smoothed and delayed

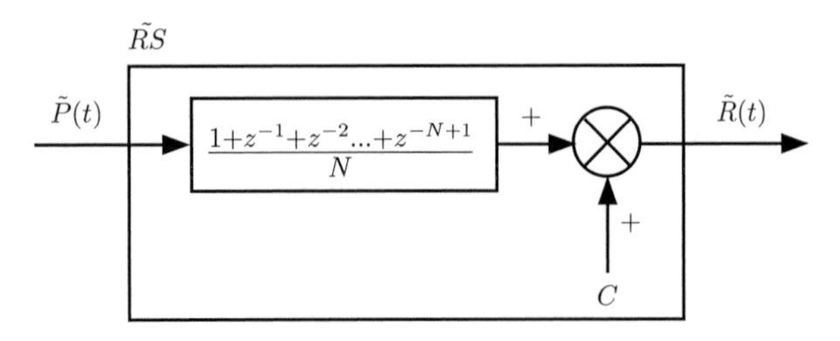

Figure 2.5: Block diagram of the rating system model when continuously rating perceived motion incongruence in a motion simulator.

version of the PMI. Without available data to support an explicit model, the response system $\tilde{RS}$ is modelled as a simple moving average filter with a window length of N seconds. A constant C is added to account for the non-zero minimum mean rating, due to spread between participants. The resulting rating system model $\tilde{RS}$ is shown in Figure 2.5. Assuming these representations of $\tilde{S}(t)$, $\tilde{PS}$ and $\tilde{RS}$, experimental CR data $R(t)$ can be used to, using linear least-squares, fit the model parameters: the 6x1 motion mismatch weight vector $\vec{W}$, the filter window length N and the constant C. The resulting model weights $\vec{W}$ show the strength of the correlation between a specific motion mismatch and the perceived motion incongruence. The perceptual system model PS can be used to minimize the motion mismatches, by implementing it as a cost function in the optimization algorithms for MCAs. The weight parameters of the simple model described here $\tilde{PS}$ could for example be used to replace the tuned weights in cost functions for MCA optimization based on adaptive [40] or model predictive control [39] algorithms.

2.3. EXPERIMENT

An experiment was performed to investigate whether a CR method can be used to measure time-varying perceived motion incongruence. For this experiment participants were exposed to a passive driving simulation in a motion-based simulator. During the simulation, different levels of motion incongruence were induced by varying the simulator MCA settings for different manoeuvres.

2.3.1. INDEPENDENT VARIABLES

The independent variables in this experiment were manoeuvre (three levels) and MCA setting (three levels), which were all embedded in a simulation trial, resulting in nine different simulation segments. The following manoeuvres were used in the simulation:

- CD: Curve Driving at 70 km/h, on a curve with a 257 meter radius and a 120 degrees deflection angle
- BA: Braking from 70 km/h to full stop and again Accelerating to 70 km/h on a straight road
- BCDA: Braking from 70 km/h to 50 km/h while entering the curve, Curve Driving at 50 km/h and Accelerating again to 70 km/h when exiting the curve, on a curve with a 131 meter radius and a 120 degrees deflection angle

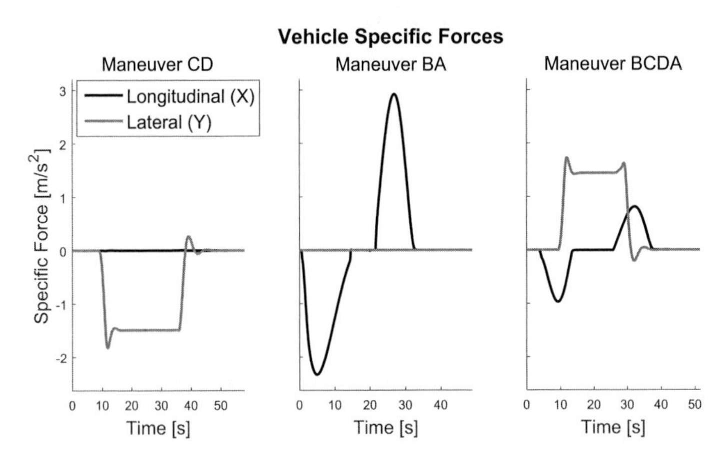

Figure 2.6: Longitudinal and lateral vehicle specific force for the three manoeuvres.

With these three manoeuvres, the simulation consists of motion incongruence in different motion channels. As shown in Figure 2.6, manoeuvres CD and BA primarily affect the longitudinal (X) and lateral (Y) specific forces, respectively, while the BCDA manoeuvre combines both forces. The MCAs were all classical washout filters [121], which map the vehicle specific force and rotational velocity vectors onto the simulator workspace. These algorithms make use of motion washout, returning the simulator to a neutral position with accelerations and rotations below human perception threshold, and tilt-coordination, tilting of the simulator cabin to simulate sustained acceleration. Tilt-rate limiting is applied to keep the rotation rate below human perception thresholds, for which values of ~3 deg/sec are often used [122].
The washout filter parameters that serve as a basis for the three MCAs used here were tuned to reproduce the above described manoeuvre motions within simulator limits, while making maximum use of tilt-coordination and not applying scaling or tilt-rate limiting. To induce specific motion mismatches, only the scaling or the tilt-rate limiting parameters were adjusted, which resulted in the following three MCAs:

- MCA_{Scal}: Scaling
  - Motion scaling (gain=0.6), which leads to scaling and small rotational errors (<4 deg/sec)
- MCA_{TRL}: Tilt-Rate Limiting
  - Rotation rate limiting to 1 deg/sec, which leads to missing or false cues, and very small rotational errors
- MCA_{NL}: No Limiting
  - Neither tilt rate limiting nor scaling is applied, which leads to large rotational errors (<8 deg/sec)

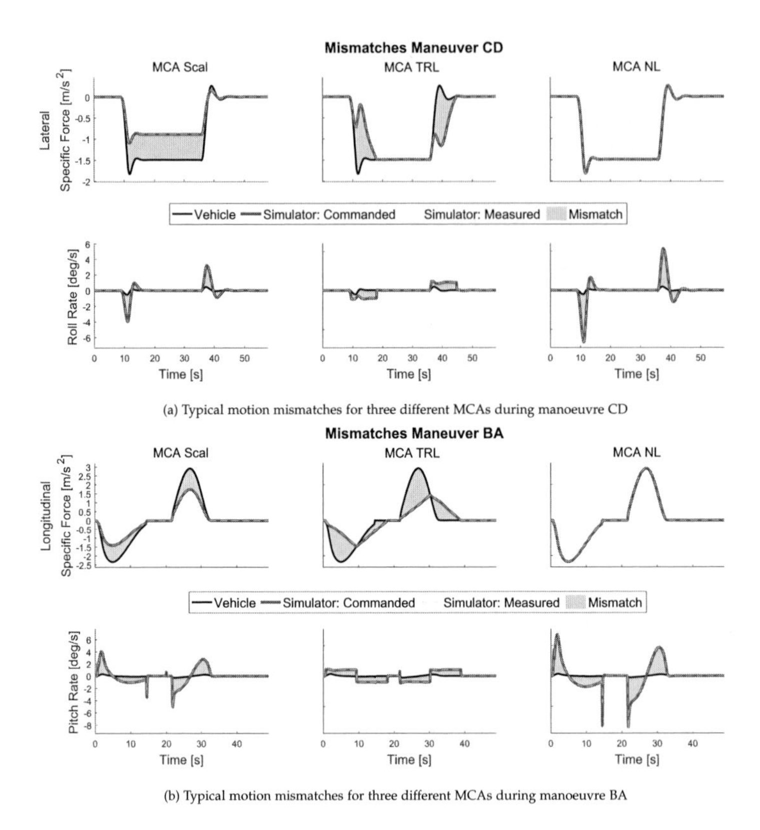

(a) Typical motion mismatches for three different MCAs during manoeuvre CD

(b) Typical motion mismatches for three different MCAs during manoeuvre BA

In Figure 2.7 the vehicle motions for each manoeuvre are shown together with the measured and commanded simulator motions resulting from the use of different MCAs. As the measured and commanded motions are very similar, the physical motion mismatches are hereby defined as the difference between vehicle and commanded simulator motion and are indicated with the light gray area in Figure 2.7. The longitudinal specific force and pitch rate for manoeuvre CD, as well as the lateral specific force and roll rate for manoeuvre BA, are zero for both vehicle and simulator motion and are not shown in Figure 2.7. A scaling error, visible in the specific force during the turn, the acceleration and the deceleration motions for MCA_{Scal}, is caused by a constant gain between vehicle and simulator motion. A missing cue, visible in the specific force at the beginning of these motions for MCA_{TRL}, is here defined as a simulator motion that has a lower amplitude than the vehicle motion but, unlike the scaling error, the gain between vehicle and simulator motion is not constant over all frequencies. A false cue, visible in the specific force at the end of these same motions for MCA_{TRL}, is similar to the missing cue, but here the variable motion gain is greater than one. False cues can also refer to simulator motion when no vehicle motion is present, such as the rotation errors visible in all rotational velocity plots in Figure 2.7.

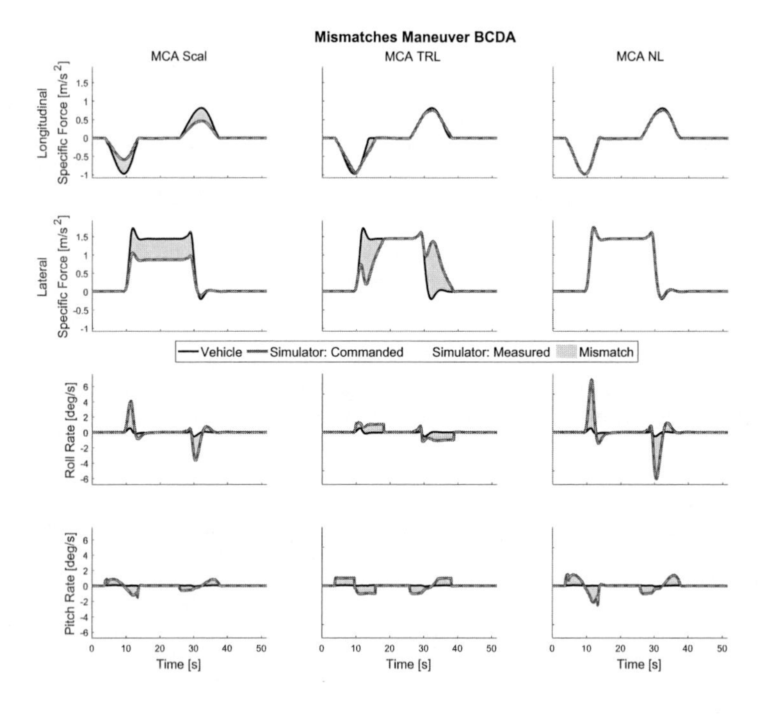

(c) Typical motion mismatches for three different MCAs during manoeuvre BCDA

Figure 2.7: The figures show vehicle motion, as calculated by CarSim, the commanded simulator motion resulting from the different MCAs, the motion that was measured in the simulator and the mismatch between vehicle and commanded simulator motion.

2.3.2. Dependent Variables

The dependent variables were the continuous motion incongruence rating (MIR) throughout the simulation trial, repeated three times, and the off-line MIR for each of the nine simulation segments.

2.3.3. Apparatus

The experiment was performed in the CyberMotion Simulator at the Max Planck Institute for Biological Cybernetics. This dynamic simulator was developed to expand the limited workspace and dexterity of traditional hexapod-based simulators. It is an 8-degrees-of-freedom serial robot derived from an industrial robot manipulator (Kuka GmbH, Germany), where a 6-axes industrial robot manipulator is mounted on a linear rail and equipped with a motorized cabin at the end effector. The cabin is equipped with two WUXGA (1920x1200 pixels) projectors (Eyevis, Germany) and interference filter stereo projection system (Infitec GmbH, Germany), which provide up to 160x90 degrees Field-of-View on the cabin inner side. The visuals and vehicle physical motions were generated using the simulation software CarSim (Mechanical Simulation, US). The rating interface, shown in Figure 2.1, consisted of a rotary knob (SensoDrive GmbH, Germany) to express the rating, and a rating bar rendered on the dashboard of the virtual vehicle for visual feedback on the current rating.

2.3.4. Participants

In total 16 participants, one female, aged between 22 and 38 years partook in the experiment. Their levels of simulator experience ranged from no simulator experience (7), participated in simulator studies before (5), to motion cueing expert (4), and all had a valid driving license.

2.3.5. Procedure and Instructions

Participants were first trained to use the rating interface and familiarize themselves with the simulation via the rating interface and congruence range trainings as described in Section 2.2.2. For the congruence range training two simulation trials were rated, after which the within-subject consistency was visually checked by the experimenter. If a low consistency was detected, a third training trial was given. After a short break the CR measurement part started, where participants were asked to observe and continuously rate three simulation trials, each including all nine combinations of manoeuvres and MCAs. After a second break the OR measurement part was started, where the participants were asked to observe nine short simulation trials, containing only one segment each, and provide one off-line rating after each trial using the rating interface. The same simulation segments were used throughout the experiment.
The simulation trial used for the congruence range training had a fixed segment order. The three trials used for the CR measurements all had a different segment order and were never the same as the training trial. For the OR measurement part each trial always consisted of the same initial acceleration and final deceleration and one of the nine segments, such that the simulation always had a natural start and ending. The order of these trials was randomized per participant. Per participant, the experiment lasted approximately 2 hours.
Throughout the experiment participants were asked how they felt regularly, such that the experiment could be stopped if early signs of simulator sickness, such as sweating and burping, were detected.

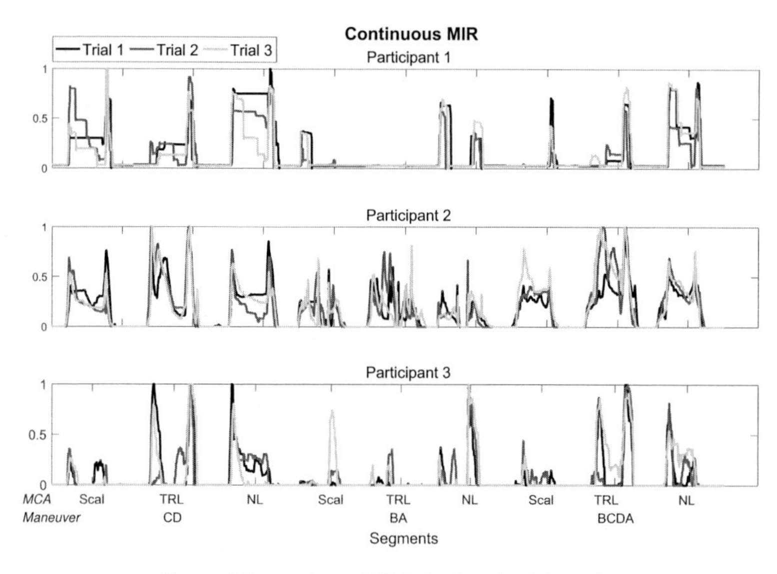

Figure 2.8: Raw continuous MIR during three simulation trials.

2.3.6. HYPOTHESIS

It is hypothesized that the continuous MIR will show sufficient consistency within- and between-subjects and that this rating will significantly correlate with the corresponding off-line ratings. It is also hypothesized that the continuous MIR will show an increase in motion incongruence during the physical motion mismatches shown in Figure 2.7. This in turn leads to the hypothesis that a simple model which makes use of these motion mismatches, described in Section 2.2.4, can explain a significant portion of the motion incongruence measured with the continuous rating.

2.4. RESULTS

2.4.1. RELIABILITY

As mentioned in Section 2.2.4, the reliability within- and between-subjects is determined by Cronbach's Alpha, using each time step sample as a separate measurement. The within-subjects reliability is calculated using the three simulation trial repetitions. The raw rating data for each of these three simulation trials from the first three participants is shown in Figure 2.8. As each simulation trial has a different sequence of the same simulation segments, for comparison, the rating data for each trial has been reordered to fit the same base sequence. The rating data in Figure 2.8 shows that participants rated the three trials consistently. The alpha for within-subject reliability had a median across all participants of 0.771 and an interquartile range between 0.727 and 0.897. Figure 2.8 also shows that there is variability between participants. For example, during the vehicle acceleration and deceleration in manoeuvre BA for MCA_{NL}, which causes rotation rate mismatches, participant 1 rated the motion as being much more incongruent than participant 2. This difference, which is also visible in the other manoeuvres for MCA_{NL}, could be explained by a difference in rotation rate perception threshold between these participants. Instead, participant 2 gave higher incongruence ratings than the other two participants during manoeuvre BA for MCA_{TRL}. This could possibly be explained by the ability of participant 2 to extract the vehicle motion more accurately from the visuals than the two other participants and therefore observing the incongruence better. Another explanation could be the preference of errors in one motion channel over another. Rotational errors might have had a stronger influence on the perceived incongruence of participant 1, while participant 2 was more focused on, or had a preference for, accurate linear acceleration. As humans have different motion sensitivities and thresholds, but also dissimilar higher level processes such as motion preferences, expectations and experiences, the observed rating differences between participants are to be expected. However, an alpha of 0.855 for between-subject reliability indicates that, in general, participants did agree on the occurrence and magnitude of the perceived motion incongruence during the simulation. From these results it is concluded that the method provides sufficient reliable measurements.

2.4.2. Validity

As mentioned in Section 2.2.4, the validity analysis is done by comparing the off-line and continuous motion incongruence rating (MIR) for each of the nine simulation segments, as well as comparing the continuous MIR to the physical motion mismatches shown in Figure 2.7. For this analysis Figure 2.9, showing the mean off-line and continuous MIR and their standard error during each segment, is used. For the correlation calculation between off-line and continuous MIR, the latter needs to be reduced to one variable. Figure 2.9 shows that, with the exceptions of the BA/MCA_{TRL} and BCDA/MCA_{NL} cases, the off-line MIR can be accurately predicted by the maximum continuous MIR during that segment. To compare the off-line and continuous MIR the latter is therefore "summarized" as the mean across all participants of the maximum rating per participant per segment. As the set of maximum ratings per segment for each participant has a smaller variance and higher mean than the set of off-line ratings per participant, for the correlation calculation, both sets are standardized to have zero mean and unit variance for each participant. Figure 2.10 shows the means over all participants of 1) the standardized off-line MIR and 2) the standardized maximum continuous MIR per segment. The Pearson correlation coefficient between the standardized mean OR and maximum CR is $r = 0.86$ ($p < 0.01$). This indicates that there is a significant linear relationship between the off-line and continuous MIR, and it can therefore be reasonably assumed that both methods measure the same perceived motion incongruence.
The time variations in the continuous MIR were hypothesized to correlate with the induced physical motion mismatches illustrated in Figure 2.7. The continuous MIR for manoeuvre CD, shown in Figure 2.9(a), clearly shows differences between the three MCAs. The false cue generated by MCA_{TRL}, starting at around 37 seconds, is clearly rated to induce the strongest motion incongruence. The second strongest perceived motion incongruence can be related to the missing cue starting at 11 seconds for the same MCA. The scaling error throughout the turn caused by MCA_{scal} is also clearly visible, but there is an unexpected increase in rating towards the end of the turn. More detailed analysis showed that this increase is visible in the ratings of seven of the sixteen participants. Finally, the peaks seen in the right graph in Figure 2.9(a) can be related to the roll rate error at the beginning and end of the turn for MCA_{NL}.
The ratings for manoeuvre BA, shown in Figure 2.9(b), are overall much lower than the

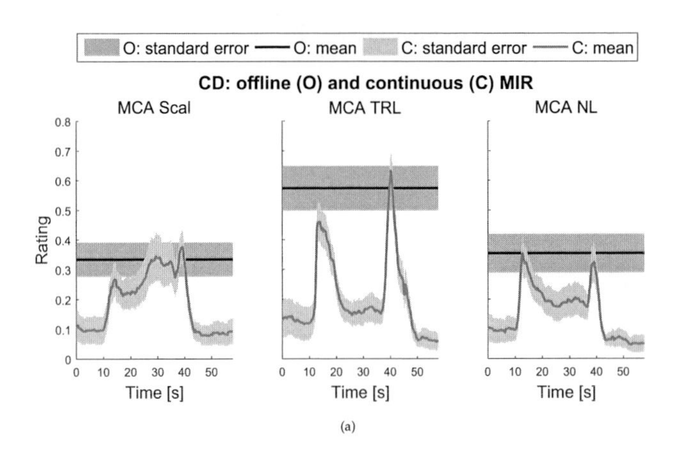

(a)

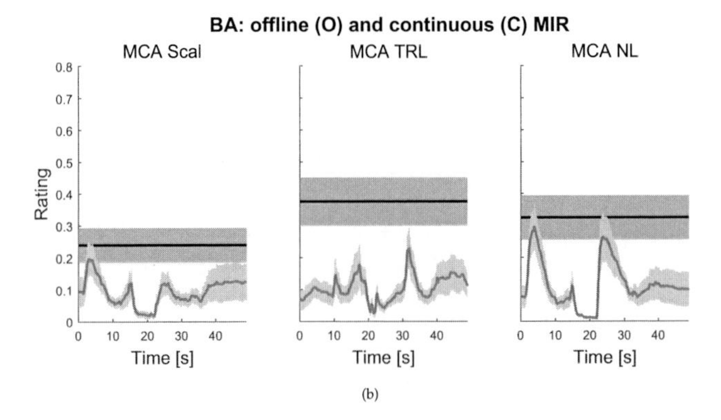

(b)

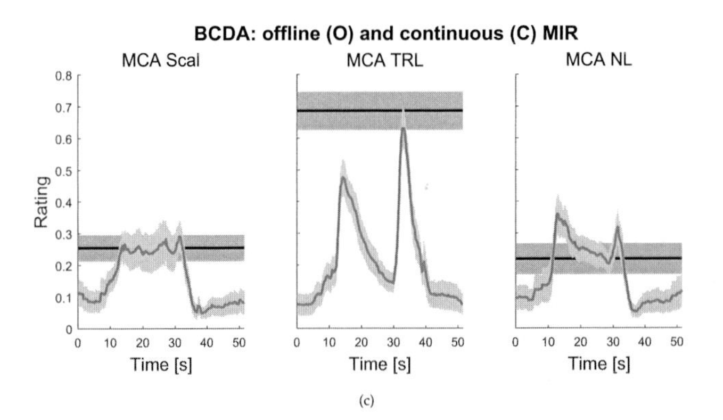

(c)

Figure 2.9: Mean values for off-line and continuous MIR and their standard error for each of the MCAs during manoeuvres CD (a), BA (b) and BCDA (c).

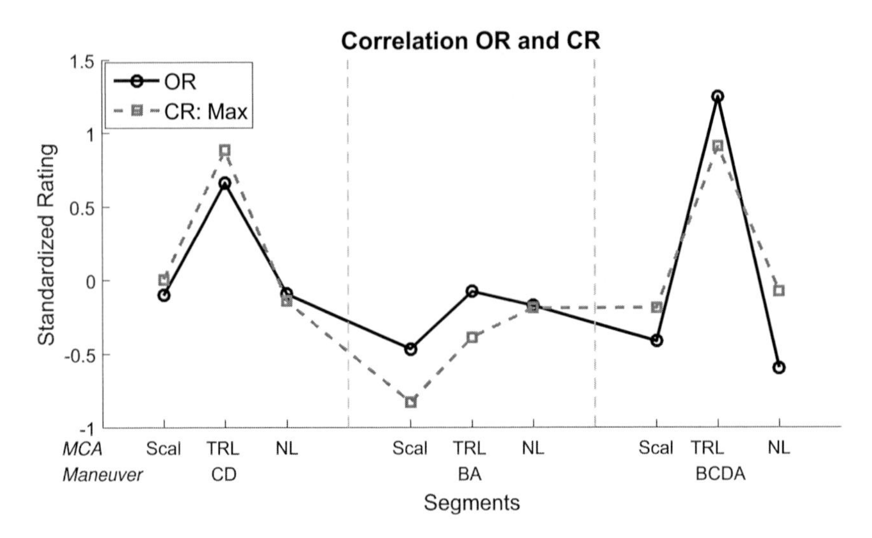

Figure 2.10: Mean standardized rating per simulation segment for the two rating methods.

ratings for manoeuvre CD. The main peaks in the continuous MIR for this manoeuvre are found when using MCA_{NL}, which causes large tilt rates at the onset of braking and accelerating. At around 20 seconds, the continuous MIR seems to approach zero, which can be attributed to the absence of any vehicle or simulator motion during the full stop, see Figure 2.7. The continuous MIR for MCA_{TRL} does not approach zero as observed for the other two MCAs and, in fact, shows a small peak, which can be attributed to the missing high-frequency motion cue at the end of the full stop. This peak, however, was only visible in the ratings of six participants. Some participants reported verbally that they did not rate this missing cue, even though it did clearly increase the motion incongruence. They reported waiting for the cue to arrive but, when realizing it would not occur, felt it was too late to rate accordingly, which could explain the difference between off-line and maximum continuous MIR for this manoeuvre.
Ratings for manoeuvre BCDA, shown in Figure 2.9(c), are very similar to the ratings for manoeuvre CD, which reveals that similar physical motion mismatches, indeed result in a very similar continuous MIR, indicating consistent rating behaviour. Again the scaling error, missing/false cues and roll rate errors caused by MCA_{Scal}, MCA_{TRL} and MCA_{NL}, respectively, result in increased continuous MIR. The main difference with the continuous MIR is seen for MCA_{scal}: the increase in rating towards the end of the turn that was visible in the CD manoeuvre is not observed in the BCDA manoeuvre. Overall, the continuous MIR can be visually correlated to the physical motion mismatches during each simulation segment rather well, suggesting that the CR method can indeed be used to measure the perceived motion incongruence.

2.4.3. Applicability

To show how the results of the novel measurement method can be used, the simple model described in Section 2.2.4 is fit to the measured continuous MIR. Such models

Table 2.1: Estimated model parameters

(a) $\tilde{PS}$

Motion Channel		$\vec{W}$	I
Specific	X	0.000	0
Force	Y	0.227	37
	Z	0.389	18
Rotational	Roll	0.047	17
Velocity	Pitch	0.003	2
	Yaw	0.012	26

(b) $\tilde{RS}$

Parameter	Value
C	0.087
N	3 [s]
δ_{avg}	1.45 [s]

may lead to a better insight in how PMI results from physical motion mismatches. Furthermore, the fitted motion mismatch weights $\vec{W}$ can be used in MCA optimization. As explained in Section 2.2.4 we do not directly measure the true perceived motion incongruence $P(t)$, but rather the motion incongruence rating $R(t)$. The latter being the output of the human response system RS, rather than the output of the human perception system PS. For this reason, not only a model of the perceptual system $\tilde{PS}$, but also a model of the human response system $\tilde{RS}$ needs to be fit. The model parameters motion mismatch weight $\vec{W}$, filter window length N and rating constant C are estimated by fitting the modelled $\tilde{R(t)}$ to the mean measured continuous MIR over all participants.

When comparing the estimated mismatch weights $\vec{W}$, it should be taken into account that the corresponding mismatches did not have equal strength in the simulation trial. The motion mismatches in longitudinal specific force, for example, were mainly present during one third of the total simulation, i.e., during the BA manoeuvre. For this reason an additional parameter, the influence factor I, is calculated to represent the percentage of $\tilde{P}(t)$ caused by mismatches in a specific motion channel:

$$I_i = 100 \cdot \frac{\Sigma(|\tilde{S}_{veh_i}(t) - \tilde{S}_{sim_i}(t)|) * W_i}{\Sigma\tilde{P}(t)} \tag{2.2}$$

Here I_i is the influence factor, W_i the weight, $\tilde{S}_{veh_i}$ the vehicle motion and $\tilde{S}_{sim_i}$ the simulator motion for the ith motion channel. The resulting estimated parameters and the influence factor per motion channel are listed in Tables 2.1(a) and 2.1(b). Additionally the average filter delay δ_{avg}, resulting from the fitted filter window length N, is shown. Figure 2.11(a) shows the resulting modelled continuous MIR, as well as the mean and standard error of the measured continuous MIR averaged over all participants. Figure 2.11(b) shows the PMI components per motion channel: the absolute difference between vehicle and simulator motion multiplied with the estimated weight vector. To determine the goodness of fit of the modelled PMI the coefficient of determination r^2 [119] is calculated with:

$$r^2 = \frac{\Sigma(\tilde{R} - \bar{R})^2}{\Sigma(R - \bar{R})^2}, \tag{2.3}$$

where $\bar{R}$ is the mean R over all time steps. The r^2 was found to be 0.79, i.e., 79% of the variations in $R(t)$ can be accounted for by the model.

The lateral specific force and yaw rate motion channels had the highest influence factors. Figure 2.11 shows that the motion mismatches in these channels only occurred

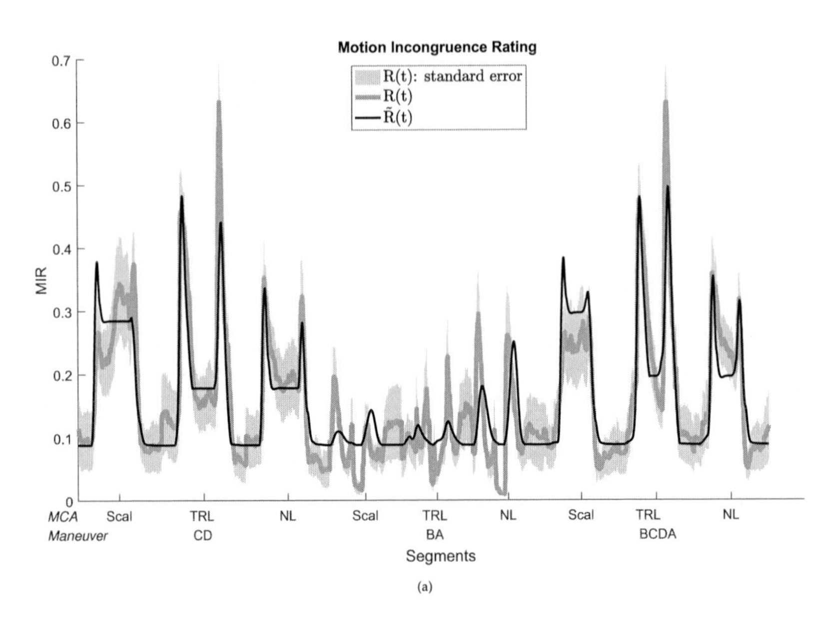

(a)

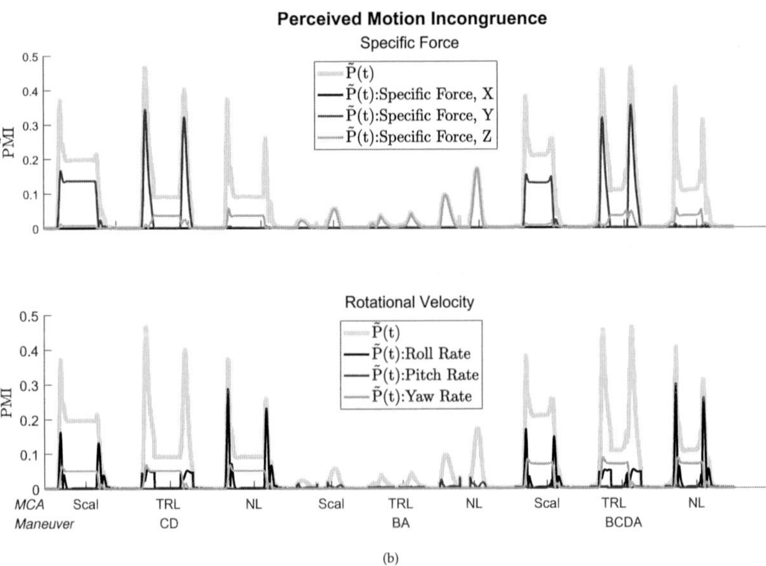

(b)

Figure 2.11: (a) measured continuous MIR $R(t)$, its standard error over all participants and the modelled continuous MIR $\tilde{R}(t)$. (b) modelled PMI $\tilde{P}(t)$ and its components.

during the curve driving manoeuvres CD and BCDA. As expected, the specific force mismatches are the main contributor for the PMI when using MCA_{scal} and MCA_{TRL}, while the roll rate mismatches are the main contributor for the clear peaks in PMI when using MCA_{NL}. The PMI measured throughout the curve for all MCAs is best modelled with the yaw rate mismatches.
During the BA manoeuvre, $\tilde{P(t)}$ is mainly based on the motion mismatches in the vertical specific force, caused by tilt-coordination. It is surprising that the motion mismatches in the longitudinal specific force did not influence $\tilde{P(t)}$ at all. The pitch rotation rate had a very small influence on $\tilde{P(t)}$.
The perceptual system model $\tilde{PS}$ together with the estimated weights $\vec{W}$ from Table 2.1(a) could now be used to minimize the perceived motion incongruence for manoeuvres and motion mismatches similar to those used in this experiment, by implementing it as a cost function in an MCA optimization algorithm. The estimated weights would then replace tuned weights normally used in the cost function. As the range of motion incongruence used in this experiment is relatively small, compared to those that can possibly be present during a vehicle simulation, the perception model $\tilde{PS}$ should be further improved using additional experimental data, before it can be used for a larger range of manoeuvres and motion mismatches.

2.5. DISCUSSION

2.5.1. SUMMARY

The experiment analysis described in this paper shows that the proposed continuous rating method can indeed be used to obtain a valid and reliable measure of time-varying perceived motion incongruence during vehicle motion simulation. Within- and between-subject reliability of the continuous motion incongruence rating (MIR) was shown to be sufficiently high to assume reliable measures. The validity of the measurement was investigated by comparing the continuous MIR first, to an off-line MIR generally used to measure overall perceived motion incongruence and second, to the physical motion mismatches in six motion channels over time. A significant correlation between the off-line MIR and the maximum continuous MIR was found, indicating that the continuous rating method indeed results in a measure of the perceived motion incongruence similar to that measured with the off-line rating method. Because also the different physical motion mismatches can be identified from the continuous MIR, it is reasonable to assume that the continuous MIR is indeed a measure for perceived motion incongruence. Finally, the applicability of the results from this method was investigated by fitting a simple model, describing the formation of perceived motion incongruence and the resulting continuous MIR from the sensory input in a motion simulator, to the measured continuous rating. This simple model could already explain a large part of the measured continuous MIR. Next to giving more insight in the importance of certain motion mismatches on the formation of perceived motion incongruence, these estimated weights for motion mismatches in six motion channels, specific force and rotational velocity, can be used in cost functions for MCA optimization.

2.5.2. SPECIFIC FINDINGS

Even though the current experiment was set up to validate the measurement method, it also already provided some interesting results on motion cueing that require further discussion. The correlation between the off-line rating per segment and the continuous rating seemed to depend on the maximum, rather than the mean continuous rating

during a manoeuvre. This finding is in accordance with earlier findings on the relation between overall and a continuous rating of video quality [123], where it is found that the off-line rating can best be predicted from the continuous rating by the peak impairment. This finding should also be taken into account when designing a cost function for MCA optimization, where currently, the overall motion incongruence is assumed to be a summation of the motion incongruence over all time steps of a certain manoeuvre [93] or the prediction horizon [55].
The correlation between the continuous ratings and motion mismatches in different motion channels showed that the false cues during the curve driving manoeuvres were rated with a higher motion incongruence than the scaling and missing cues during these manoeuvres. This is in accordance with general knowledge on false cues [51], but has now been measured directly for the first time.
The missing cue in longitudinal specific force at the end of the full stop in the "Braking, Accelerating" manoeuvre, unlike the missing cues during the "Curve Driving" manoeuvre, was not clearly rated as highly incongruent with the continuous rating. The off-line rating for this segment, however, did show high overall incongruence. The missing cue makes one feel like the car never came to a full stop, while the visuals do not show any vehicle motion any more. The timing of this cue is difficult to deduce from the visual motion, but, from experience, participants do expect the cue to appear at the end of the full stop. Because of this, some participants had reported to be too late in rating this perceived motion incongruence during the continuous rating, but said they did take it into account during the off-line rating. This could explain the difference between off-line and continuous rating, which is an interesting topic for further research. Depending on the application of the measurement method, it can be useful to clearly instruct the participants to rate any motion incongruence, even if their rating is delayed. It is often more important to obtain the incongruence rating than it is to avoid time delays, as the latter can be detected and removed during the data analysis.
The "Braking, Accelerating" manoeuvre, was clearly rated to be less incongruent for all MCAs than the curve driving manoeuvres, even though the objective motion mismatch magnitude, shown in Figure 2.7, was similar. An explanation for this can be that participants are less capable of perceiving longitudinal vehicle acceleration, derived from the changes in velocity observed in the simulator visuals, than they are at extracting vehicle yaw rate from these visuals during curve driving, due to differences in the optic flow for these two degrees-of-freedom, as explained in [96]. Less accuracy in the visually perceived vehicle motion, results in a larger range of simulator motions still being perceived as congruent.
A simple model, mapping sensory input to a motion incongruence rating can already produce a good fit, explaining 79% of measured mean continuous ratings. The fit was best for the curve driving manoeuvres, where the model shows that the rating was most likely caused by the lateral specific force mismatches.
The peak in motion incongruence due to the false cues for the curve driving manoeuvres, when using an MCA that includes tilt rate limiting, is modelled to be much lower than the peak that was measured with the continuous rating, while the peaks related to the missing cues are a much better fit. Both these peaks are caused by motion mismatches in the same motion channel, i.e., lateral specific force. As mentioned before, false cues are in general perceived as more incongruent than missing cues, but with the current implementation of the model this difference cannot be emphasized. It is therefore advised for future work that the motion mismatches defined as false cues are assigned a different weight than mismatches defined as missing cues.

The fit for the "Braking, Accelerating" manoeuvre is much less accurate than for the curve driving manoeuvres. A surprising finding is that the longitudinal specific force and the pitch rate are modelled to have, respectively, no and very little impact on the perceived motion incongruence. Instead, the motion mismatch in vertical specific force is modelled as the main influence on perceived motion incongruence during this manoeuvre. The latter mismatch exists due to tilt-coordination used in washout filters, where linear acceleration is simulated via the gravitational vector by tilting the simulator cabin. The motion mismatch in vertical specific force is thus related to the pitch angle. This implies that instead of responding to the perceived pitch rate, participants might have responded to the resulting pitch angle. This is consistent with results presented in [92], where it was concluded that subjects relied strongly on mismatches in attitude to determine the goodness of the motion cueing, at least during the tested longitudinal acceleration manoeuvre.
The estimated weight of zero for longitudinal specific force can be related to the aforementioned less accurate perception of this motion from simulator visuals, leading to a large range of simulator motions still being perceived to be congruent with the visual motion. In [124] it is concluded that simulator jerk has a large influence on perceived motion. In future work it is therefore advised to investigated if additional sensory inputs, such as rotational angle and linear jerk, would result in a better fit for the "Braking, Accelerating" manoeuvre.
The human response system during the rating task was modelled as a moving average filter and an added constant offset. The estimated window length for the filter resulted in an average delay of 1.45 seconds. This delay seems reasonable as it is in the same range as delays found in previous research where a continuous rating method was applied: [125] reports delays between 1.5 and 2 seconds, [111] reports delays between 0.5 and 0.7 seconds and [126] reports delays between 0.9 and 1.2 seconds. The constant offset should have accounted for the non-zero minimum mean rating, due to spread between participants, at points in the simulation where instead a minimum rating of zero was expected. A minimum rating is expected, for example, while the car is stopped and no simulator motion is present. The non-zero minimum mean continuous ratings found at these points in the simulation, however, are much lower than the estimated constant offset. This could be an indication that the constant offset value was not estimated correctly, or that the constant offset fulfils a different role in the model than the role that was intended. Some participants verbally reported that they rated the absence of road rumble. It is possible that such ratings were accounted for in the model by an inflated value of the constant offset. In future work it is therefore advised to include the absence of road rumble as one of the motion mismatches in the perception system.

2.5.3. Method considerations

This subsection gives a short overview of several aspects of the rating method that should be taken in consideration when using it to measure perceived motion incongruence.

- Time resolution: the major advantage of this method is that perceived motion incongruence is measured continuously over time, resulting in much more information than the currently existing methods that only provide overall measurements for a certain simulation trial. Even though the rating is continuous, it cannot be assumed to be instantaneous. Due to processing delays and filtering, which can also differ somewhat between participants, the time resolution of the mean motion incongruence rating is limited.

- Passive driving: the main drawback of the presented method is that it involves a passive driving simulation, where participants are not asked to control the vehicle. In [68] it is shown that, at least in some cases, perceptual thresholds differ between passive and active driving tasks, which can be an indication that perceived motion incongruence will also differ. The perceptual thresholds, however, seem to increase, which could lead to less sensitivity to motion incongruence during active driving. If this is the case, measuring perceived motion incongruence during a passive rather than an active driving task might actually lead to a more sensitive measurement. Another drawback of passive driving simulations is that only the incongruence between visual and physical motion stimuli can be measured. Perceived visual-physical motion incongruence, however, is not the only aspect of motion quality in a simulator. Motion quality depends on both the congruence between all motion stimuli, including stimuli such as proprioceptive feedback from control devices during active driving, and on how well these stimuli combined represent the actual vehicle motion. For motions such as the high-frequency vehicle motion at the end of a full stop manoeuvre, which is not easily derived from visual information, incongruence between the proprioceptive stimuli during active driving and the physical motion might be a more important measure. This, however, cannot be measured with passive driving simulations. In [113] an experiment is described where continuous ratings are used to measure strain during driving in both active and passive driving simulations. Here continuous ratings were taken during a simulation while performing an active driving task, followed by continuous rating while passively observing a repetition of this simulation. To include the effect of performing an active driving task on perceived motion incongruence, it would be interesting to use a similar experiment set up in future work.

- Direct measurement: an advantage of the method is that it provides direct measurements of perceived visual-physical motion incongruence rather than indirect measurements such as control behaviour or performance. Critical aspects of motion cueing can therefore more easily be identified.

- Memory workload: an advantage of the continuous rating method as compared to off-line rating is that the memory workload is reduced, as participants do not need to evaluate an entire trial and compare it to another one. Instead the perceived motion incongruence is only compared to the one instant of maximum incongruence that is unchanged throughout the experiment. This decreased memory workload also allows for rating of longer simulation trials, as compared to off-line rating. Longer rating trials in turn might help participants reaching a higher level of immersion into the simulation.

- Measurement scale: another advantage of the method is that it allows for measurements on an interval scale, i.e. containing information about order as well as having equal intervals, rather than an ordinal scale, i.e. only containing information about order, such as the often used Cooper-Harper scale [127] or paired comparison method [128]. However, unlike the Cooper-Harper scale, the scale in this study is derived from a specific experiment, which has the drawback that comparison between experiments is more difficult.

- Participant engagement: a negative effect of passive driving simulations is that participants can loose concentration due to the lack of activity. In the continuous rating method this is much reduced by requesting participants to actively rate throughout the simulation.
- Realistic simulation: The experiment described in this paper was deliberately set up to create very clear and specific cueing errors, such that the method could be validated. For future work it would be interesting to perform an experiment using a more realistic environment by using a dedicated driving simulator and realistic visuals and manoeuvres.

2.6. CONCLUSION

This paper describes a first experiment using a continuous rating method to measure time-varying perceived motion incongruence in a motion-based simulator. Results show that participants with different backgrounds and expertise in motion cueing and motion simulation are able to continuously rate perceived motion incongruence during passive driving simulations in a consistent manner. The correlation between retrospective off-line and continuous rating methods suggests that both methods indeed measure the same underlying variable, i.e., the perceived motion incongruence. This result is strengthened by the similarities between the continuous rating and the presented physical mismatches between vehicle and simulator motion. The continuous rating could therefore be used to determine the relative importance of short-duration motion mismatches such as scaling errors, missing and false cues. A simple model, mapping a selected set of sensory inputs to the motion incongruence rating, was fitted to the measured continuous rating data. The estimated model parameters showed the relative importance of each of the selected sensory inputs on the formation of perceived motion incongruence. Using this novel measurement method more complex models can be designed, which can significantly increase knowledge on perceived motion incongruence and that can also be used to further improve simulator motion cueing.

3

Comparison between Filter- and Optimization-Based Motion Cueing Algorithms for Driving Simulation

In this chapter a Motion Cueing Algorithm (MCA) based on a real-time Classical Washout Filter (CWF) is compared to an off-line optimization-based MCA. An experiment was performed where participants were requested to use the method developed in Chapter 2 to rate the Perceived Motion Incongruence (PMI) throughout a vehicle motion simulation with a high level of realism in the Daimler Driving Simulator. Results show a significantly better PMI for the optimization-based MCA throughout the simulation. For this thesis, a more important result is that participants were able to rate consistently, also in a realistic vehicle simulation. The data obtained with this experiment are used in Chapters 4-6 for the development of PMI prediction models.

This chapter is based on the following publication:
Cleij, D., Venrooij, J., Pretto, P., Katliar, M., Bülthoff, H. H., Steffen, D., Hoffmeyer, F.W. and Schöner, H. P. (2016). "Comparison between filter- and optimization-based motion cueing algorithms for driving simulation." in *Transportation Research Part F: Traffic Psychology and Behaviour*, vol. 61, no. 1, pp. 53-68

3.1. Introduction

Motion cueing is the process of converting a desired physical motion, obtained from, e.g., a vehicle model, into motion simulator input commands. This conversion is done by a motion cueing algorithm (MCA). In past decades, many different types of MCAs have been introduced [85]. The vast majority of them are variations of the filter-based approach, which relies mainly on scaling down and filtering the physical motions such that the commanded motion lies within the limited motion envelope of a simulator.
Recently, several optimization-based MCAs have been developed [56, 93, 129, 130]. The most important difference with filter-based approaches is that an optimization-based MCA produces an optimized output, in which simulator constraints are explicitly accounted for, instead of a filtered output for which it is not guaranteed it lies within the simulator's operational capabilities. In some optimization-based MCAs the motions of the simulator platform are optimized which are then converted to simulator control commands by low-level controllers [56, 129]. In others, the simulator control commands are optimized directly [93, 130].
It is clear that filter-based and optimization-based MCAs are fundamentally different algorithms, but it is not readily apparent which provides better motion cueing, if at all, and under which conditions. One can compare the algorithms' output, i.e., the commanded simulator motions, but that does not provide a direct answer to the question how the motion cueing quality of the algorithms is actually perceived by simulator occupants. The study presented in this paper aimed at providing some answers to that question by performing an experimental comparison.
The filter-based and optimization-based motion cueing approaches were compared in a driving simulation experiment, executed on the Daimler Driving Simulator (DDS) of Daimler AG in Sindelfingen, Germany. In the experiment, an optimization-based algorithm, developed by the Max Planck Institute for Biological Cybernetics in Tübingen, was compared against Daimler's filter-based MCA, using a newly developed motion cueing quality rating method [106]. The two algorithms will be referred to, in this paper, as MCA_{OPT} and MCA_{FIL} respectively. The goal of the comparison is to investigate whether optimization-based MCAs have, compared to filter-based approaches, the potential to further improve the quality of motion simulations.

3.2. Motion cueing algorithms

3.2.1. Filter-based motion cueing

Filter-based motion cueing algorithms consist of a combination of gains and filters, which transform (desired) vehicle motion into simulator set-points in real-time. Typically, the filters have a high-pass characteristic to prevent low-frequency accelerations from consuming a considerable part of the simulator's motion space. The gain functions can be linear or nonlinear and are adjustable, with the aim of reaching a good motion representation in a wide range of manoeuvres, preferably with a constant set of parameters. Special requirements of the manoeuvre that is to be simulated, like tight curves or turns, might be considered separately by the algorithm in order to provide good motion cueing quality while keeping the simulator within its operational limits. Well-known characteristics of the filter-based approach are tilt-coordination (where low-frequency components of the linear acceleration are reproduced by tilting the simulator platform) and motion washout (the ever-present push to return to the initial position). Such MCAs are commonly referred to as washout filters.
The MCA_{FIL} for non-professional driver applications is based on a classical washout

algorithm. Scaling factors and filters are used in all six degrees of freedom (DOF) to calculate the motion cues. Lateral, vertical and yaw excitations are dynamically limited by high-pass filters. A modified tilt-coordination algorithm provides an impression of steady-state acceleration in longitudinal direction and maximizes the use of the linear rail.
The main goal of the MCA_{FIL} is to provide linear motion cues within the envelope of the motion system even during worst-case manoeuvres. The algorithm takes the outputs of the vehicle simulation (accelerations and rotation angles and rates in 6 DOF) and calculates the commands for the motion system at 500 Hz. When used in driver-in-the-loop studies, the algorithm operates in real-time such that the driver is free to choose velocity, acceleration, deceleration and manoeuvres like lane change or overtaking other cars.

3.2.2. OPTIMIZATION-BASED MOTION CUEING

Optimization-based motion cueing optimizes simulator motions or control commands through an optimization algorithm. An often-used approach is Model Predictive Control (MPC). MPC is a control methodology that optimizes the current control signal based on a process model and a future reference trajectory of finite length, while taking constraints into account [131]. The optimization is governed by an objective function which quantifies the difference between (desired) vehicle motion and simulator motion. The optimization is constrained by the simulator's actuator limits. As a result, the optimized simulator control inputs and states always lie within the simulator's operational capabilities. MPC-based algorithms utilize predictions of future reference signals, using a 'prediction horizon' of a certain length, to compute the current control action. The advantage of this is that the current control action is optimized while taking future simulator states and control actions into account [132].
In general, an MPC-based MCA finds a sequence of controls u and states x which minimizes the following objective function:

$$J(\boldsymbol{x},\boldsymbol{u}) = \sum_{k=1}^{N}\left(\|\boldsymbol{u}_k\|_P^2 + \frac{1}{2}\left(\|\boldsymbol{y}(\boldsymbol{x}_k,\boldsymbol{u}_k) - \hat{\boldsymbol{y}}_k\|_R^2 + \|\boldsymbol{y}(\boldsymbol{x}_{k+1},\boldsymbol{u}_k) - \hat{\boldsymbol{y}}_{k+1}\|_R^2\right)\right) + \sum_{k=1}^{N+1}\|\boldsymbol{x}_k - \boldsymbol{x}_0\|_Q^2 \tag{3.1}$$

subject to the constraints:

$$\begin{aligned} \boldsymbol{x}_{k+1} - F(\boldsymbol{x}_k,\boldsymbol{u}_k) &= \boldsymbol{0} \\ \boldsymbol{u}_{min} \leq \boldsymbol{u}_k &\leq \boldsymbol{u}_{max} \\ \boldsymbol{x}_{min} \leq \boldsymbol{x}_k &\leq \boldsymbol{x}_{max} \end{aligned} \tag{3.2}$$

where N – number of time steps, $\boldsymbol{x}_0$ – the "neutral" state of the simulator, $\boldsymbol{y}(\boldsymbol{x}_k,\boldsymbol{u}_k)$ – physical motion signal at the head point in the simulator as a function of its state and input, $\hat{\boldsymbol{y}}_k$ – physical motion signal at the head point in the vehicle (reference value), P,Q,R – symmetric positive-definite weighting matrices for penalizing control input, deviation from the neutral state and error in the physical motion signal, respectively. F is the function that describes discrete-time dynamics of the system, u_{min}, u_{max}, x_{min}, x_{max} are the lower and upper bounds of the inputs and states. The physical motion signal is defined as:

$$\boldsymbol{y} = \begin{bmatrix} \boldsymbol{f} \\ \boldsymbol{\omega} \end{bmatrix} \tag{3.3}$$

Table 3.1: Comparison of algorithm characteristics

	MCA_{FIL}	MCA_{OPT}
Type	Filter-based	Optimization-based
Real-time capable	Yes	No
Driver-in-the-loop applications	Suitable	Not suitable
Sampling rate	500 Hz	50 Hz
Future Reference	Not applicable	Entire trajectory
Accounting for simulator limits	Manual tuning	Constrained optimization
Tuned	Yes	No

where $\boldsymbol{f}$ – specific force, $\boldsymbol{\omega}$ – rotational velocity at the head point.
The MCA_{OPT} is described in more detail in [39, 93]. For the experiment described in this paper the process model was adjusted to incorporate and cope with the parallel, rather than serial, model of the Daimler simulator. In the current paper, a trajectory-based optimization was performed, which means that the information of the entire trajectory was provided to the algorithm at the start of the optimization: i.e. in Equation 3.1, N is the total number of trajectory samples and $\hat{\boldsymbol{y}}$ is obtained from a recording of the manoeuvre that was to be simulated (instead of a prediction of the future reference). In theory, this should lead to the best cueing quality, as the maximum amount of available information (i.e., a 'perfect prediction') is provided to the optimization algorithm. A clear disadvantage of this approach is that it makes the algorithm only suitable for simulation of pre-recorded manoeuvres. It is possible to use prediction methods to obtain real-time predictions of the future reference signal, which makes the algorithm suitable for driver-in-the-loop simulations, e.g., [56]. It is to be expected that this would result in a lower simulation quality compared to the trajectory-based optimization approach used in the current study [87].
The weighting matrices used in the objective function are: $P=0.1$, $Q=0$, $R=diag(1,1,1,10,10,10)^2$. As the value for Q was zero, there was no penalty for deviation from the neutral state, which implies that the algorithm did not exhibit washout behaviour. The weighting factor 10 for the rotational velocities is chosen as an approximate ratio of standard deviation of specific force components in m/s^2 and standard deviation of rotational velocity components in rad/s for typical car manoeuvres.
The optimization was constrained by the actuator limits. For safety reasons, the bounds were set at 95% of the actual position, velocity and acceleration limits of each actuator. In addition, the simulator state was constrained by the condition that the initial and final position of the simulator should be upright (zero degrees of roll, pitch and yaw) and the initial and final velocity of simulator should be zero. The optimization was performed using CasADi toolbox [133] and Ipopt solver [134].

3.2.3. Algorithm comparison

It is important to note that the two MCAs described above are very different algorithms, each with their own characteristics, see Table 3.1.
The MCA_{FIL} is a robust, real-time algorithm, suitable for driver-in-the-loop simulations. The MCA_{OPT} performs its optimization based on perfect knowledge of the entire driving manoeuvre, making it unsuitable for real-time driver-in-the-loop applications, but suitable for passive simulations.

The MCA_{OPT} does not run in real-time due to the high computational load associated with the optimization. The optimization of the trajectory used in this study – with a duration of approximately 5 minutes – took a few hours on a regular PC. After the optimization, the output of the MCA_{OPT}, which provided data at a sampling rate of 50 Hz, was resampled (interpolated) to 500 Hz, in order to run synchronously with the output of the MCA_{FIL}.
During the optimization, the MCA_{OPT} utilized exact knowledge on the desired motion for all future time steps (trajectory-based optimization). Such an optimization would not be possible if the knowledge about the future is limited, as is the case in real-time driving scenarios with a driver in the loop. In that case, prediction algorithms would be required to obtain an estimate of the future reference trajectory. The effect of using (different approaches to) real-time prediction on simulation quality remains a topic to be addressed in future studies.
Furthermore, the optimization of the MCA_{OPT} is constrained by the simulator's actuator limits. As a result, the optimized simulator control inputs and states always lie within the simulator's operational capabilities. This is not guaranteed for the output of the MCA_{FIL}, where simulator limits are typically accounted for by tuning the algorithm's parameters. The MCA_{OPT} did not need tuning for the experiment described in this paper. The implementation of the MCA_{OPT} used in the current study did not account for perceptual factors like drivers' motion sensitivity and thresholds [39]. It is to be assumed that the implementation of such features will further improve the cueing quality of the MCA_{OPT}.
Due to the above differences, this study is not to be considered as a competitive comparison between MCA alternatives, but rather as an attempt to gain insight in the potential that an optimization-based approach has to offer with respect to well-established filter-based approaches.

3.3. METHODS

3.3.1. RESEARCH QUESTIONS

The primary research question of this study is whether optimization-based MCAs have the potential to further improve the quality of motion simulations compared to filter-based approaches. At the start of the study it was unknown whether there would be any measurable differences between the two algorithms, and if so, what can be learned from these differences to further improve motion cueing.
In order to measure the quality of the motion cueing, a quality rating method developed at the Max Planck Institute for Biological Cybernetics in Tübingen was utilized. The method was described and evaluated in [106]. As the rating method was only recently developed, a secondary research question was whether the method provides reliable and repeatable results within and between participants.

3.3.2. APPARATUS

The experiment was conducted in the Daimler Driving Simulator (DDS), an electrical hexapod platform mounted on a 12 m long linear axis [17] (Figure 3.1(a)). For this experiment, the car's longitudinal axis was aligned with the simulator's linear axis by rotating the cabin in the dome (Figure 3.1(b)). This adjustment provides a relatively large motion space for the reproduction of longitudinal accelerations and deceleration with the disadvantage that the space for lateral motion is limited. The driver's cabin was a standard Mercedes-Benz C-Class model (W204) equipped with an additional

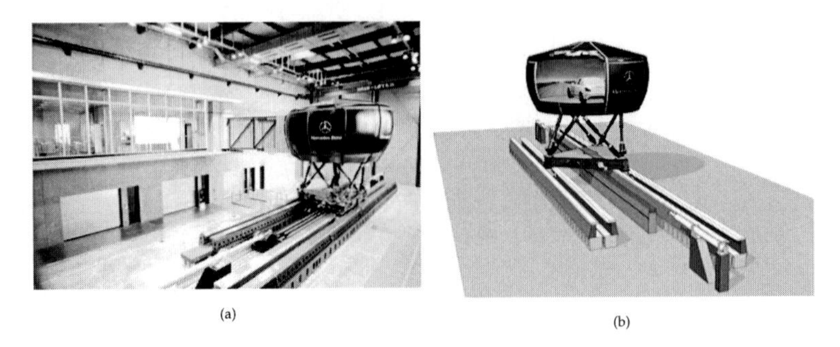

(a) (b)

Figure 3.1: Exterior of the Daimler Driving Simulator (DDS) (a) and the car orientation during the experiment (b).

Table 3.2: Overview of experiment procedure

TRAINING (1 trial pair, 10 min)	Familiarization with rating device and procedure.
EXPERIMENT PART 1 (1 trial pair, 10 min)	Motion mismatch rating: CR followed by OR for each trial
EXPERIMENT PART 2 (2 trial pairs, 20 min)	Motion mismatch rating: CR followed by OR for each trial

display showing the rating bar (described below).

3.3.3. Participants

In total 18 participants, 9 females, aged between 21 and 40 ($mean = 29.3$, $std = 5.7$) took part in the experiment. All had previous experience in driving simulators, but no or limited knowledge of motion cueing. They were expert drivers with a minimum mileage of 10,000 km per year ($mean = 17.222$, $std = 7.496$). Two participants did not complete the experiment due to motion sickness symptoms and their data were excluded from the analysis.

3.3.4. Experimental procedure

In the experiment, participants were presented with four pairs of evaluation trials, of which the first pair was used for training purposes. Each trial consisted of the playback of an identical pre-recorded simulated drive. While the visuals remained unaltered, the vehicle motions of the simulated drive were processed by either the MCA_{FIL} or the MCA_{OPT}, generating two different simulator trajectories. These trajectories were repeatedly presented in random order at each trial pair. In total, participants rated each trajectory four times, of which the last three were included in the data analysis. During the playback, the participants did not need to take any actions on the steering wheel or pedals. Instead, they were asked to concentrate on the movements of the simulator and rate the perceived motion mismatch, i.e., the perceived mismatch between the motion felt in the simulator and the motion one would expect from a drive in a real car, taking the simulator visuals as a reference. The experiment lasted approximately 1 hour, of

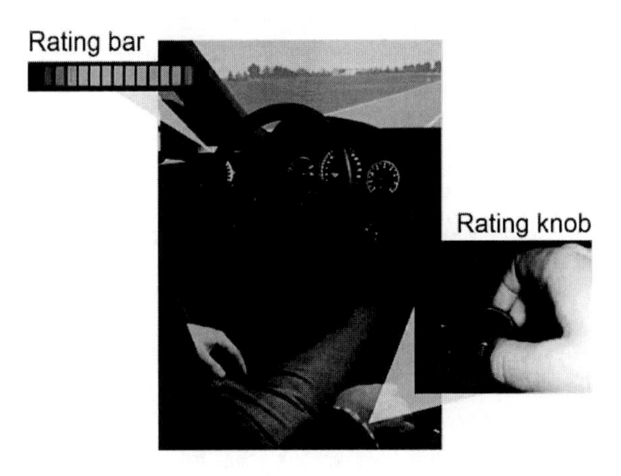

Figure 3.2: Location of rating bar and rating knob in the car cabin. The rating bar has 15 coloured marks, where green/red marks indicated low/high perceived mismatch.

which 45 minutes in the DDS (Table 3.2). Throughout the experiment participants were asked how they felt regularly, such that the experiment could be stopped if early signs of simulator sickness, such as sweating and burping, were detected.

3.3.5. RATING PROCEDURE

The ratings were provided using the built-in rotary COMAND-knob of a Mercedes C-Class. By rotating it, participants controlled a rating bar with 15 coloured markers, visible on a small screen located to the left of the steering wheel (Figure 3.2). A rotation to the left reduced the number of visible marks (lower motion mismatch); a rotation to the right increased the number of visible marks (higher motion mismatch). There was always at least one green mark visible. The quality of the motion cueing was rated in two ways:
A continuous rating (CR) method was used to measure time-varying aspects of the perceived mismatch between real and simulated drive. For the CR, participants were asked to continuously assign a value (magnitude) to the instantaneous perceived motion mismatch via the rotary knob during the playback. If no mismatch was perceived they were asked to provide a rating of zero.
After each trial the participants were asked to provide an overall rating (OR), by indicating the perceived motion mismatch of the entire playback. The OR resulted in a single rating for each trial. The rating method is described in more detail in [106].

3.3.6. STIMULI

The recorded simulated drive that was used in this experiment was performed by a human driver. The drive consisted of different manoeuvres combined into one realistic drive in both rural and city surroundings of about 5 minutes. The chronological list of manoeuvres that occurred during the drive is shown below.

- InitAcc: initial acceleration along a rural road up to the speed of 100 km/h

- RuralCurves: drive over a rural road consisting of a large-radius left, right and left curve, during which a constant speed was maintained.
- OverTake: double lane change manoeuvre at constant speed to avoid a car parked on the right-hand side of the road.
- SlowDown50: upon entering an urban area the speed is initially reduced from 100 to 70 km/h and then from 70 to 50 km/h.
- TrafLightDec: driving through a gentle curve and decelerating to a full stop in front of a red traffic light.
- TrafLightWait: standing still in front of the red traffic light for 6 seconds.
- TrafLightAcc: accelerating from stand-still to 50 km/h after the traffic light turns green
- City1: multiple gentle curves through the city at a constant 50 km/h speed.
- Roundabout: decelerating to 20 km/h, driving through a four-exit roundabout, exiting at the second exit and accelerating back to 50 km/h.
- City2: multiple curves through the city at a constant 50 km/h speed.
- TurnLeft: decelerating to 20 km/h, driving through a 90-degrees left turn and accelerating back to 50 km/h.
- City3: multiple gentle curves through the city at a constant 50 km/h speed.
- FinalDec: deceleration to a full stop at a red traffic light.

3.4. Results

To compare the two MCAs, first the results and reliability of the overall and continuous ratings are shown in Section 3.4.1. To determine which parts of the drive are responsible for these differences, the continuous rating is analysed per manoeuvre in Section 3.4.2. In Section 3.4.3 a comparison between the continuous rating and the motion cueing errors over this time interval is shown as an indication of what caused the perceived mismatch during these manoeuvres. To minimize the cueing errors, both MCAs made use of different cueing mechanisms which are described in Section 3.4.4. Finally in Section 3.4.5 the implementation of these mechanisms to generate the linear acceleration in the driver frame of reference in the Daimler Driving Simulator is explained.

3.4.1. Rating results and reliability

Using the method described above, participants rated the perceived motion mismatch between 0 (no motion mismatch) and 14 (strong motion mismatch). Note that a higher rating value implies a lower cueing quality.
To test for significant differences between mean ratings the parametrized paired t-test (test statistic = t) and repeated measures ANOVA (test statistic = F) were used. Because the mean ratings were not always normally distributed, the Lilliefors test for normality and generalized ESD (extreme Studentized deviate) test were used to check the normality and outlier assumptions of these tests. If these tests were not passed, the non-parametric Wilcoxon signed-rank test (test statistic = z) and the Friedman test (test statistic = χ^2) were used instead of the paired t-test and ANOVA respectively.

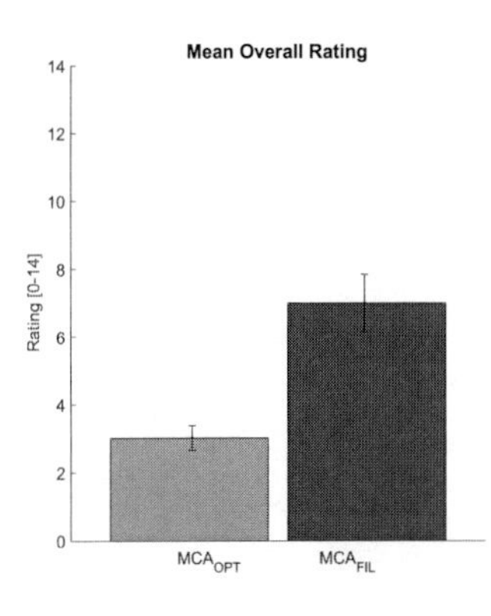

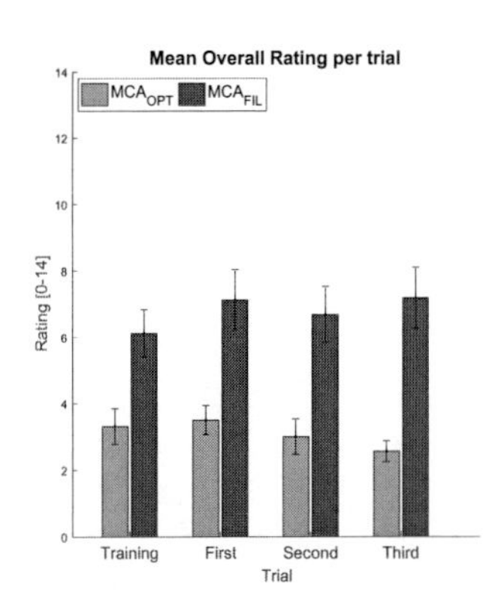

Figure 3.3: Mean overall rating across three evaluation trials (left) and mean overall rating per trial (right). Error bars indicate the standard error.

The average results obtained for the overall rating (OR) are shown in Figure 3.3. The mean overall rating ($MCA_{FIL} = 7.0$, $MCA_{OPT} = 3.0$) across all participants differs significantly between MCAs: $z = 3.2154$, $p < 0.01$. This indicates that participants felt less motion mismatch with the MCA_{OPT} than with the MCA_{FIL}. The overall rating does not change significantly for either of the MCAs over the three evaluation trials (MCA_{FIL}: $F(47) = 0.91123$, $p > 0.05$, MCA_{OPT}: $\chi^2(47) = 2.4615$, $p > 0.05$).
The average results obtained for the continuous rating (CR) are shown in Figure 3.4, showing the mean continuous rating across all participants. Before computing the means, the CR raw data were standardized per trial pair by subtracting the minimum rating and dividing by the rating range. The mean values of the continuous ratings ($MCA_{FIL} = 4.0$, $MCA_{OPT} = 1.5$) also differ significantly; $t(15) = 3.6708$, $p < 0.001$. Participants were consistent when rating the perceived motion mismatch continuously (Cronbach's alpha (α), $mean = 0.83$, $std = 0.07$). As shown in Table 3.3, from the 16 participants only two did not pass the statistical test for consistency ($\alpha < 0.7$) [118]. This result is in line with previous findings in which the same method was used to determine MCA perceived quality [106].

3.4.2. Rating results per manoeuvre

Differences between MCAs rating on specific manoeuvres were analysed by comparing the mean continuous rating of the two MCAs per manoeuvre. The continuous rating for manoeuvres 'InitAcc' and 'FinalDec' was not recorded fully and was therefore excluded from the analysis. To control for the false discovery rate of multiple comparisons the Benjamini-Hochberg correction was applied to all corresponding p-values before testing for significance. The resulting test statistics and the significantly different means per MCA are shown in Table 3.4.

[1] This column indicates participant number as assigned during the experiment, where numbers 4-7 are missing.

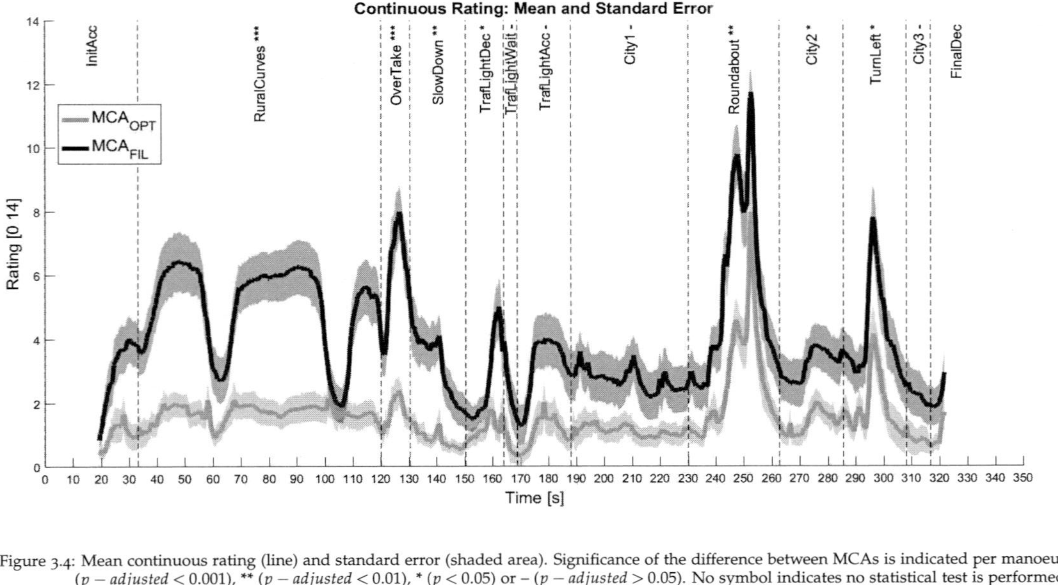

Figure 3.4: Mean continuous rating (line) and standard error (shaded area). Significance of the difference between MCAs is indicated per manoeuvre with *** ($p-adjusted < 0.001$), ** ($p-adjusted < 0.01$), * ($p < 0.05$) or – ($p-adjusted > 0.05$). No symbol indicates no statistical test is performed.

Table 3.3: Consistency test (Cronbach's Alpha) for participants' continuous rating. [* Two participants did not reach 0.7].

Participant [1]	Cronbach's Alpha	Participant	Cronbach's Alpha
0	0.894	12	0.888
1	0.828	13	0.84
2	0.883	14	0.683 *
3	0.868	15	0.875
8	0.822	16	0.671 *
9	0.897	17	0.886
10	0.884	18	0.727
11	0.843	19	0.854

Table 3.4: Test statistics for the differences in mean continuous rating per manoeuvre. *** p-adjusted<0.001, ** p-adjusted<0.01, * p-adjusted<0.05 and otherwise p-adjusted>0.05

Manoeuvre	Test statistic (means MCA_{FIL}/MCA_{OPT})
RuralCurves ***	t = 4.3771 (5.1/1.7)
OverTake ***	z = 3.9762 (6.1/1.7)
SlowDown **	t = 3.4159 (3.2/0.9)
TrafLightDec *	t = 2.8780 (2.7/1.2)
TrafLightWait	z = 1.7721
TrafLightAcc	z = 2.0919
City1	z = 1.5270
Roundabout **	t = 3.3070 (5.5/2.6)
City2 *	t = 2.5093 (3.2/1.4)
TurnLeft *	t = 2.8131 (4.1/1.9)
City3	z = 0.9526

Table 3.5: Correlation coefficients between continuous rating and absolute motion errors

Set	Longitudinal				Lateral				Combination			
MCA	MCA_{OPT}		MCA_{FIL}		MCA_{OPT}		MCA_{FIL}		MCA_{OPT}		MCA_{FIL}	
Correlation/Error	r	ϵ_{max}	r	ϵ_{max}	r	ϵ_{max}	r	ϵ_{max}	r	ϵ_{max}	r	ϵ_{max}
Longitudinal Acc	0.84	0.09	**0.86**	**2.22**	0.60	0.10	0.43	0.74	0.70	0.82	0.45	1.92
Lateral Acc	0.86	0.07	0.48	0.29	0.62	0.23	**0.92**	**1.41**	0.92	1.62	**0.90**	**3.61**
Vertical Acc	0.87	0.13	0.72	0.02	0.90	0.08	0.79	0.03	0.86	0.26	0.84	0.16
Roll Rate	0.68	0.25	0.23	0.53	0.61	1.98	0.38	1.58	**0.93**	**6.27**	0.70	1.53
Pitch Rate	0.78	1.95	0.72	0.70	0.68	1.05	0.45	0.31	0.85	4.27	0.57	0.79
Yaw Rate	0.85	0.16	0.43	1.12	0.69	3.20	0.81	5.06	0.84	9.42	0.88	26.10
Roll Angle	0.65	0.96	0.26	0.86	**0.91**	**8.39**	0.88	3.23	0.80	14.16	0.85	4.88
Pitch Angle	**0.92**	**7.55**	0.82	2.60	0.50	5.43	0.42	1.36	0.60	6.47	0.44	3.09
Yaw Angle	0.81	17.79	0.71	13.33	0.87	18.50	0.72	43.00	0.46	58.83	0.55	77.90

3.4.3. COMPARISON BETWEEN RATING AND CUEING ERRORS

To determine which motion cueing errors were responsible for the measured perceived mismatch the data is split in different Manoeuvre Sets. For these sets only the manoeuvres that were rated significantly different between the two MCAs are used.

- Set Longitudinal (X): 'SlowDown', 'TrafLightDec'
- Set Lateral (Y): 'RuralCurves', 'OverTake', 'City2'
- Set Combination (XY): 'Roundabout', 'TurnLeft'

The mean of the continuous ratings over all participants was compared to the errors in 9 motion channels: linear accelerations in x, y and z direction, rotational rate and angle in roll, pitch and yaw. In Table 3.5 the correlations (Pearson correlation coefficient r) between MCA continuous rating and the absolute error in different motion channels is shown per Manoeuvre Set. Correlations also occur between some of the motion channels themselves, such as yaw and lateral acceleration, which can make it difficult to determine exactly which error is rated. Nonetheless the correlation together with the maximum size of the absolute error (ϵ_{max}) can give a good indication of whether a certain rating is likely to be caused by the error in a specific motion channel. The highlighted values in Table 3.5 show the highest correlation coefficients and corresponding maximum error for each column of the table. From Table 3.5 one can see that the mean continuous rating correlates with errors in different motion channels for each MCA. For the longitudinal manoeuvres the MCA_{OPT} rating correlates best with the pitch angle error, while the MCA_{FIL} rating correlates best with the longitudinal acceleration error. For the lateral manoeuvres a similar observation can be made; here the MCA_{OPT} rating correlates best with the roll angle error, while the MCA_{FIL} rating correlates best with the lateral acceleration error. It is also notable that the maximum pitch angle error is relatively high for the MCA_{OPT}, but no clear correlation between this error and the rating is found. For the manoeuvres with both strong longitudinal and lateral acceleration errors the MCA_{OPT} correlates best with the roll rate error, while the MCA_{FIL} correlates best with the lateral acceleration error. Additionally a strong correlation between lateral acceleration error and the rating for the MCA_{OPT} is also found.
Overall the rating for the MCA_{FIL} correlates best with the linear acceleration error, while the rating for the MCA_{OPT} correlates best with the rotational error.

3.4.4. MOTION CUEING MECHANISMS

The MCAs utilize very different approaches to deal with the limited motion space of the simulator. While these mechanisms are explicitly programmed for the MCA_{FIL}, the mechanisms used by MCA_{OPT} are only revealed when analysing the resulting optimized simulator motions. In the following paragraphs these different mechanisms are described.

Global scaling of the vehicle motions is an often used mechanism to guarantee that the simulator motions stay within the simulator motion space, while the visual and vestibular motions remain coherent. The MCA_{FIL} relies to a large extent on a global scaling of the vehicle motion, resulting in significantly reduced motion strengths. Because the MCA_{OPT} uses an optimization at each time step, motion scaling is only applied at those time steps where this is required (i.e., local scaling), resulting in virtually no global scaling.

Washout of the simulator motions is an often used mechanism to keep the simulator within its motion limits. This mechanism returns the simulator to its neutral position, such that simulator excursions in all simulator degrees of freedom are still possible. This mechanism is used by the MCA_{FIL} via high pass filtering of the accelerations that are produced via the simulator's linear rail and hexapod translations. While the MCA_{OPT} has a similar mechanism implemented, currently this was not used (as Q=0 in Eq. 1).

Tilt-coordination is an often used mechanism to simulate sustained linear accelerations while keeping the simulator within the simulator workspace. The simulator is slowly tilted such that a component of the gravitational acceleration can be used to simulate the desired longitudinal or lateral acceleration. This mechanism is explicitly implemented in the MCA_{FIL} where filters are used to obtain the low frequent vehicle accelerations that can be simulated with tilt-coordination. Even though this mechanism is not explicitly implemented in the MCA_{OPT}, the optimization can result in similar behaviour, trading tilt rate errors for improved performance in linear acceleration simulation. The use of tilt-coordination by both MCAs is clearly visible in Figure 3.7 where the green line indicates the acceleration in the driver reference frame that is produced via tilt-coordination.

Prepositioning is a mechanism used by the MCA_{OPT} to extend the motion space of the simulator. Knowing future motions, the MCA_{OPT} positions the simulator in such a way that the available excursion for this future motion is maximized. Instead of moving the simulator to the neutral position as done with the washout mechanism, prepositioning results in moving the simulator to extreme positions of the simulator. For example, several seconds before the manoeuvre 'OverTake', the hexapod is slowly moved to a large lateral offset to the right (1.1m) (MCA_{OPT} at t=5.5[s], Figure 3.5). This prepositioning doubles the available leftward excursion needed for the linear accelerations of the future motion. A similar prepositioning strategy can be identified for the manoeuvre 'Roundabout'. Here, the available clockwise yaw excursion of the hexapod is maximized by very slowly prepositioning the hexapod to a counter-clockwise yaw angle of 21 degrees (MCA_{OPT} at t=120[s], Figure 3.6 top) before entering the roundabout. In this case the mechanism starts about 110 seconds before the start of the roundabout

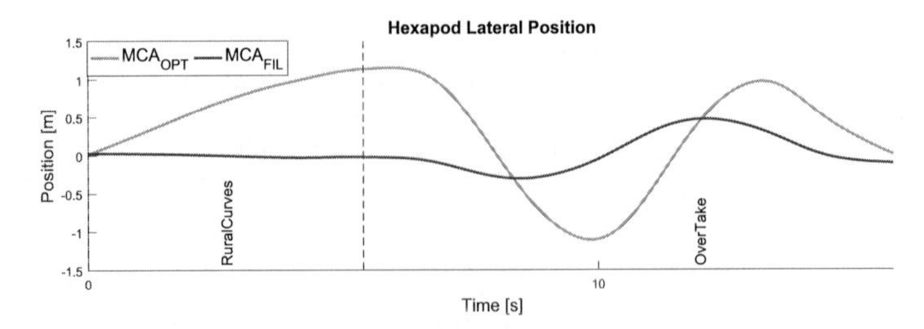

Figure 3.5: Hexapod lateral position for the end of the manoeuvre 'RuralCurves' and the manoeuvre 'OverTake' for both MCAs.

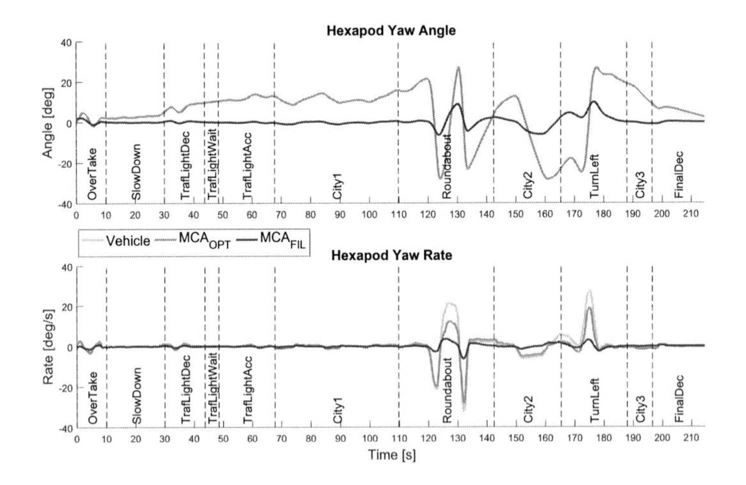

Figure 3.6: Yaw angle (top) and yaw rate (bottom) during several manoeuvres for both MCAs.

manoeuvre. For the manoeuvre 'LeftTurn', the available yaw excursion is maximized about 10 seconds before the turn starts (MCA_{OPT} at t=161[s], Figure 3.6 top). Here the algorithm traded an error in yaw rate just before the turn (MCA_{OPT} between t=161[s] and t=171[s], Figure 3.6 bottom) for additional yaw excursion possibilities while driving through the turn (between t=171[s] and t=179[s]).

Velocity buffering , which is only utilized by the MCA_{OPT}, is the final mechanism that will be discussed here. In velocity buffering the simulator's velocity is utilized to maximize the simulator's future acceleration capabilities. It can be interpreted as the velocity equivalent of the prepositioning mechanism. By giving the simulator a velocity in one direction, the duration for which the simulator can then accelerate in the opposite direction is increased. In the experiment, velocity buffering was most clearly observed during manoeuvres 'InitAcc' and 'TrafLightWait'. As shown in the top plot of Figure 3.7, the MCA_{FIL} did not result in simulator motion during the ve-

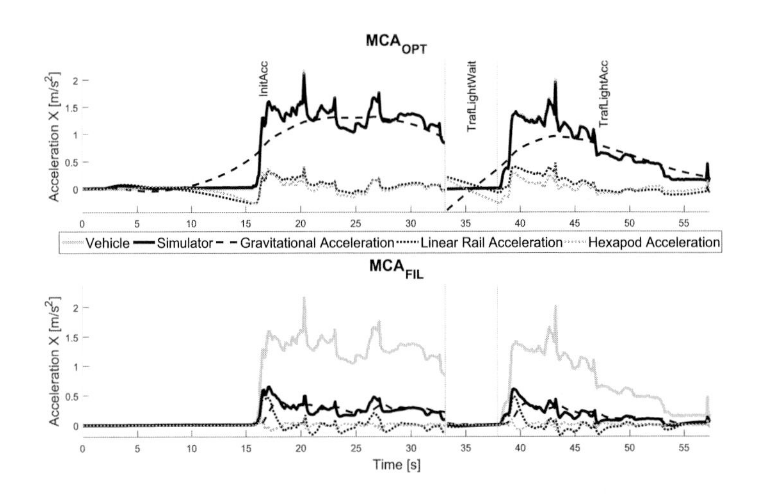

Figure 3.7: Contributions of different acceleration sources to the longitudinal acceleration in the driver frame of reference for manoeuvres 'InitAcc', 'TrafLightWait' and 'TrafLightAcc' using the MCA_{OPT} (top) and the MCA_{FIL} (bottom).

hicle's standstill (between $t = 0\,[s]$ and $t = 15.5\,[s]$ and between $t = 33\,[s]$ and $t = 38\,[s]$), but the MCA_{OPT} generated a backwards simulator motion via the linear rail (blue line) and hexapod translation (red line) during the vehicle's standstill. At the same time the simulator pitch angle was slowly increased (green line) to counteract the generated backward motion, resulting in a linear acceleration (black line) of zero in the driver frame of reference. The pitch rate was constant at $0.9\,[deg/s]$ for the manoeuvre 'TrafLightWait' and increased from 0 to 1.5 deg/s for the manoeuvre 'InitAcc'. Upon accelerating (at $t = 15.5\,[s]$ and $t = 38\,[s]$), the initial forward vehicle acceleration was simulated by first slowing down the backwards motion (green and red lines) before the simulator obtained forward velocity, effectively extending the duration at which the forward acceleration could be sustained. This mechanism was mainly used to improve the high frequent initial acceleration cue provided by the hexapod translation (during both manoeuvres) and the linear rail (mainly during manoeuvre 'InitAcc'). Effective velocity buffering requires accurate knowledge on future accelerations in order not to exceed actuator position limits.

3.4.5. LINEAR ACCELERATION SIMULATION IN DAIMLER SIMULATOR

There are multiple ways in which the Daimler simulator, consisting of a linear rail and a hexapod, can be used to generate linear acceleration in the driver frame of reference. Investigating the sources of the linear accelerations resulting from either MCA gives a better insight in how the different cueing mechanisms and the capabilities of the Daimler simulator are exploited by these MCAs. Table 3.6 shows the contributions of the different sources of acceleration as a percentage of the required vehicle acceleration (first value) and as a percentage of the generated total simulator acceleration (second value) in the driver frame of reference. As the MCA_{OPT} results in very small accelera-

Table 3.6: Contributions of different acceleration sources to the longitudinal and lateral acceleration in the driver frame of reference.

Acceleration	Longitudinal (X)		Lateral (Y)	
MCA	MCA_{OPT}	MCA_{FIL}	MCA_{OPT}	MCA_{FIL}
Gravitational acceleration (GA)	99.5%/100%	19.3%/80.3%	82.0%/87.6%	24.2%/90.7%
Linear rail acceleration (LRA	41.3%/41.5%	12.4%/51.4%	7.8%/8.3%	0.2%/0.8%
Hexapod X acceleration (HXA)	17.1%/17.2%	2.4%/9.9%	2.5%/2.7%	0%/0.2%
Hexapod Y acceleration (HYA)	4.3%/4.3%	0.2%/0.9%	17.6%/18.8%	4.6%/17.4%
Hexapod Z acceleration (HZA)	0.2%/0.2%	0.0%/0.0%	0.1%/0.1%	0.0%0.0%
Hexapod roll rate (HRR)	0.8%/0.8%	0.1%/0.3%	2.5%/2.7%	1.1%/4.1%
Hexapod pitch Rate (HPR)	2.4%/2.4%	0.7%/2.8%	0.5%/0.5%	0.0%/0.0%

tion errors (i.e. vehicle and simulator accelerations in the driver frame of reference are almost equal), the two percentages are very similar for this MCA. The gravitational acceleration contributes to both linear and longitudinal acceleration in the driver frame of reference via the tilt-coordination mechanism. When the hexapod yaw angle is zero, an acceleration of the hexapod over the linear rail simulates longitudinal acceleration in the driver frame of reference. When the hexapod yaw angle differs from zero, this source produces lateral acceleration in the driver frame of reference. The hexapod translation capabilities can also be used to simulate short duration longitudinal and lateral linear accelerations in the driver frame of reference. In combination with a hexapod tilt angle, translation accelerations in one direction can contribute to the linear acceleration in a different direction in the driver frame of reference. Finally, because the hexapod coordinate system is positioned $1.6\,[m]$ below the driver's eye point, also the roll and pitch rate of the hexapod cause small linear accelerations in the driver frame of reference.
As can be derived from Table 3.6, the sum of the contributions of all these sources is not necessarily 100%. This can be explained by the occurrence of counter acting accelerations produced by different sources resulting in a simulator acceleration of zero in the driver frame of reference. As shown in Figure 3.7 (top) this occurs, for example, during prepositioning (from $t = 0\,[s]$ to $t = 15.5\,[s]$) and velocity buffering (from $t = 33\,[s]$ to $t = 38\,[s]$) where the linear rail (blue line) and hexapod (red line) accelerations counter act the gravitational acceleration (green line). The total acceleration produced by the individual sources is thus higher than the resulting acceleration in the driver frame of reference.
Both the MCA_{FIL} as well as the MCA_{OPT} use the gravitational acceleration as their main source for simulating both longitudinal and lateral acceleration in the driver frame of reference. However, the angular rates produced by the MCA_{OPT} are much larger than those produced by the MCA_{FIL} (means of 0.47 and $0.39\,[deg/s]$ versus 0.14 and $0.09\,[deg/s]$ in roll and pitch respectively). Table 3.6 shows that this source is responsible for 100% of the total longitudinal simulator acceleration in the drivers reference frame when using the MCA_{OPT}. To keep the corresponding tilt rate low, the tilt angle is slowly increased before the motion starts as can be seen from the green line between for example $t = 30\,[s]$ and $t = 33\,[s]$ in the top plot of Figure 3.8. At the same time the resulting false rotational cue is minimized by the hexapod (red line) and linear rail (blue line) accelerations in opposite direction. The MCA_{FIL} also counter acts the excess gravitational acceleration from tilt-coordination with linear rail accelerations, but does so only at the end of the manoeuvre as can be seen around $t = 15\,[s]$ and $t = 22\,[s]$ in the bottom plot of Figure 3.8. The MCA_{FIL} is designed to use tilt-coordination for

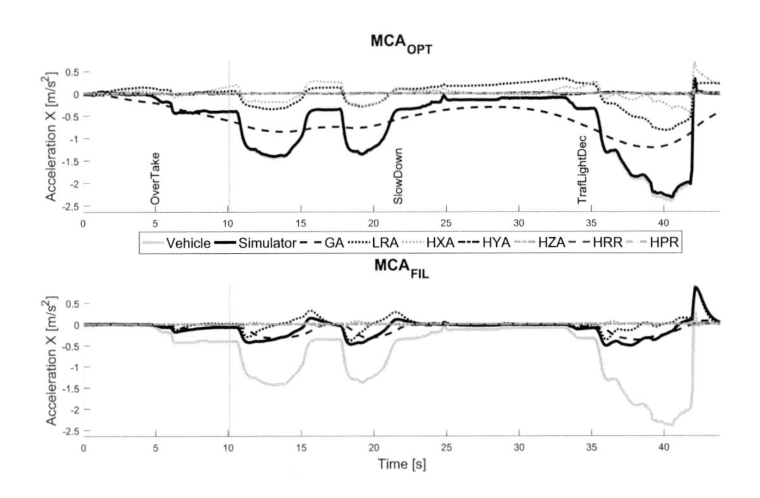

Figure 3.8: Contributions of different acceleration sources to the longitudinal acceleration in the driver frame of reference for the manoeuvres 'OverTake', 'SlowDown' and 'TrafLightDec' using the MCA_{OPT} (top) and the MCA_{FIL} (bottom).

sustained accelerations, while using the linear rail and hexapod translations for high frequent motions. This is clearly visible in Figure 3.8 (bottom) during the manoeuvre 'SlowDown' where the linear rail (blue line) is used for the initial and final high frequent part of the motion while the low frequent part of the motion is produced by the gravitational acceleration (green line).In Figure 3.9 (bottom) instead of the linear rail, the hexapod lateral acceleration (yellow line) is used for the very low amplitude high frequent part of the motion throughout the curves. Even though the MCA_{OPT} does not explicitly implement such a division in the frequency domain, the optimization algorithm comes up with a very similar solution. This effect is especially well visible in Figure 3.9 (top), where all high frequent variations of the lateral accelerations are generated via hexapod lateral acceleration (yellow line). The benefit of not implementing a hard division for high and low frequencies explicitly is that the linear rail can also be used for low frequent motions as seen clearly by the blue line in Figure 3.8 (top) during the manoeuvre 'TrafLightDec'. Table 3.6 also shows an interesting difference between the usage of the linear rail accelerations as a source for both lateral and longitudinal acceleration in the driver reference frame. In the currently used Daimler simulator configuration, the linear rail can be used for longitudinal acceleration when the hexapod yaw angle is equal to zero. Table 3.6 shows that both MCAs produce around half of the longitudinal simulator motions in the driver reference frame using this source. Due to the use of linear rail accelerations for low frequent motions as well as the use of the prepositioning and velocity buffering mechanisms, the linear rail accelerations account for a much larger part of the total vehicle acceleration when using the MCA_{OPT} than when using the MCA_{FIL}.
When applying a yaw angle to the hexapod, the linear rail acceleration can also be used to simulate lateral vehicle acceleration which, as Table 3.6 indicates, is only used by the

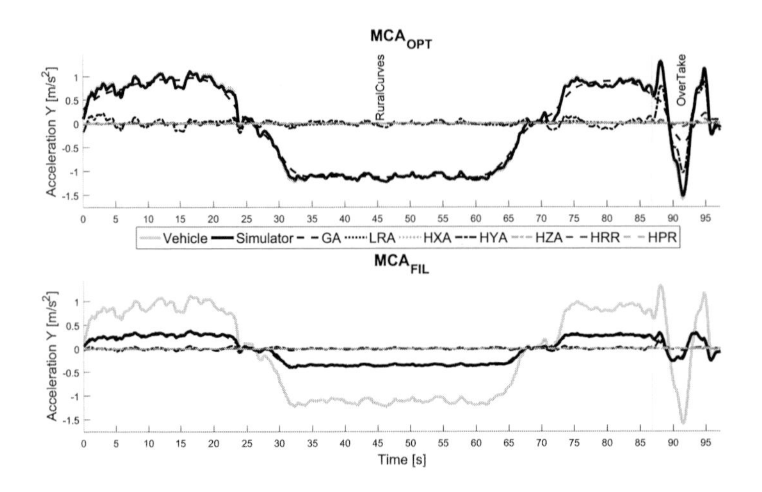

Figure 3.9: Contributions of different acceleration mechanisms to the lateral acceleration in the driver's coordinate system for the manoeuvres 'RuralCurves' and 'OverTake' using the MCA_{OPT} (top) and the MCA_{FIL} (bottom).

MCA_{OPT}, where 7.8% of the lateral vehicle acceleration is produced in this way. As shown in Figure 3.10, the MCA_{OPT} uses this mechanism effectively during the roundabout manoeuvre where the linear rail acceleration (blue line) contributes maximally at $t = 15\,[s]$ with $0.5\,\left[m/s^2\right]$ to the lateral acceleration in the drivers reference frame. Since the yaw angle of the hexapod is always smaller than $90\,[deg]$, using this mechanism for lateral acceleration creates an additional parasitic longitudinal acceleration. Figure 3.11 between $t = 14\,[s]$ and $t = 17\,[s]$ shows that the MCA_{OPT} uses pitch tilt-coordination (green line) to counteract the parasitic longitudinal acceleration (blue line).

3.5. DISCUSSION

The perceived motion mismatch for the MCA_{FIL} and MCA_{OPT} were rated using two measurement methods, both showing a significantly larger perceived motion mismatch when using the MCA_{FIL} than when using the MCA_{OPT}. This indicates that the MCA_{OPT} indeed has the potential to further improve motion cueing in simulators such as the DDS. The rating methods both showed very similar ratios between the MCA_{FIL} and the MCA_{OPT} (OR: 2.32, CR:2.67), implying that the mean of the continuous rating was a good indicator for the overall rating of the complete drive. The consistency of both the overall and the continuous rating between trials indicates that learning, habituation or fatigue effects did not impact these ratings significantly, and participants were able to provide consistent estimates during the whole experiment.

A more in detail analysis of the continuous rating shows that the difference between MCAs is mainly caused by the manoeuvres 'Overtake' and 'RuralCurves', where the largest significant difference between the mean ratings was found. It is especially surprising that during the 'RuralCurves' manoeuvre, where the MCA_{FIL} simply scales down the motion, such a large difference between MCAs is found. The often used

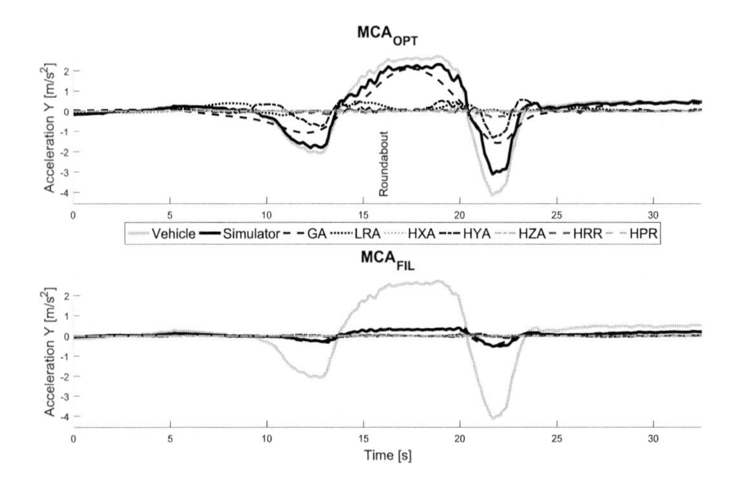

Figure 3.10: Contributions of different acceleration mechanisms to the lateral acceleration in the driver's coordinate system for the manoeuvre 'Roundabout' using the MCA_{OPT} (top) and the MCA_{FIL} (bottom).

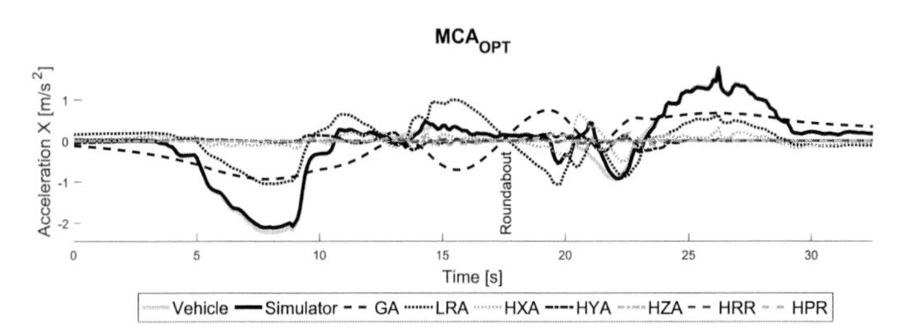

Figure 3.11: Contributions of different acceleration mechanisms to the longitudinal acceleration in the driver's coordinate system for the manoeuvre 'Roundabout' using the MCA_{OPT}.

mechanism of scaling can thus have a larger impact on the perceived mismatch in motion simulation than often assumed. The continuous rating also showed that for both MCAs the manoeuvre 'Roundabout' is rated as the worst of all manoeuvres presented during the drive. It is notable that even with full knowledge of the future motions the MCA_{OPT} still resulted in a strong perceived motion mismatch during this manoeuvre. However, also here an optimization-based approach has the potential to significantly improve motion cueing compared to a filter-based approach.
The extensive use of global scaling mechanism by the MCA_{FIL} and the lack of any global scaling in the MCA_{OPT} is the largest difference between the two MCAs. This resulted in larger linear acceleration errors for the MCA_{FIL} than for the MCA_{OPT}. Instead, the MCA_{OPT} made much more use of tilt-coordination mechanism, resulting in higher rotation errors. The effect of these choices is also visible in the continuous rating results. The MCA_{OPT} rating showed a higher correlation to rotation errors, while the MCA_{FIL} showed a higher correlation to the errors in linear accelerations. These correlations indicate that both error types were perceived by the participants. The mean continuous rating, however, was significantly higher for the MCA_{FIL} than for the MCA_{OPT}. This indicates that, a higher tilt rate that results in false rotational cues, would have been more acceptable than the degree of global scaling as done by the MCA_{FIL} to avoid these false cues.
One advantage of global scaling is that the resulting simulator motions are less strong, but still in coherence with the desired vehicle motions. Currently the MCA_{OPT} managed to minimize the acceleration errors so well, that low coherence did not seem to be an issue during most of the drive. If the acceleration errors increase, for example when using a limited prediction horizon, unpredictable scaling of the linear accelerations might become a problem. In this case global scaling could also become necessary for the MCA_{OPT}. The correlation analysis also indicates that the participants rated rotational angles instead of rotational rates during the lateral and longitudinal manoeuvre sets. This finding indicates that not only the tilt rate, but also a tilt angle can be perceived as a false cue and should be taken into account when optimizing a motion cueing algorithm.
During the 'TrafLightWait' manoeuvre there were no vehicle motions. The MCA_{OPT}, however, uses this time for velocity buffering and prepositioning which results in significant simulator motion. The corresponding continuous rating of the MCA_{OPT} for this manoeuvre does not show a significant increase in perceived mismatch, indicating that this motion was not perceived or did not bother the participants. This could be explained by the tilt rate of $0.9\left[deg/s\right]$, which is often found to be below the human threshold for rotational rate, and a resulting acceleration of zero. In practice, however, simulator motion can often be felt through the generation of parasitic rumble-like motions. In the Daimler simulator special attention was given in the design for this purpose, resulting in very smooth simulator motions. Using simulator motions for prepositioning or velocity buffering during standstill is therefore only recommended for high quality motion simulators. For both these mechanisms knowledge of the future vehicle trajectory is necessary. Reducing the prediction horizon of an MPC-based algorithm will therefore also reduce the possibilities of using these mechanisms.
The trade-off made by the MCA_{OPT} to reduce the yaw rate before the 'LeftTurn' manoeuvre in order to increase the available yaw rate while driving through the turn could be responsible for the slight increase in continuous rating at this point. During the 'LeftTurn' manoeuvre continuous rating also shows a clear perceived motion mismatch. Further research could determine if such trade-offs are beneficial and the

MPC cost function should be adjusted correspondingly.
For the manoeuvres 'City1' and 'City3' no significant difference between the MCAs was found. The maximum vehicle acceleration during these manoeuvres was $0.36\left[m/s^2\right]$ in lateral direction, compared to $1.45\left[m/s^2\right]$ during manoeuvre 'City2', where significant differences between MCAs were found. The high scaling factor of the MCA_{FIL} thus does not seem to reduce the perceived motion mismatch significantly for accelerations below $0.36\left[m/s^2\right]$. It is likely that the resulting absolute errors were close to the human perceptual threshold for lateral physical acceleration and were thus not perceived as incoherent with the visual motions.

3.6. CONCLUSIONS

The results of the experiment lead to the following conclusions:

- The results show that participants were able to rate the perceived mismatch consistently over the various repetitions. This holds for both the overall rating (Figure 3.3 right plot) as for the continuous rating (Table 3.3).
- The rating results (Figure 3.3 right plot and Figure 3.4) show that optimization-based cueing algorithms such as MCA_{OPT} indeed have the potential to improve motion cueing compared to filter-based approaches such as MCA_{FIL}.
- The continuous rating (Figure 3.4) shows that avoiding global scaling of the vehicle motions, as can be done with optimization-based cueing algorithms, has a large impact on the perceived motion mismatch especially during rural road simulations. The rating also showed that simulating a roundabout manoeuvre is difficult even for optimization-based approaches such as MCA_{OPT}. However, a significant improvement can still be made compared to filter-based-approaches such as MCA_{FIL}.
- The comparison of the continuous rating with the motion errors (Table 3.5) indicates that a certain amount of detectable rotation errors could be preferred over large scaling errors.
- Several mechanisms were identified that contributed to the differences observed between the two algorithms: global scaling, washout, tilt-coordination, prepositioning and velocity buffering. Additionally an overview was given (Table 3.6) of how the two MCAs used the different sources of linear acceleration in the Daimler Driving Simulator.

It should be noted that the two MCAs are very different algorithms, each with their own characteristics. This study is therefore not to be considered as a competitive comparison between MCA alternatives, but rather as an attempt to gain insight in the potential that an optimization-based approach has to offer. The results show that there exists a potential to further improve the quality of the motion simulation with optimization-based methods, deserving of further research.
Regarding the rating method, the results show that the rating method provides reliable and repeatable results within and between participants, which further confirms the reliability and utility of the method. Ongoing research investigates the dynamics and limitations of the rating behavior.

In future experiments it could be investigated how the quality of the MCA_{OPT} degrades if the prediction horizon is decreased (i.e., no longer using trajectory-based optimization) or if the prediction is imperfect (i.e., no longer using the pre-recorded trajectory but a predicted trajectory). Also, it could be studied how the objective function can be adapted to further improve the quality of the MCA_{OPT}. Finally, it would be interesting to investigate whether the tuning of the MCA_{FIL} can be further improved based on the more detailed analysis of the results presented in this paper.

II

Modelling Perceived Motion Incongruence

4
Cueing Error Detection Algorithm

This chapter presents an algorithm for automatically detecting different types of cueing errors within one motion channel. The algorithm parameters are fit on a first dataset and validated using a second dataset obtained with experiments described Chapter 2 and reference [135], respectively. The results show that the algorithm can differentiate between scaled, missing and false cues consistently. Literature shows that these cueing error types generally have a different effect on the Perceived Motion Incongruence (PMI). The algorithm described here is used to include this knowledge in Perceived Motion Incongruence (PMI) prediction models presented in Chapters 5 and 6, by allowing for different weights to be assigned to each cueing error type.

4.1. Introduction

Motion Cueing Algorithms (MCAs) for vehicle simulation are used to map the vehicle motions onto the simulator workspace. Optimizing such an MCA requires minimizing the perceived differences between motion cues perceived in a real vehicle and those perceived in the simulator. Differences between simulator and vehicle motion, i.e. cueing errors, are inherent to motion simulation due to the limited motion space of the simulator. Tuning experts therefore attempt to tune an MCA by minimizing only those cueing errors that are considered to be the most detrimental to cueing quality. Mathematically predicting cueing quality should therefore also take into account that not all cueing error types are equally detrimental. Currently, however, no algorithm exists that can mathematically1 distinguish different cueing error types based on recorded simulator and vehicle motions alone. In this chapter a Cueing Error Detection Algorithm (CEDA), that uses only simulator and vehicle motion signals as its inputs, is presented.
In [51] guidelines are provided for how to use expert knowledge to manually tune a classical washout filter MCA. This expert knowledge includes, for example, the fact that errors due to motion limiting are much more detrimental to the cueing quality than cueing errors that result from pure downscaling. In [51] also an extensive overview is given on the varying influence of ten different types of cueing errors on the cueing quality. In [136] the different influence of missing, scaled and false cues on the perceived cueing quality was confirmed. Here a model, based solely on the euclidean distance between simulator and vehicle motions in each motion channel, was fitted to measured subjective ratings of cueing quality. The differences between the model and subjective ratings could largely be explained by an increased weight on false and missing cues compared to overall scaling errors. It was proposed that adding a CEDA to this model, such that different cueing error types could be weighted separately, would improve the model fit.
Optimization-based MCAs, such as described in [137], attempt to optimize the cueing quality by minimizing the euclidean distance between simulator and vehicle motions for each motion channel. Much of the expert knowledge of different cueing error types is thus discarded in such optimizations. Using a cueing quality prediction model that includes separately weighted cueing error types, could add this expert knowledge to the optimization and thus help to improve such MCAs.
In literature, cueing errors are often subdivided in scaled, missing and false cues. In [51] ten different cueing errors are categorized on the basis of their source and divided in two main groups: false cues and scaled/missing cues. A false cue is described as motion that is presented in the simulator that either does not occur in the real vehicle, or that is of opposite direction as compared to the vehicle motion, or a simulator motion containing a high frequency disturbance of a sustained motion cue. Scaled or missing cues, described as relatively less detrimental to cueing quality, indicate a simulator motion that is reduced or absent as compared to the vehicle motion.
In [138] a subdivision of four cueing error types in car simulations is made: false cues, missing cues, phase errors and scaling errors. False cues are described as motion cues in the wrong direction as compared to the vehicle motion or simulator motion cues without predefined stimuli. Missing cues are described as missing stimuli in the simulator motion that were present in the vehicle motion and phase errors as noticeably delayed stimuli. Finally, scaling errors are described as the noticeable amplitude difference between simulator and vehicle accelerations.
In [139] and [138] cueing errors are divided in only two subgroups: scale and shape errors. Where the scale errors are described as a pure scaling between simulator and

vehicle motion cues and shape errors include all other cueing errors, where the former are expected to be less detrimental to cueing quality than the latter. The importance of similarity in simulator and vehicle motion signal shapes was also noted [140] and [141], where low correlation between simulator and vehicle motion is used to indicate poor cueing quality.
Another interesting signal characteristic that tends to influence the effect that cueing errors have on the cueing quality is the frequency content of the motion. The subdivision of the main error groups described in [51] is partly based on the frequency of the error signal, where high frequency errors caused by reaching simulator motion limits, are often more detrimental to the cueing quality than low frequency cueing errors. In [142] cued motions themselves are subdivided in high-frequency onset cues, transient cues, and low-frequency sustained cues. In both [142] and [51] it is mentioned that the effect that cueing errors have on the cueing quality depends on the type of the corresponding cued motions. From perceptual research in motion cueing we also know that the motion frequency is an important factor in human self-motion perception. For example, both our visual and vestibular systems can detect motions only at limited frequency ranges, which is known to result in varying motion perception thresholds depending on the frequency of the motion [143]. Furthermore, Valente Pais [101] showed that the range of amplitude differences between the visual and physical motion cues that still results in a coherent perception of the motion, the so-called coherence zone, depends on the frequency of the motion. Even though the exact relation between (reduced) cueing quality and the cueing error frequency is not yet fully known, it is therefore useful to already account for the motion frequency characteristics in the design of a CEDA.
Here we propose an automatic wavelet-based CEDA that can distinguish between the three most common cueing errors, i.e., scaled, missing and false cues, given simulator and vehicle motion signals. Wavelet-based methods are used to determine if the vehicle and simulator motion signals locally differ in shape and thus can be classified as a false or missing cue or if their shape is the same and a purely scaled motion cue is observed. Although currently not implemented, using wavelet-based methods also makes it easier to extend the algorithm in the future with detection of cueing errors based on the frequency characteristic of the motion cue, by performing the analysis on only a subset of the motion frequencies.
In Section 4.2 the CEDA is further explained. This section contains the algorithm goal and requirements, mathematical definitions of scaled, missing and false cues, a short overview of the algorithm steps, an explanation of the used shape and gain measures, and finally an overview of the resulting tunable algorithm parameters. In Section 4.3 two datasets, used to tune and validate the algorithm and each containing different cueing error types, are presented. In Sections 4.3.3 and 4.4 the desired algorithm output and the algorithm results are presented, respectively. A discussion of these results is given in Section 4.5, followed by a conclusion in Section 4.6.

4.2. ALGORITHM

4.2.1. GOAL

The main reason for developing a CEDA is to automatically detect and localize different types of cueing errors that occur in a certain simulation segment. In Figure 4.1(a) a simplified example is given of typical vehicle and simulator motions and the corresponding cueing errors that can occur due to simulator motion washout. Figure 4.1(b) shows the euclidean distance between simulator and vehicle motions, i.e., the absolute

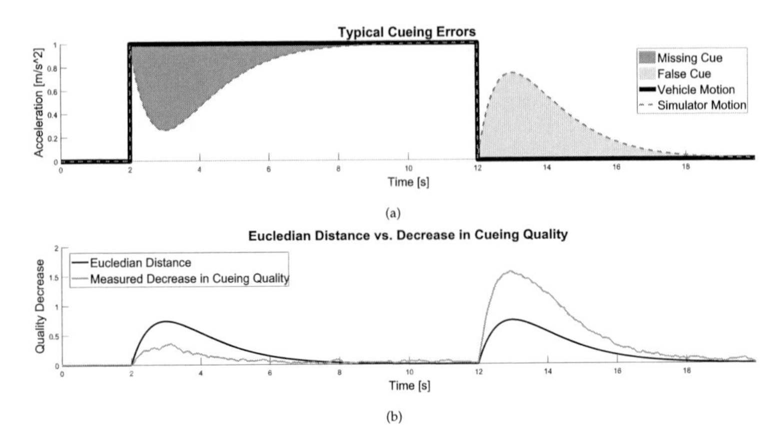

Figure 4.1: Simplified vehicle and simulator motions as the result of motion washout with corresponding cueing errors (a) and the fictional corresponding decrease in cueing quality and the actual euclidean distance between simulator and vehicle motions (b).

cueing error, as well as a fictional example of the decrease in perceived cueing quality that such cueing errors can cause, i.e., as mentioned in [139] false cues are often more detrimental to the cueing quality than missing cues. To model, and eventually predict, the decrease in cueing quality, not only the euclidean distance (which is equal in this case), but thus also the cueing error type information is needed. The goal of the CEDA is to provide this cueing error type information. In Figure 4.2(a) the desired output of the CEDA, given the simulator and vehicle motions, is shown. In Figure 4.2(b) a simple linear model of cueing quality decrease, using the additional cueing error type information in the CEDA output, is shown. Here the model weighs missing cues (MC) with 0.5, while a higher weight of 2 is attributed to false cues (FC). This model has a higher explanatory power than when only euclidean distance information is used.

4.2.2. Requirements

The main requirement of the algorithm is that it should use only the simulator and vehicle motions to generate three time signals that each contain only one cueing error type. This requirement distinguishes the algorithm presented here from the shape and scaling error calculations used in [139, 144]. Here the classical washout filter MCA parameter for global scaling was used to distinguish between these error types, defining any cueing error that is the result of the scaling parameter as scale errors and any additional difference between simulator and vehicle motions as shape errors. While this can be an effective way to distinguish between scale and shape errors when classical washout filter MCAs are used, it cannot be used in the absence of a global scaling parameter, such as is the case in model predictive control-based MCAs as described in [137].

Implementing a CEDA as part of a Perceived Motion Incongruence (PMI) model for optimization purposes imposes additional model requirements. While not all requirements are known at this stage, the algorithm should at least not be unnecessarily complex, such that fast computation can be achieved. For this reason, the algorithm de-

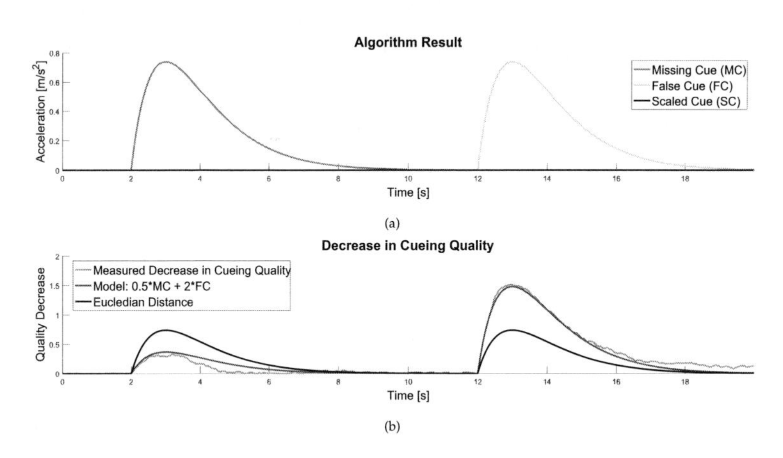

Figure 4.2: Desired algorithm output (a) and simple linear model using the CEDA output fitted to the fictional decrease in cueing quality (b).

scribed here will only be able to distinguish a limited number of different cueing errors that are known to have a different effect on the cueing quality. Additionally, as optimizing smooth functions is easier, the algorithm should be smooth, i.e., have continuous derivatives. As will be explained further in the next sections, the algorithm described here makes use of threshold comparisons and absolute value calculations. As these calculations generally involve non-smooth functions, here smooth approximations of these functions are used instead. For smooth threshold comparisons the following function is used:

$$F_{th}(X, th_x, b) = 0.5 + 0.5\tanh\left(\frac{X - th_x}{b}\right), \tag{4.1}$$

where X is the vector or matrix for which the threshold signal is calculated, th_x is the threshold for X and b is the steepness of the threshold function. Multiplying this threshold function with the signal X results in a signal where all values below threshold are set to zero.

The absolute value calculations used in the algorithm described here are instead approximated with the smooth function [145]:

$$|X|_s = \sqrt{X \cdot X^* + \eta} \tag{4.2}$$

where $\eta = 1e^{-6}$, which is also used in other functions of the algorithm to ensure smoothness, and $*$ indicates the conjugate. In the following sections the resulting algorithm is further explained.

NON BINARY SHAPE AND GAIN MEASURES

In the next sections error detection based on close to binary threshold functions will be introduced. This directly results in detected cueing errors that can instantly switch from one type to another, which might not always be an accurate description of human perception. More smooth transitions between cueing error types can be generated by adjusting the steepness parameter b of the threshold functions. In Figure 4.3 the effect of changing b on the threshold function is shown.

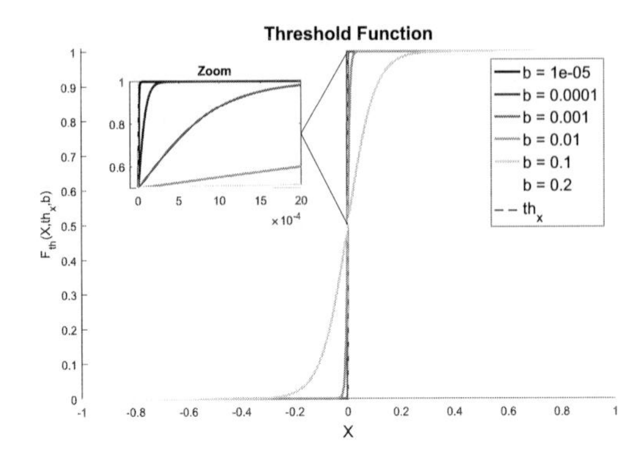

Figure 4.3: Threshold function for different values of the steepness parameter b.

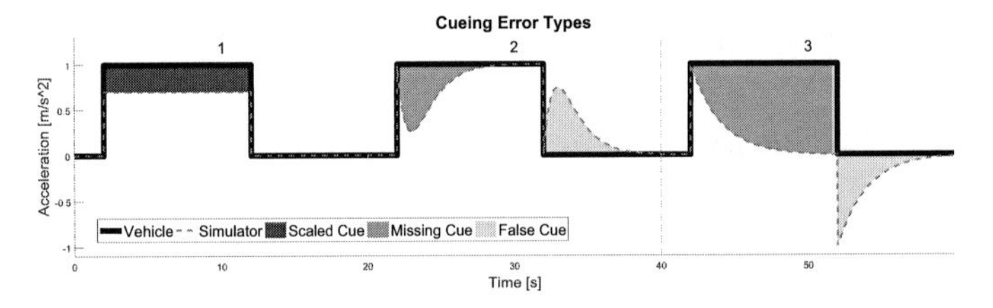

Figure 4.4: Examples of the three types of cueing errors, scaled, missing and false cues, considered in this study.

4.2.3. Mathematical Cueing Error Definitions

In literature many different types of cueing errors have been discussed [51],[139] and [138]. To comply with the aforementioned requirement for simplicity, the algorithm introduced here only focuses on a small selection of common cueing errors (scaled, false and missing cues), known to affect the cueing quality differently. While many written definitions of these cueing errors exist, it is unclear how these different cueing errors can be defined mathematically.

In this section scaled, missing and false cues are defined mathematically, based on the many definitions of these cues found in literature. In Figure 4.4 examples of these three cueing error types are shown. In green the scaled cue, resulting from downscaling of the simulator compared to the vehicle motion, is indicated. In blue and red examples of missing and false cues are shown, respectively. Here the simulator and vehicle motion signals differ in shape. During the missing cue the simulator motion is also smaller than the vehicle motion, while during the false cue the simulator motion is larger in magnitude than the vehicle motion.

One important cueing error characteristic thus seems to be the shape similarity between simulator and vehicle motion signals. In [141] the correlation between certain objective indicators of the differences vehicle and simulator motion signals and the subjective rating of the resulting simulation was analyzed. The results showed that the Pearson correlation between simulator and vehicle motion signals mainly correlates with the subjective rating of the MCA, i.e., the higher the Pearson correlation, the higher the

Table 4.1: Mathematical cueing error definitions

Cueing Error	Description	Mathematical definition
Scaling	similar shape	$SM \geq th_{SM} \cap GM \geq 1$
Missing	different shape and less motion in simulator than in vehicle	$SM < th_{SM} \cap 0 < GM < 1$
False	different shape and more motion in simulator than in vehicle, or simulator and vehicle motion of opposite sign	$(SM < th_{SM} \cap GM \geq 1) \cup GM < 0$

ratings. In [138] and [139] the importance of shape similarity is mentioned and a distinction between scaling and shape errors is made to analyze MCA performance. In [140] the cross-correlation between the simulator and vehicle motion signals is used as an additional parameter, next to the euclidean distance between simulator and vehicle motion, in the cost function of the optimization algorithm, such that this correlation can be directly optimized. In the CEDA described here, the shape similarity characteristic is therefore used to distinguish between the scaled and missing and false cues.
To further differentiate between missing and false cues, the algorithm described here uses the ratio between the amplitudes of simulator and vehicle motions. False cues encompass all cueing errors due to simulator motion in the wrong direction or due to simulator motion when no vehicle motion is expected. Missing cues, instead, are defined as shape errors due to simulator motion with a lower amplitude than the vehicle motion. A summary of the three different cueing error types and their mathematical description, based on measures of the gain GM and shape similarity SM between the simulator and vehicle motion signals, is defined in Table 4.1. As will be further explained in Section 4.2.5, the measure for shape similarity SM is chosen to be the time-varying mean semblance over a range of frequencies. A value of SM smaller than a certain threshold th_{SM} indicates a perceived difference in signal shape, while a value above this threshold indicates that both time signals have equal shape and any error that is detected must therefore be a scaling error, or, in the case of a negative gain between simulator and vehicle motions, a false cue. In Table 4.1, GM is a measure for the gain between simulator and vehicle motions, which is also further explained in the next subsection.
As it is still unknown when two signals are perceived as similar in shape and when not, the algorithm parameters are tuned based on clear examples of scaled, missing and false cues present in a reference dataset (Dataset 1), which is described in Section 4.3. The tuned algorithm is subsequently tested on known cueing errors in a second dataset described in the same section.

4.2.4. Algorithm Overview

The overview of the different steps made in the cueing error detection algorithm are shown in the scheme in Figure 4.5. The algorithm inputs are vectors containing the simulator (y_{sim}) and vehicle (y_{sim}) motion time signals for a specific motion channel and the output consists of three vectors each containing the time signal of one type of cueing error, i.e., scaled (ϵ_{sc}), missing (ϵ_{mc}) or false cue (ϵ_{fc}).

The upper path, Steps 1-3, shows the steps taken to calculate the shape measure SM. In Step 1 the time (t) and scale (s) dependent semblance S is calculated. From this the relevant semblance S_{rel} is calculated in Step 2, which, after averaging and smoothing in Step 3, leads to the shape measure. These three steps are explained further in Section 4.2.5. The bottom path, Step 4, shows the calculation of the gain measure GM, which is further explained in Section 4.2.5.
In Step 5, using the gain and shape measures, binary time signals for all cueing errors (sc_{bin}, mc_{bin}, fc_{bin}) are computed by applying a set of rules further explained in Section 4.2.5, to detect the different cueing error types. These binary signals are unity when a specific cueing error type occurs and zero otherwise. Multiplying these binary signals with the total cueing error, i.e., the absolute difference between vehicle and simulator motion ϵ, isolates the portions of this error attributable to the different error types (Step 6) and with that gives the final algorithm output.

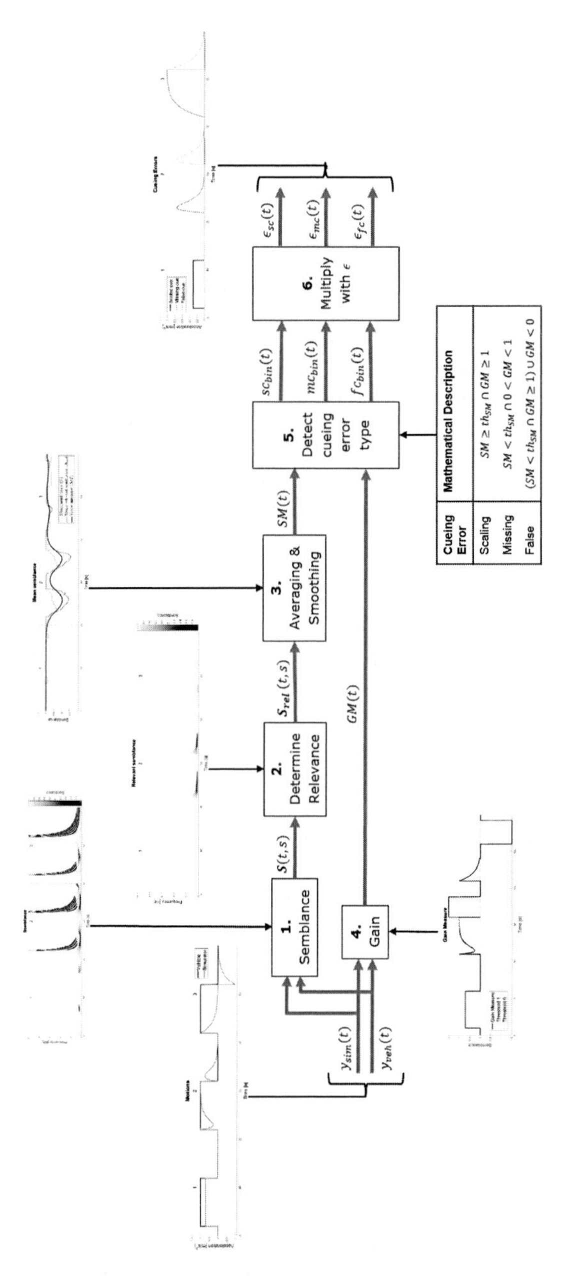

Figure 4.5: A scheme of the main steps in the algorithm.

Table 4.2: Shape and gain measures. Where SM indicates the shape measure, th_{SM} the shape measure threshold and GM the gain measure.

Variable	Symbol	Value of 1
Shape	SV	$SM > th_{SM}$
Gain	$GV0$	$GM \geq 0$
	$GV1$	$GM > 1$

4.2.5. Shape and Gain Measures

Table 4.1 shows that the shape measure needs to be compared to the threshold th_{SM} and the gain measure to zero and to unity. These comparisons result in three variables, hereby called the shape variable SV, the gains variable $GV0$, for the comparison of the gain measure with zero, and the gain variable $GV1$, for the comparison with unity. These variables are calculated at each simulation time step and can have values of zero or one, depending on the simulator and vehicle motion signals that are compared. Values between zero and one are also possible, but are not considered for now and will be further elaborated on in subsection 4.2.2. To comply with the smoothness requirement, all threshold comparisons indicated in this section are performed using the smooth threshold function F_{th} described in Equation 4.1, using a value for b that is close to zero (here $1e-5$).
In Table 4.2 the shape and gain variables are listed and the condition under which they have a value of unity is shown. Under all other conditions the value of the variable is set to zero. Following the comparisons presented in Table 4.1, if we can calculate the appropriate shape (SM) and gain (GM) measures, the cueing error types can be defined as the following smooth functions:

$$\epsilon_{sc} = \epsilon \cdot \left(GV0 \cdot (1 - SV) + 1 - GV0\right), \quad (4.3a)$$
$$\epsilon_{mc} = \epsilon \cdot (GV0 - GV1) \cdot (1 - SV), \quad (4.3b)$$
$$\epsilon_{fc} = \epsilon \cdot GV0 \cdot SV, \quad (4.3c)$$

where ϵ indicates the absolute difference between simulator and vehicle motion and ϵ_{sc}, ϵ_{mc} and ϵ_{fc} the cueing errors due to scaled, missing and false cues, respectively.

Shape Measure

The purpose of SM is to provide a value smaller than the chosen threshold th_{SM} when the shapes of the two signals differ significantly, and provide a value higher than this threshold when any difference between the two signals is a pure scaling error. In [146] the basis for such a measure is described. In [146] the shape difference between two time signals is quantified in both the time and frequency domains using this wavelet-based semblance. The current study uses a similar measure to determine whether simulator and vehicle motion signals have a similar shape. The measure used in this study only differs from [146] in how it deals with the noise sensitivity of the wavelet-based semblance, as will be explained in the next paragraph.

Semblance To calculate the semblance between the simulator y_{sim} and vehicle motion y_{veh}, first the continuous wavelet transform CWT with scale parameter s (equivalent to frequency) and time shift parameter τ of both motion time signals is calculated. This transform decomposes a signal into its time and scale (frequency) components and is calculated with:

$$CWT(t,s) = 2\int_{-\infty}^{\infty} y(\tau)\Psi^*\left(\frac{\tau - t}{s}\right) d\tau, \tag{4.4}$$

where y represents the analyzed time signal and Ψ^* indicates the complex conjugate of the selected wavelet function. The scaling factor of two before the integral in equation 4.4 is used to obtain equal wavelet transform and time signal amplitudes and should only be applied when using analytic wavelets [147], i.e., complex wavelets with zero magnitude in the negative frequencies. Due to the zero magnitude at negative frequencies, the wavelet transform of a signal using an analytic wavelet eliminates the magnitude of that signal at negative frequencies, resulting in a wavelet transform magnitude at half the amplitude of the signal amplitude. To obtain equal amplitudes, the wavelet transform using an analytic wavelet should thus be scaled with a factor of two. The choice of the wavelet function depends on the signal to be analyzed and the information one needs from it, i.e., magnitude only or both magnitude and phase. As the semblance requires the calculation of phase, a complex-valued wavelet is needed for this study. Additionally, the shape of the wavelet function should be chosen to correspond to the signal features that one wants to analyze. In this study, many different types of signal features, such as abrupt changes due to simulator hitting a limit, as well as slow changes as can be seen in many scaling errors, need to be analyzed. While one could argue that it is best to use different wavelets for each feature type, to reduce the complexity of the algorithm, here only one wavelet function best resembling most signal features, the complex Morlet wavelet, is used.
The complex Morlet wavelet, which is approximately analytic, is described by:

$$\Psi(t) = \frac{1}{\sqrt{(\pi f_b)}} e^{2\pi i f_c t} e^{-\frac{t^2}{f_b}}, \tag{4.5}$$

where f_b and f_c are the wavelet bandwidth parameter and center frequency, respectively. The set of scales for which the wavelet transform is calculated is determined by:

$$s_j = s_0 2^{j\delta j}, j = 0,1,...,J, \tag{4.6}$$

where s_0 indicates the smallest scale, which should be chosen to be larger than the sampling time of y, δj is the spacing between the scales and J the number of scales minus 1. The largest scale should be chosen as less the length of y.
To show the signal information that can be represented with the wavelet transform, in Figure 4.6 the magnitude and phase of the wavelet transform of a time signal composed of three sine waves with different amplitudes and frequencies of 0.5, 8 and 3 Hz, using the complex Morlet wavelet, is shown. Figure 4.6(a) shows the time signal and Figure 4.6(b) the magnitude of its wavelet transform. An increased magnitude is visible during the sine waves, with a peak magnitude at the specific frequency of each sine wave. The peak magnitude is equal to the amplitude of the time signal. Finally, Figure 4.6(c) shows the phase of the wavelet transform, where the phase data for wavelet transform magnitudes lower than 0.1 are hatched out for clarity. This plot shows that the phase during all sine waves at magnitudes higher than 0.1 is regular, i.e., no phase shifts occur during the sine waves.
Depending on the choice of the wavelet parameters f_b and f_c, the time or frequency

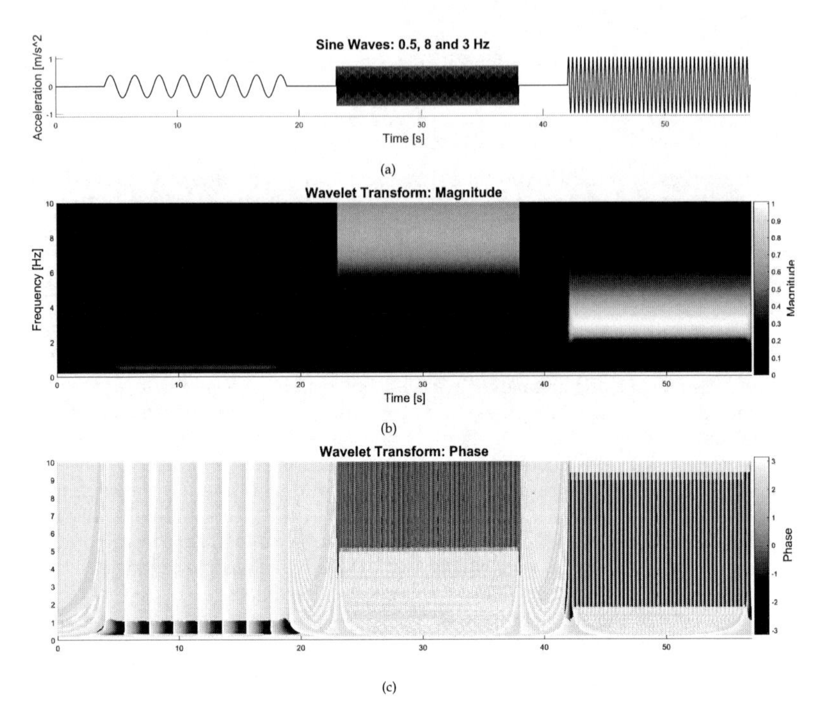

Figure 4.6: Wavelet transform magnitude (b) and phase (c) plots of sine waves with different frequencies and amplitudes (a) using the complex Morlet wavelet ($f_b = 0.8$, $f_c = 0.8$). The phase with a wavelet transform magnitude below 0.1 is hatched out for clarity.

resolution can be improved: increasing f_b or f_c improves the time or frequency localization, respectively. For Figure 4.6 parameters specifically tuned for the algorithm ($f_b = 0.8$ and $f_c = 0.8$) are used, resulting in a much better time than frequency resolution.

In Figure 4.7 the wavelet transform magnitude and phase for the cueing error examples from Figure 4.4 are shown. Figures 4.7(a) and 4.7(b) show the simplified vehicle and simulator motions during three turns, respectively. The wavelet transform magnitude plots (4.7(c) and 4.7(d)) show that most power is in the lower frequencies, except for the start and end of each turn, where also the high frequencies have power. The wavelet transform phase plots (4.7(e) and 4.7(f)) show that at lower frequencies the phase between the vehicle and simulator motion differs only during the second and third turn when missing and false cues occur. It is this difference in phase that is captured with the semblance measure.

The semblance [146] is based on the cross-wavelet transform $XWT_{veh,sim}$:

$$XWT_{veh,sim}(t,s) = CWT_{veh} \times CWT_{sim}^{*}, \tag{4.7}$$

where CWT_{veh} and CWT_{sim} indicates the wavelet transform of the vehicle motion y_{veh} and simulator motion y_{sim}, respectively, and * again indicates the complex conjugate.

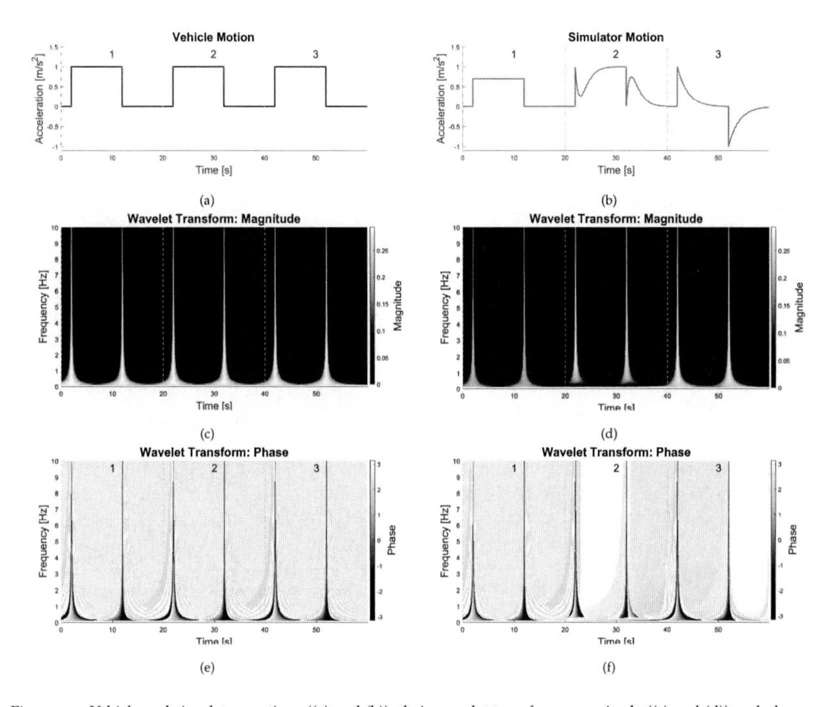

Figure 4.7: Vehicle and simulator motions ((a) and (b)), their wavelet transform magnitude ((c) and (d)) and phase ((e) and (f)) plots, using the complex Morlet wavelet ($f_b = 0.8$, $f_c = 0.8$), for typical cueing errors. The phase with a wavelet transform magnitude below 0.1 is hatched out for clarity.

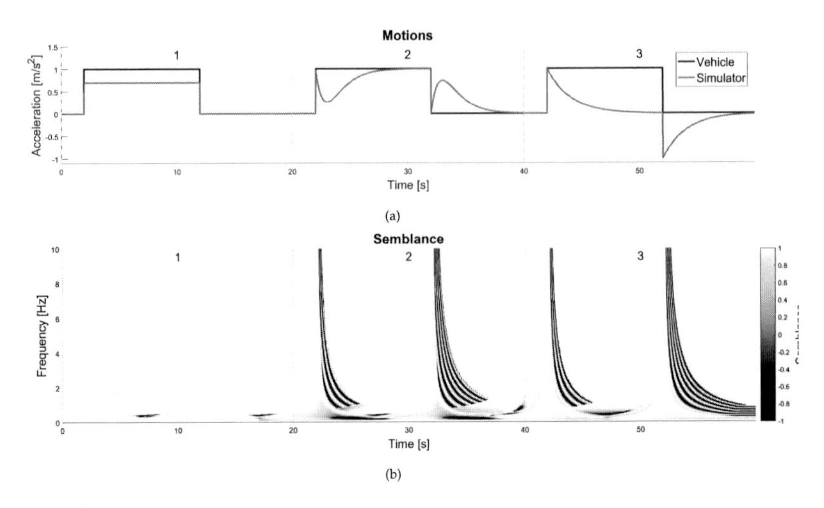

Figure 4.8: The semblance ((b)) between the vehicle and simulator motions ((a)) of typical cueing errors.

For the semblance, only the phase information contained in this cross-wavelet transform is important. The local phase γ between vehicle and simulator motion is calculated with:

$$\gamma_{veh,sim}(t,s) = \tan^{-1}\left(\frac{\Re\left(XWT_{veh,sim}\right)}{\Im\left(XWT_{veh,sim}\right)}\right) \tag{4.8}$$

Finally, the semblance as a function of time and scale $S(t,s)$, ranging between 1 and -1, is calculated as the cosine of this local phase.

$$S(t,s) = cos(\gamma) \tag{4.9}$$

In Figure 4.8 the semblance $S(t,s)$, resulting from the wavelet transforms of y_{veh} and y_{sim} shown in Figure 4.7, is presented. Figure 4.8 shows that the semblance between vehicle and simulator motions is reduced when false or missing cues occur (turns two and three), but stays close to unity during the scaling error (turn one).

Relevant semblance The semblance measure is very sensitive to noise, making it difficult to distinguish between semblance measures related to motion signals at a certain time and frequency with perceivable amplitude, i.e., relevant semblance, and semblance measures related to non-perceivable noise signals, i.e., non-relevant semblance. In [146] the sensitivity to noise is reduced by multiplying the semblance with the amplitude of the cross-wavelet transform, which makes the resulting measure, described in [146] as D, dependent on the motion amplitudes that are used. This measure D is unsuitable for the application described here, where the shape measure SM needs to be compared to a fixed threshold. Instead, this issue is solved by only taking into account the semblance of motion signals that are above the human absolute and discriminatory perception thresholds reported in literature [143, 148].
To determine whether motions are above the human perception thresholds, wavelet

transforms of the original vehicle (y_{veh}) and simulator (y_{sim}) motion signals, are compared to these perception thresholds at each time and frequency. These comparisons produce a logical matrix LM of similar size as the semblance S. This matrix consists only of values one and zero, indicating significant and insignificant motion at a particular time and scale, respectively:

$$LM = \begin{cases} 1, & \text{if } LM_{abs} = 1 \cap LM_{disc} = 1 \\ 0, & \text{otherwise,} \end{cases} \tag{4.10}$$

where LM_{abs} is the logical matrix related to the absolute motion perception threshold th_{abs}. This threshold refers to the minimal motion amplitude in a specific motion channel which can be detected by the human perceptual system. LM_{abs} is calculated with:

$$LM_{abs} = \begin{cases} 1, & \text{if } |CWT_{veh}|_s > th_{abs} \cup |CWT_{sim}|_s > th_{abs} \\ 0, & \text{otherwise} \end{cases} \tag{4.11}$$

The logical matrix LM_{disc} is related to the discriminatory motion perception threshold th_{disc}. This threshold refers to the minimal difference between two motion signals in the same motion channel which can be detected by the human perceptual system. LM_{disc} is calculated with:

$$LM_{disc} = \begin{cases} 1, & \text{if } |CWT_{veh} - CWT_{sim}|_s > th_{disc} \\ 0, & \text{otherwise} \end{cases} \tag{4.12}$$

Multiplying this logical matrix LM with the semblance $S(t,s)$ then gives the relevant semblance $S_{rel}(t,s)$ that is used to determine the difference between scaled and false or missing cues. In Figure 4.9 $S_{rel}(t,s)$ for the example cueing errors of Figure 4.4 is shown. Figure 4.9 now clearly shows that $S_{rel}(t,s)$ during the scaled cue (first turn) equals one, while during the missing and false cues (turns two and three) the semblance is smaller than one. At time steps where no relevant semblance is present in any frequency, $S_{rel}(t,s)$ is set to zero. Here $S_{rel}(t,s)$ is not important for the algorithm's outcome, as there is no error or the error is below all known human perception thresholds.

Averaging and smoothing For the shape measure SM this semblance needs to be mapped onto a measure over time only. Here we use a simple averaging over the relevant semblance to obtain the time-dependent semblance $\bar{S}_{rel}(t)$:

$$\bar{S}_{rel}(t) = \frac{\sum_{i=1}^{N} S(t,s(i)) \times LM(t,s(i))}{\sum_{i=1}^{N} LM(t,s(i))}, \tag{4.13}$$

The final step to calculate the shape measure SM is to smooth the mean relevant semblance to avoid high-frequency variations, which are unlikely to be perceived as such during motion simulation. The mean relevant semblance $\bar{S}_{rel}$ is therefore filtered with a zero phase Butterworth filter with order n and cut-off frequency ω_c, here tuned using Dataset 1, resulting in the shape measure SM. The mean semblance $\bar{S}$, mean relevant semblance $\bar{S}_{rel}$ and the final shape measure SM are shown in Figure 4.10. Here, averaging over only the relevant semblance S_{rel}, as compared to the total semblance S, results in a more pronounced measure of the shape differences during Turn 2 and, moreover, avoids the noise related detection of a shape difference during Turn 1. The final shape

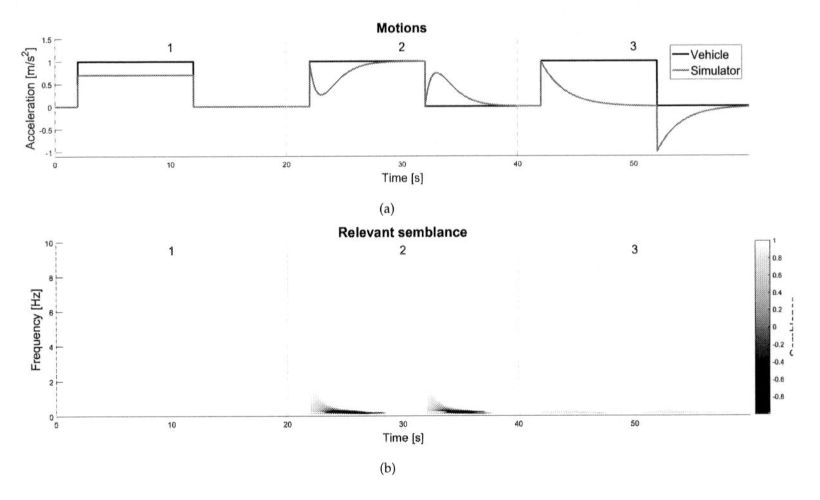

Figure 4.9: The relevant semblance ((b)) between the vehicle and simulator motions ((a)) of typical cueing errors.

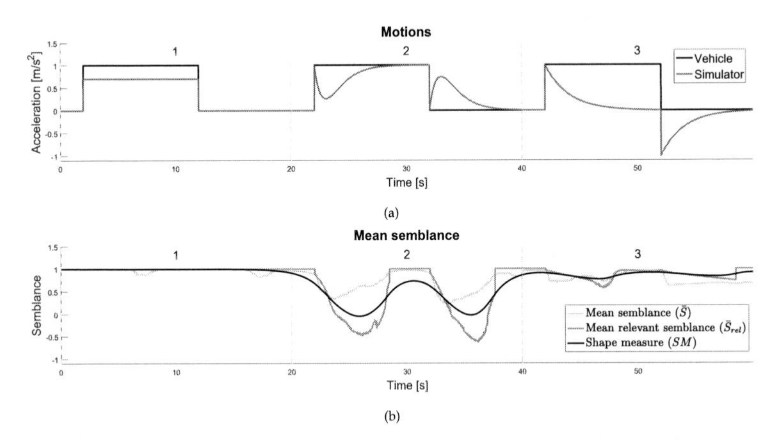

Figure 4.10: The mean semblance over frequency and the shape measure ((b)) of simplified vehicle and simulator motions ((a)).

measure SM, shown in Figure 4.10, correctly indicates no shape difference, i.e., $SM = 1$, during the scaled cue (Turn 1) and clear shape differences, i.e., $SM < 1$, during the missing and false cues in Turn 2 and Turn 3.

Shape Measure Threshold It is expected that the human perception system will not detect shape differences between vehicle and simulator motion signals for which the shape measure SM only deviates marginally from unity. A shape difference is therefore defined to occur only when the shape measure SM comes below a certain shape measure threshold th_{SM}. At which value for SM a shape difference can indeed be perceived, however, is currently not known. In this chapter, therefore, the th_{SM} is treated as a tunable parameter.

GAIN MEASURE

As mentioned before, also a gain measure GM is needed to distinguish between the different error types. To avoid divisions by zero and incorporate perceptual thresholds, the gain measure is slightly more complicated than a simple ratio between vehicle and simulator motion signal amplitudes.
The vehicle and simulator motions are first compared to the absolute motion perception threshold th_{abs} and the vehicle and simulator motions are set to zero when they are below this threshold using the smooth threshold function, resulting in yth. To avoid division by zero, the gain between the resulting motion signals (yth_{sim} and yth_{veh}) is calculated with a smooth gain function:

$$G_{sim,veh} = \frac{yth_{veh} * yth_{sim}}{yth_{veh}{}^2 + \eta}, \tag{4.14}$$

With this smooth gain function, the gain $G_{sim,veh}$ is zero when either yth_{veh} or yth_{sim} is zero. When $yth_{veh} \neq 0$ and $yth_{sim} = 0$ we indeed expect $G_{sim,veh}$ to be zero, however, when $yth_{veh} = 0$ and $yth_{sim} \neq 0$, we instead expect $G_{sim,veh}$ to be infinite. The final gain measure GM therefore equals $G_{sim,veh}$, except for the case when $yth_{veh} = 0$ and $yth_{sim} \neq 0$, where $GM > 1$. The exact value of GM in this case is not important for the algorithm outcome, as only the comparisons $GM > 1$ and $GM \geq 0$ are made.
In Figure 4.11 the resulting gain measure GM, for plotting purposes only limited to $[-1.5, 1.5]$, for the typical cueing errors of Figure 4.4 are shown. Figure 4.11 shows that only during the false cues (around 35 and 55 seconds) the value of the gain measure is larger than unity or smaller than zero.

4.2.6. ALGORITHM OUTCOME

Using the shape and gain measures combined with the cueing error equations described in Equation 4.3, the scaled, missing and false cues between the vehicle and simulator motions can be detected. The algorithm produces three separate time signals of the same length as the original cueing error, each containing only one type of cueing error. Applying the algorithm to the cueing error ϵ shown in Figure 4.4 results in the cueing errors ϵ_{sc}, ϵ_{mc} and ϵ_{fc} presented in Figure 4.12. This figure shows that a scaled cue occurs during the first turn, while the second and third turn contain a missing cue initially and a false cue at the end.

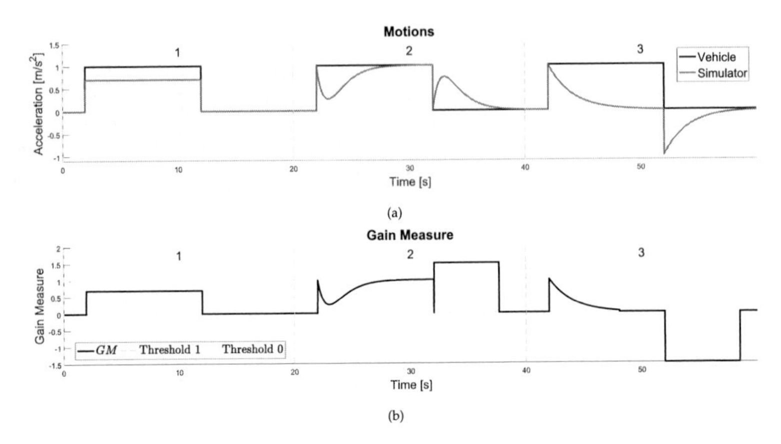

Figure 4.11: The simplified vehicle and simulator motions ((a)) and corresponding gain measure ((b)). The gain measure is limited to $[-1.5, 1.5]$ for plotting purposes only.

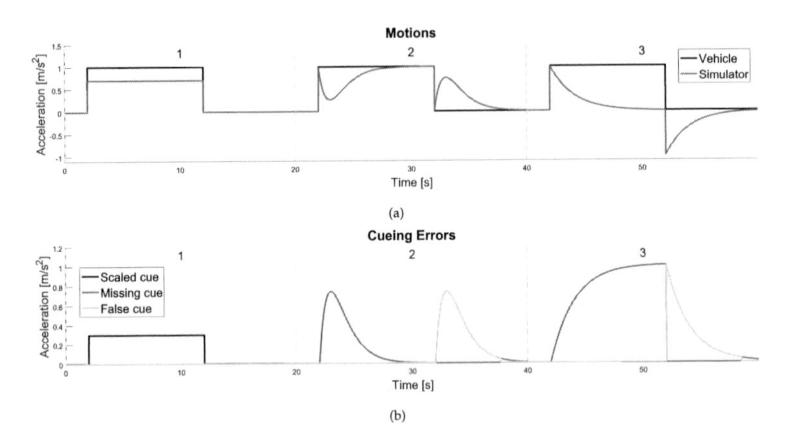

Figure 4.12: The algorithm output ((b)), in terms of scaled, missing, and false cue occurrences, for the vehicle and simulator motions ((a)) of example cueing errors.

4.2.7. Algorithm Parameters

The algorithm described in this section has several parameters which are listed in Table 4.3. All parameters, apart from the absolute and discriminatory perception thresholds, were tuned such that the algorithm output closely resembles the in Section 4.3.3 true cueing errors in Dataset 1, presented in Section 4.3. The final parameter values are also shown in Table 4.3.

The scale parameters s_0, J and δj were found via optimization, such that the combination of all wavelet transform parameters resulted in a frequency range between 0.1 and 10 $[Hz]$. The algorithm was tested on typical motion profiles for linear acceleration in the horizontal plane, i.e., longitudinal (a_x) and lateral (a_y) acceleration. The perceptual motion thresholds th_{abs} and th_{disc} are therefore also only shown for these accelerations. In [143] an absolute linear acceleration threshold of $0.04m/s^2$ in both longitudinal and lateral direction was found for motion frequencies in the range $1.0 - 14\ [rad/s]$ or $0.159 - 2.228\ [Hz]$. For low frequencies, i.e., in the range $0.096 - 0.159\ [Hz]$, the absolute thresholds were found to increase with decreasing frequency. For simplicity, in this study the lowest absolute threshold of $0.04\ \left[m/s^2\right]$ is applied for all frequencies.

A lot less research can be found on the discriminatory thresholds for linear motion perception. In fact, these thresholds were only investigated by Naseri and Grant in [148]. Assuming that Weber's law holds, they estimated that for sinusoidal motion with a frequency of 0.4 or 0.6 $[Hz]$ the Weber fraction was 0.05 and 0.02, respectively. Again for simplicity, the value for the Weber fraction resulting in the lowest threshold is chosen. In case future research would provide a more detailed function describing the relation between absolute and discriminatory motion thresholds and motion frequency over the frequency range used in this study, these functions can easily be implemented in the algorithm.

Table 4.3: Algorithm parameters

Symbol	Description	Effect on error detection	Value used in this study
Thresholds			
b	Steepness parameter of the slope from 0 to 1 for all but the shape and gain thresholds	A lower value reduces the threshold steepness.	$1e-5$
b_s	Steepness parameter of the slope from 0 to 1 for the shape and gain thresholds	A lower value reduces the threshold steepness and can result in ϵ_{all} being classified as partly one and partly another cueing error.	$1e-5$
th_{abs}	Absolute perceptual threshold	This value should be based on literature. Increasing it results in less sensitivity of the shape similarity measure to low amplitude motions.	$0.04 \left[m/s^2\right]$ for both a_x and a_y [143]
th_{disc}	Discriminatory perceptual threshold	This value should be based on literature. Increasing it results in less sensitivity of the shape similarity measure to low amplitude errors.	for a_x and a_y $\lvert CWT_{veh}\rvert \times wfrac$ is used, with Weber fraction $wfrac = 0.02$ [148].
th_{SM}	Shape measure threshold, above which scaling errors are assumed	Increasing it causes higher sensitivity to shape differences and consequently results in a higher detection rate for missing and false cues.	0.95
Wavelet Transform			
s_0	Smallest scale	Increasing it results in decreased influence of lower frequency motions on the shape similarity measure	$8.13[s]$

J	Number of scales	Increasing it results in increased influence of higher frequency motions on the shape similarity measure	$100\,[scales]$
δj	Spacing between scales	Increasing it results in less accurate shape similarity measure	$0.0664\,[scales]$
f_b	Wavelet bandwidth	Increasing it results in a higher frequency resolution, but lower time resolution of the semblance	$0.8\,[Hz]$
f_c	Wavelet center frequency	Increasing it results in a higher frequency resolution, but lower time resolution of the semblance	$0.8\,[Hz]$
Smoothing			
ω_c	Butterworth low-pass filter cut-off frequency	Increasing it decreases smoothness of the similarity measure	$0.1\,[Hz]$
n	Butterworth low-pass filter order	Increasing this reduces the amplitude of the similarity measure at frequencies higher than w_c	1

Table 4.4: Segment numbers for all nine segments in Dataset 1.

Maneuver/MCA	**MCA1:** scales down linear acceleration	**MCA2:** limits tilt-rate	**MCA3:** does not limit motions
M1: Curve driving at constant speed	1	2	3
M2: Braking to full stop and accelerating to 70 $[km/h]$	4	5	6
M3: Curve driving at variable speed	7	8	9

4.3. Motion Profiles

For the tuning and validation of the algorithm, two experiment datasets, here referred to as Dataset 1 and Dataset 2, were used. In this section the cueing errors occurring in these datasets are discussed in more detail.

4.3.1. Dataset 1

Dataset 1 is derived from the motions used in the experiment described in Chapter 2. This dataset includes typical motion cueing errors caused by washout filters in both longitudinal and lateral acceleration during three maneuvers and each performed with three different MCAs. Hence, there are nine numbered segments in total, which are summarized in Table 4.4. More information on the maneuvers and MCAs can be found in Chapter 2. In Figures 4.13 and 4.14 the vehicle and simulator motion signals for lateral and longitudinal accelerations in Dataset 1 for all nine segments are shown, respectively.

As can be seen in these figures, MCA1 (Segments 1, 4 and 7) introduces a scaling difference, while the MCA2 (Segments 2, 5 and 8) introduces a shape difference between vehicle and simulator motion. MCA3 (Segments 3, 6 and 9) instead results in close to no difference between the vehicle and simulator motion. Furthermore, the longitudinal acceleration contains slightly higher frequency motions and smaller errors between simulator and vehicle motion than the lateral acceleration.

4.3.2. Dataset 2

Dataset 2 is derived from the motions used in the experiment described in [135] and includes 7 different types of cueing errors with varying levels of intensity in the lateral acceleration during 22 identical curves. In Figure 4.15 the vehicle and simulator motions for the lateral acceleration in Dataset 2 for all 22 segments are shown.

This dataset shows a wide variety of pure scaling and shape differences between vehicle and simulator motion in lateral acceleration only. In Segments 13 and 18 no difference between vehicle and simulator acceleration is introduced. Pure scaling differences between vehicle and simulator motion are found in Segments 2, 6, 10, 16, 20 and 21, of which only 6 and 21 contain larger amplitude simulator than vehicle motion. Segments 5, 7, 12 and 19 all show similar shape differences between vehicle and simulator motion, but with different intensity. Different delays are introduced in Segments 4, 17 and

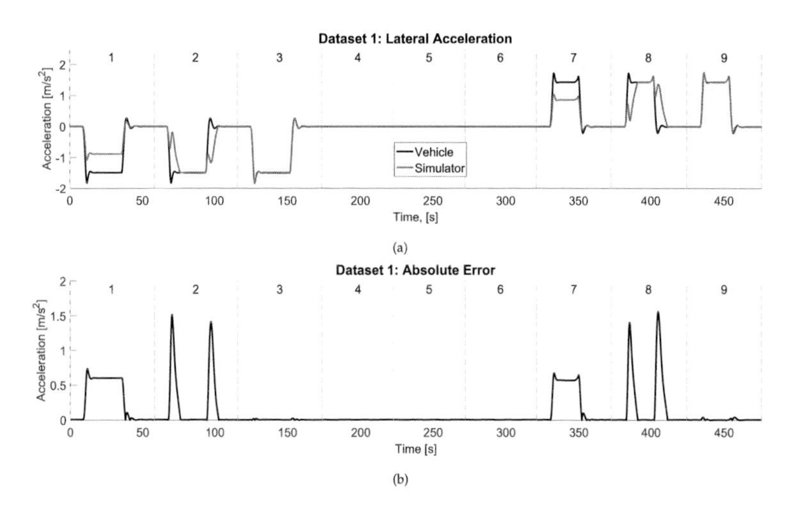

Figure 4.13: Lateral accelerations (a) and absolute errors between vehicle and simulator motions (b) for Dataset 1.

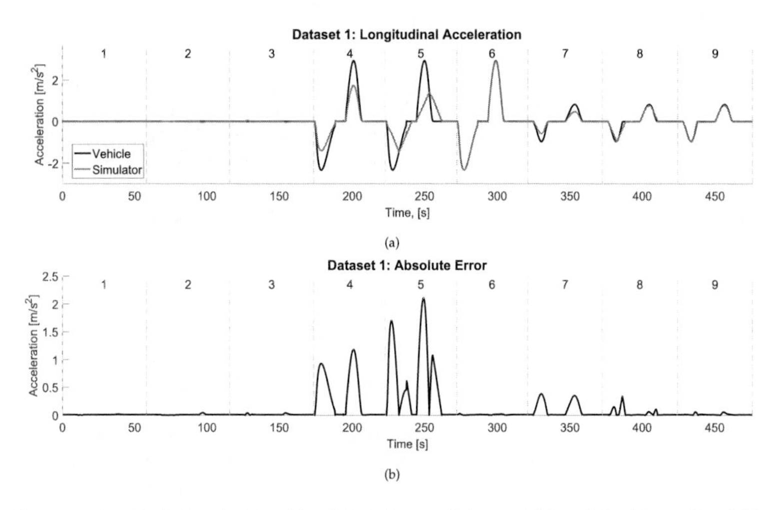

Figure 4.14: Longitudinal accelerations (a) and absolute errors between vehicle and simulator motions (b) for Dataset 1.

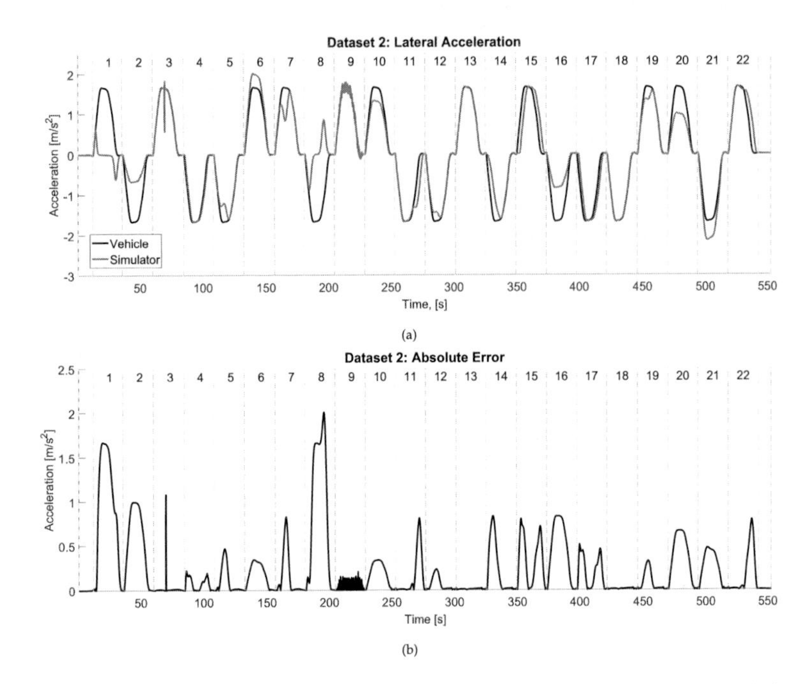

Figure 4.15: Longitudinal accelerations (a) and absolute errors between vehicle and simulator motions (b) for Dataset 2.

15 and in Segments 1 and 8 only the high-frequency part of the vehicle motion is still present in the simulator motion. Very high-frequency errors between vehicle and simulator motion are introduced in Segment 3, due to reaching a simulator motion limit, and Segment 9, introducing simulator noise. Finally, in Segments 11 and 22 additional simulator motion is introduced at the end of the curve, while in Segment 14 some of the vehicle motion at the beginning of the curve is missing from the simulator motion.

4.3.3. Error Detection Truth Data

The algorithm defines only three different cueing errors types based on shape and amplitude differences. The cueing error types in lateral and longitudinal acceleration in Dataset 1 are similar for each segment, with the latter often having much smaller error amplitudes. In Dataset 1 the algorithm should detect the errors caused by MCA1 (Segments 1, 4 and 7) as pure scaling errors and detect no or very small errors for MCA3 (segments 3, 6 and 9). For MCA2 (Segments 2, 5 and 8) errors at the beginning of either the acceleration, deceleration or curve driving motions should be detected as a missing cue, while for the same maneuver errors towards the end should be detected as false cues.

In Dataset 2 the algorithm should clearly distinguish between the scaling errors in Segments 2, 6, 10, 16, 20 and 21, no errors in Segments 13 and 18 and shape errors in all other segments. Delays between simulator and vehicle motion, i.e., Segments 4, 17 and

Table 4.5: True Cueing Errors.

Cueing Error	**Segments Dataset 1** *lateral & longitudinal acceleration*	**Segments Dataset 2** *lateral acceleration*
Scaling error	1, 4, 7	2, 6, 10, 16, 20, 21
Missing Cue	start of 2, 5, 8	5, 7, 12, 14, 19 start of 1, 4, 8, 15, 17 partly in 3, 9
False Cue	end of 2, 5, 8	11, 22 end of 1, 4, 8, 15, 17 partly in 3, 9

15, and the simulator motions containing only high frequent motions, i.e., Segments 1 and 8, are expected to result in an initial missing cue and a false cue at the end. In Segments 5, 7, 12, 19 and 14 only a missing cue should be detected, while in Segments 11 and 22 only a false cue should be detected. The error due to hitting a simulator motion limit, i.e., Segment 3, is expected to mainly cause a missing cue and possibly some small false cues due to additional oscillations. Finally, the simulator noise error, i.e., Segment 9, is expected to be detected as a mixture of missing and false cues as the simulator motion oscillates around the vehicle motion. In Table 4.5 an overview of the true cueing errors is given.

4.4. RESULTS

In this section results of the different algorithm steps are shown when applied to the motion profiles introduced in Section 4.3. The figures show the following results per motion profile in the subplots from top to bottom:

(a) Vehicle $y_{veh}(t)$ and simulator $y_{sim}(t)$ acceleration
(b) Semblance $S(t,s)$ over a frequency range of $0.1 - 10\,[Hz]$
(c) Relevant semblance $S_{rel}(t,s)$ over a frequency range of $0.1 - 10\,[Hz]$
(d) Mean semblance, mean relevant semblance $\bar{S}_{rel}(t)$, the shape measure $SM(t)$ and shape measure threshold th_{SM}
(e) Gain measure GM [1] and thresholds at zero and one.
(f) The detected separate cueing errors ϵ_{sc}, ϵ_{mc} and ϵ_{fc}

4.4.1. DATASET 1

In Figure 4.16 the algorithm results for each step are shown for the lateral acceleration in Dataset 1. Figure 4.16(f) shows the final result of the algorithm: scaled cues are detected during Segments 1 and 7, missing cues occur at the beginning and false cues at the end of the turn in Segments 2 and 8 and very small scaling errors are detected during Segments 3 and 9. Since no lateral motion is present during Segments 4, 5 and 6, no error is detected here.

Between the false and missing cues in Segments 2 and 8 the two motion signals are

[1] For plotting purposes GM was limited to the range $[-0.5, 1.5]$, whereas in the algorithm the unlimited value is used.

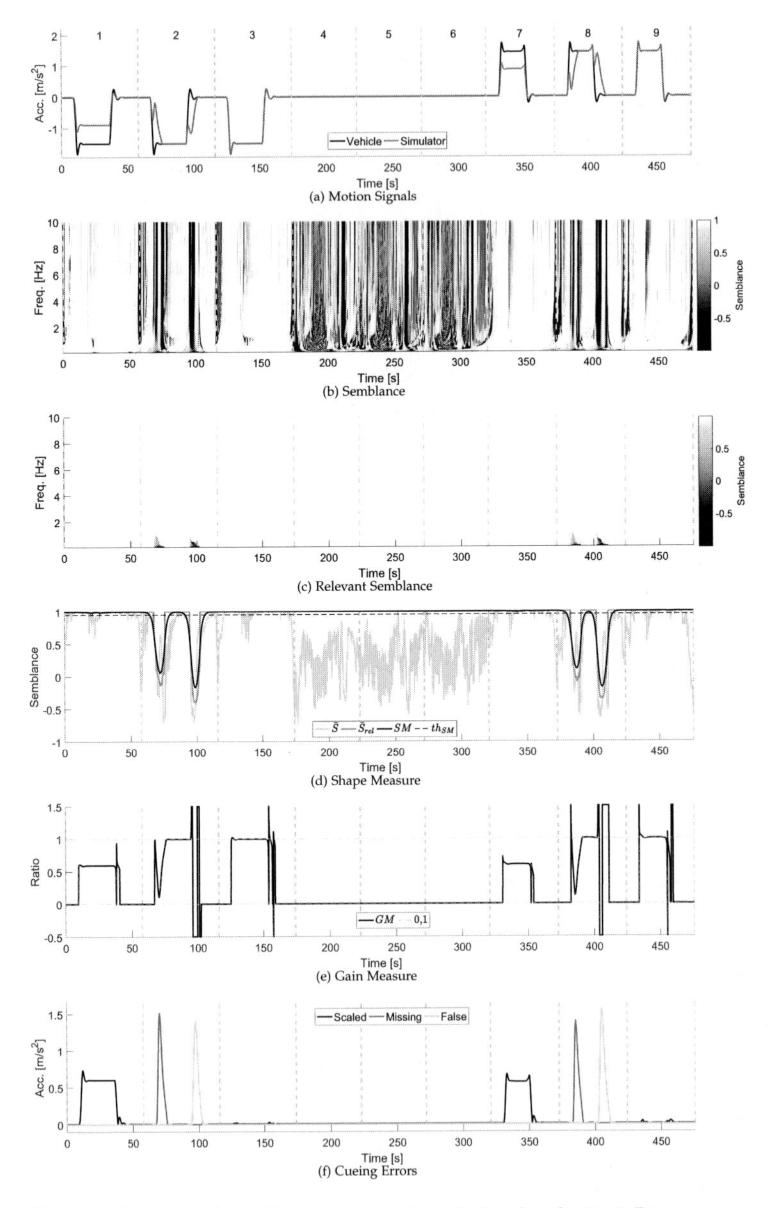

Figure 4.16: Results of each algorithm step when applied to the lateral acceleration in Dataset 1.

very similar, which can also be seen from the increasing mean semblance at this point in Figure 4.16(d). The difference between the semblance (Figure 4.16(b)) and relevant semblance (Figure 4.16(c)) plots shows that much of the semblance below one at high frequencies is due to motion with amplitudes below human thresholds. The gain measure (Figure 4.16(e)) shows high variations, especially during Segments 4, 5 and 6, where both simulator and vehicle motion hover around $0 \left[m/s^2\right]$. The gain measure for the scaled cues in Segments 1 and 7 shows a relatively constant scaling factor of 0.6 between vehicle and simulator motions, while the scaling factor for the missing cue in Segments 2 and 8 varies within the range $0 - 1$. The false cues in these segments show a jump from > 1 to < 0 back to > 1 due to the change in sign in the vehicle acceleration. In Figure 4.17 the algorithm results for each step are shown for the longitudinal acceleration in Dataset 1. Figure 4.17(f) again shows the final result of the algorithm: scaled cues are indeed detected in Segments 4, 7 and 9. Missing cues are found at the beginning and false cues at the end of the acceleration and deceleration maneuvers in Segments 5 and 8. However, in Segment 8 during the initial acceleration phase a small scaled cue is found. Again the difference between the semblance (Figure 4.17(b)) and relevant semblance (Figure 4.17(c)) shows motion with amplitudes below human threshold cause the decreased semblance visible in Segments 1 till 3.
Overall, all scaled, missing and false cues, except for the small false cue at the start of Segment 8, were detected correctly by the algorithm.

4.4.2. DATASET 2

Figure 4.18 shows the algorithm results in each step for the lateral acceleration in Dataset 2. Figure 4.18(f) shows the final result of the algorithm: scaled cues are found in Segments 2, 6 10, 16, 20 and 21, while missing cues are detected in Segments 5, 7, 12, 14 and 19, and false cues in Segments 11 and 22. The scaled cue in Segment 16 is detected as a missing cue at the very start of the turn. Missing cues are found at the beginning and false cues at the end of the curve in Segments 1 and 8.
Similar results are found during time delay errors in Segments 4, 15 and 17. The high-frequency limiting and noise errors in Segments 3 and 9, respectively, are partly detected as a missing and partly as a false cue. While in Dataset 2 most high-frequency motions were below the human perception thresholds, the relevant semblance in Figure 4.18(c) shows that during Segments 3 and 9 the high-frequency motion is above this threshold.
Unlike Dataset 1, the error type in Dataset 2 sometimes changes while the total error is still above zero, creating some steep peaks in the resulting error type signal. This is clearly visible in Segments 1 and 8 in Figure 4.18(f). Depending on the use of the algorithm such steep variations can be troublesome. Additionally, it might seem unrealistic that a human indeed perceives such abrupt changes between error types. One way of smoothing these changes between different error types is by using non-binary shape and gain measures. This is achieved by increasing the threshold steepness parameter b_s.
In Figure 4.19 the algorithm final result for Segment 1 is shown for the binary ($b_s = 10^(- 5)$) and non-binary $b_s = 0.1$ gain and shape measures. Currently it is not known which value for parameter b_s results in a cueing error transition that best resembles human perception. It is proposed that when using the cueing error detection algorithm in a PMI model, this parameter is only increased if it significantly improves the model fit to measured rating data.

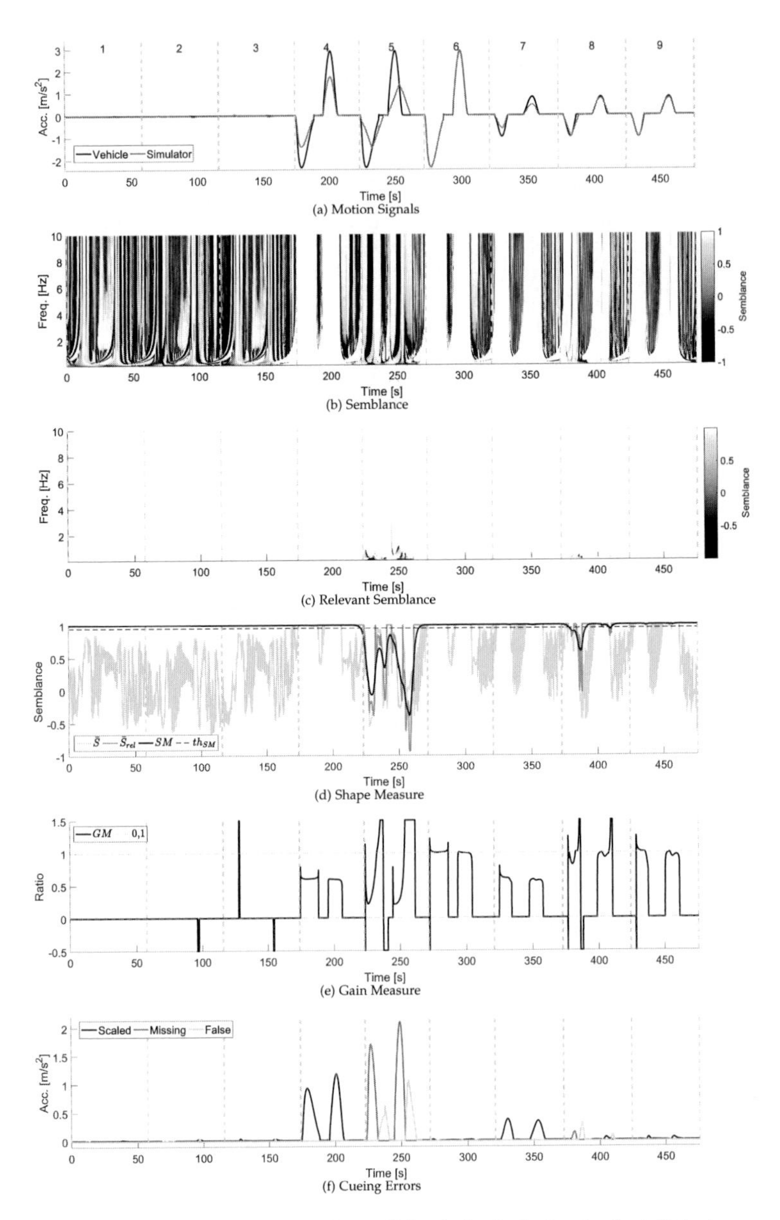

Figure 4.17: Results of each algorithm step when applied to the longitudinal acceleration in Dataset 1.

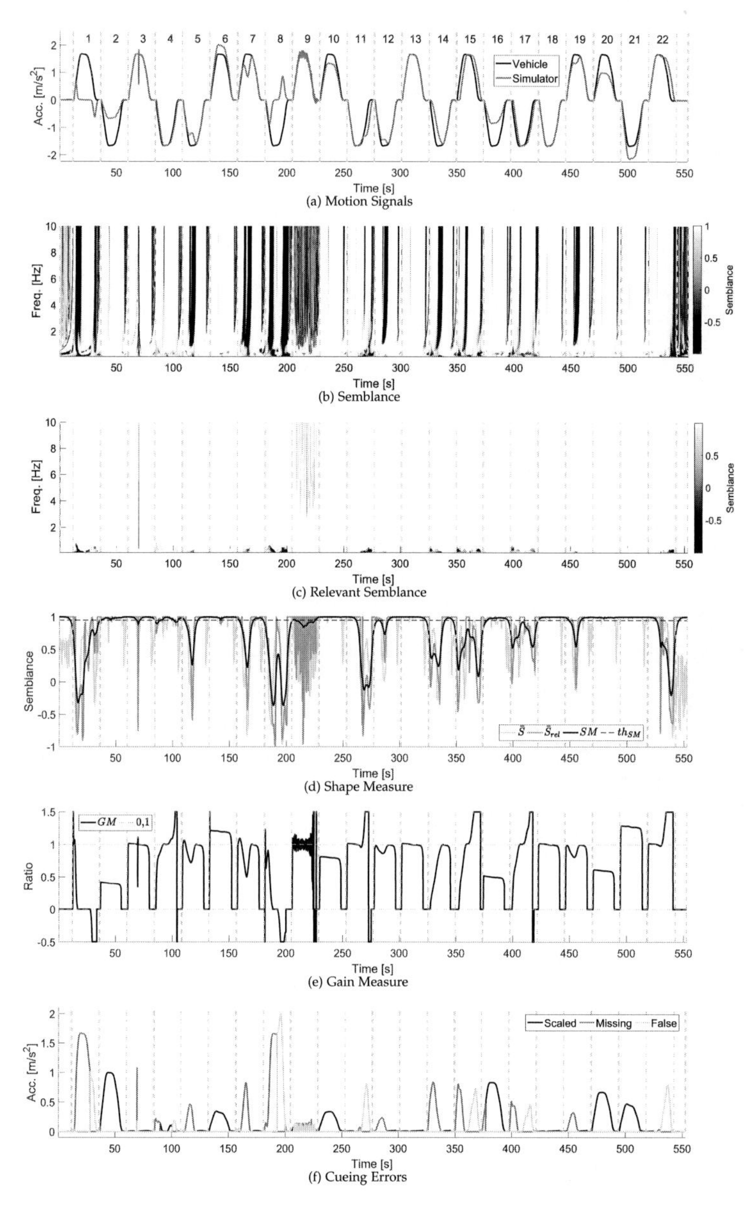

(a) Motion Signals

(b) Semblance

(c) Relevant Semblance

(d) Shape Measure

(e) Gain Measure

(f) Cueing Errors

Figure 4.18: Results of each algorithm step when applied to the lateral acceleration in Dataset 2.

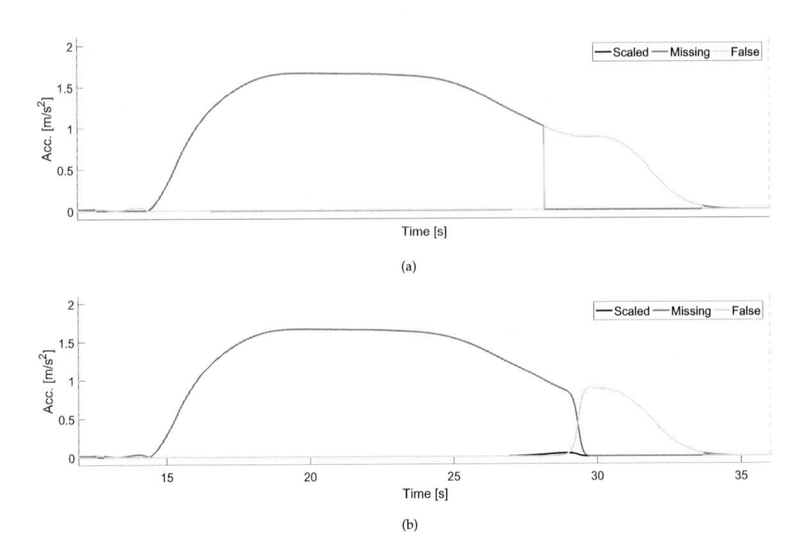

Figure 4.19: Binary and non-binary algorithm output for the lateral acceleration in Segment 1 of Dataset 2, with $b_s = 1e-5$ and $b_s = 0.1$, respectively.

4.5. Discussion

In this chapter a Cueing Error Detection Algorithm (CEDA), based on a wavelet-based semblance measure, is presented. The goal of the algorithm is to split the total cueing error into three common cueing error types: scaled cues, missing cues and false cues. Such results can then be used for improved modelling or analysis of motion cueing quality. With the aim of eventually using this algorithm for cueing quality optimization, the algorithm was designed using only smooth functions. The algorithm parameters were tuned using Dataset 1, previously presented in Chapter 2, and also applied to Dataset 2 obtained from the experiment described in [135]. Both datasets were designed to contain specific cueing error types, based on their description in literature. These true cueing error types therefore served as the truth data, and thus desired output, for testing the algorithm.

Almost all cueing errors in both datasets were correctly detected. Only in Dataset 2, during Segment 16, a scaled cue was initially detected as a missing cue. This slight deviation from the true cueing error type seems to be caused by a combination of mean semblance smoothing and a large false cue just before the start of the scaled cue. To correct the negative effect of smoothing, the cut-off frequency of the smoothing filter could be increased, however, this would result in more oscillations between cueing error types in, for example, the noise error in Segment 9 of Dataset 2. Overall, the algorithm performed well in detecting different cueing error types.

The algorithm has several parameters that require tuning. While the parameter values presented here show good results for both Datasets 1 and 2, as well as the simplified example motions, it is possible that these parameter values can be further optimized to provide good results over a wider range of cueing errors. In future research it is therefore recommended to obtain more cueing error data to improve the tuning.

For example, an experiment including one maneuver with different degrees of shape similarity between the vehicle and simulator motion could be performed. Answers from participants on the shape similarity between vehicle and simulator motions could be used to tune the shape measure threshold. Additionally, more datasets with known cueing errors and different maneuvers can be generated to improve the tuning of parameters, such as wavelet center and bandwidth frequency and smoothing filter cut-off frequency and order.

In the algorithm presented here, the wavelet-based semblance was used to determine shape similarity between two signals. This semblance was shown to be an efficient measure to distinguish between scaled and missing or false cues. Another measure that was considered for this study is the more often used wavelet-based coherence [149, 150]. This measure, however, requires smoothing in time and frequency to avoid it being equal to unity [149], which reduces the resolution of the measure and can also introduce lags. This measure is therefore less suited for use in the algorithm presented here. While the significance testing presented in [149] can be used to improve the resolution for some applications, its dependency on the total motion signal amplitude makes it unsuitable for the application presented here. The amplitude dependency would cause similar cueing errors being classified differently in case a larger motion is added to the same time signal.

The semblance measure does not require smoothing, but has a high sensitivity to noise due to its lack of amplitude information. In this study it was shown that the effect of noise on the algorithm results could be significantly reduced via amplitude filtering using the original wavelet transforms, i.e., only taking into account the relevant semblance.

The wavelet transform resolution in time and frequency depends on the choice of the center and bandwidth parameters of the selected wavelet. For the algorithm described in this chapter these wavelet parameters were tuned for optimal algorithm results for Dataset 1, which resulted in a high time resolution, but a comparatively low frequency resolution. The algorithm results presented here do not appear to be negatively influenced by this power spread in frequency. However, for future research it can be useful to investigate whether techniques, such as a prior significance test on the wavelet transforms, improve the results when applying the algorithm to a wider range of cueing errors. It should again be noted that the significance testing described in [149] cannot be directly applied here.

Currently the relevant semblance is determined by comparing the time signals and their wavelet transforms to absolute and discriminatory human perception thresholds. Research on motion perception such as [143] has shown that the perceptual thresholds depend on motion frequency. Because of the frequency information that is readily available with this method, such frequency-dependent perceptual thresholds can easily be implemented in the algorithm when they are available in a more general form. Currently, however, the non-frequency dependent thresholds are used.

The algorithm is designed to only detect three of the most common cueing errors, causing other error types to be classified as one of these three. Time delay errors, for example, were detected as consecutive combined missing and false cues. It can be debated whether a separate class of delayed cues would better represent human perception, or if humans actually do perceive delayed cues as a combination of missing and false cues. It is possible that this, for example, depends on the size of the delay and the main frequency of the maneuver. A small delay in a slalom maneuver might be perceived as a delay, while a large delay in a large radius turn might be perceived

as a combination of a missing and a false cue. An experiment testing this hypothesis could clarify if detection of separate time delay error is useful.
High-frequency cueing errors, such as the limiting and noise errors in Dataset 2, were classified as oscillating, missing and false cues. Using the frequency information present in the wavelet transforms, a separate class of high-frequency cueing errors could be made. As, for example, the limiting error is known to be detrimental to cueing quality [51], it is recommended to also extend this approach to include detection of this error type in future research.

4.6. CONCLUSION

Cueing errors of different types can affect the perceived motion cueing quality to different degrees, e.g., false cues have been shown to be more detrimental to cueing quality than scaled cues. To capture this difference in a cueing quality prediction model, an algorithm to detect and extract different cueing error types is required. In this chapter, a Cueing Error Detection Algorithm, detecting scaled, missing and false cues from simulator and vehicle motion data, was presented.
The algorithm, which uses the wavelet-based semblance to distinguish between different types of cueing errors, was tuned to a first dataset such that all present cueing errors were detected. When applying this tuned algorithm to a second dataset, as well as when applying it to simplified example motions, boh for which the algorithm was not tuned to, yielded the same level of performance, and most cueing errors were detected.
While the algorithm presented here only detects three of the most common cueing error types, the availability of frequency information makes it possible to extend the algorithm to also detect error types such as limiting or noise errors. The use of smooth functions throughout the algorithm also makes it suited for optimization purposes in the future.

5
Modelling Perceived Motion Incongruence

This chapter presents a complete design process for the time-varying Perceived Motion Incongruence (PMI) prediction models that are central to this thesis. The process consists of several steps, including the selection of process parameters, model inputs and filter orders, as well as model parameter estimation and model analysis. Three models of increasing complexity are designed by applying this process, using part of the dataset from the experiment described in Chapter 2. The prediction capabilities of these models are verified using the rest of this dataset. The results show that all models have considerable PMI prediction capabilities for data obtained within one experiment. Increasing the complexity of a model, such as including the Cueing Error Detection Algorithm (CEDA) described in Chapter 4, improves these capabilities.

5.1. Introduction

One of the main challenges in vehicle motion simulation is to determine which simulator motions result in the best perceived motion quality. Addressing this challenge requires knowledge on how motion cueing quality is affected by simulator motion. For purposes such as automated motion cueing analysis or optimization specifically, a mathematical representation of this relation is required. In this study, therefore, the design process of models that predict the time-varying motion quality due to vehicle and simulator motion differences is proposed.

Over the years, researchers have proposed different methods to assess cueing quality. A commonly used method is to perform human-in-the-loop experiments where Motion Cueing Algorithms (MCAs) with different tunings are compared via subjective ratings. In [39] this was done using a magnitude estimation method, while in [36] questionnaires were used.

Because human-in-the-loop experiments are time consuming, methods have been developed to instead tune cueing algorithms off-line. The most recent of such methods is the objective motion cueing test (OMCT) [84], which is based on criteria like the Sinacori-Schroeder [50] and the Advani-Hosman [151] criteria. The OMCT can be used to tune MCA parameters, based on desirable frequency responses of the complete simulation system for several motion channels over specific frequency ranges. While this test has been proven to be successful in assessing simulator fidelity, it is not designed for on-line or off-line mathematical optimization, as it lacks a single measure of cueing quality that can be optimized.

Instead, optimization algorithms often simplify this cueing quality measure to incorporate only the euclidean distance between vehicle and simulator motions in linear acceleration and rotational velocity [40, 136], where weights for each motion channel are for simplicity often only set to account for the differences in unit [57]. While this simplifies the computation significantly, it also removes all expert knowledge from the MCA tuning process, such as described in [51], from the optimization. Such expert tuning knowledge is often based on human-in-the-loop simulation experiments as well as a basic understanding of the human perceptual system. One of the main challenges in MCA optimization is to combine the optimization requirements for simplicity of the cost function and the complexity of the human perception system.

In an attempt to include human perception in the cost function, simplified models of the vestibular system, the main sensory organ for detection of specific force (otholiths) and rotational velocity (semicircular canals), are sometimes implemented to map the actual vehicle and simulator motions to perceived motions [38, 56, 152]. More elaborate models that determine perceived vehicle and simulator motions via visual vestibular cue integration have also been used [37]. While this provides information on one aspect of human perception, it fails to address how errors in different motion channels are combined into a single measure of simulation quality. Additionally, these models do not explain why different cueing error types, such as scaled and false cues, deter the cueing quality disproportionately. The difficulty of using such models to explain simulation quality was shown in [92], where comparing perceived motions, calculated by running the actual motions through a perception model, during a take-off run could not directly explain why an MCA was judged as good or bad.

In this chapter therefore the design process for a model that maps vehicle and simulator motions to a single time-varying motion cueing quality rating is introduced. The method focuses on a bottom-up approach and does not include any human sensory system models, but rather focuses on fitting the model output to the measured per-

ceived motion incongruence rating [136]. The resulting model can be used to predict the decrease in perceived cueing quality from vehicle and simulator inputs without the use of human-in-the-loop experiments. By extending the non-linear part of the model with the cueing error detection algorithm described in Chapter 4, the varying influences of different cueing errors on the cueing quality can also be captured by this model.
The design process introduced here, is tested on data from the experiment presented in Chapter 2. As the resulting model is only based on one limited dataset, it is not expected to describe all complexities of the human perceptual system related to vehicle motion simulation. The proposed design process can however be used on larger datasets to obtain a more accurate model for the prediction of perceived motion incongruence during a vehicle motion simulation.
This chapter is structured as follows. Section 5.2 introduces the considered input-output mapping and provides an overview of the general model structure which can be used to predict time-varying perceived motion incongruence. To determine which inputs are actually represented in the measured output, find appropriate model orders, and estimate the corresponding model parameters, a system identification process is proposed in Section 5.3. To validate this process, Section 5.4 introduces three models of different complexity, the datasets to which they are fitted and the methods that are used for the analysis of the models. In Section 5.5 the models' structural, explanatory and prediction analysis results for each of the three models are compared. A discussion on these results is provided in Section 5.6, followed by a conclusion on the proposed models and system identification process in Section 5.7.

5.2. THE MODEL

In this study a design process is proposed for a model that can be used to assess the motion cueing quality in a vehicle simulation. The main purpose of the model is to map vehicle and simulator motion signals onto a measure of the time-varying Perceived Motion Incongruence (PMI). Such a model could also be used to predict motion cueing quality off-line and possibly even serve as a cost calculation algorithm for MCA optimization. In the next subsections first the inputs and output of the model are further explained, after which an overview of the general model is provided.

5.2.1. INPUTS

During the design of an MCA, usually only differences between actual and simulated physical vehicle motions in the six motion channels for linear acceleration (or specific force) and rotational velocity are used to determine the quality of an MCA. In [136], however, it was suggested that including derivatives and integrals of these motion channels could possibly also improve the model fit.
For the current study therefore all eighteen motion channels, first derivatives and first integrals of the six motion channels, are considered as possible inputs to the modelling process. The inputs described in here are not necessarily part of the final model, but are used as an initial set of possible inputs from which relevant inputs can be selected via an input selection process as described in Section 5.3.
Before adding all eighteen motion channels as model inputs, it is necessary to determine whether including them makes sense from a perceptual point of view. To this end an overview is made on how differences between actual and simulated vehicle motions in each motion channel could be perceived. In case differences between vehicle and simulator motions in a specific channel cannot be perceived, this channel should not be

used as a model input.
In this study it is assumed that the actual physical vehicle motions are estimated using a combination of the unrestricted visual motion cues and the participant's experience with car driving. Visual motion cues can be derived from optic flow [153–155], while orientation can be derived from the geometric layout of the visual environment [156]. Even though not all motion channels can be estimated with equal accuracy, for simplicity, perfect estimation of the actual vehicle motions is assumed in this study.
The ability to perceive differences between actual and simulated vehicle motions is thus assumed to depend on the gravito-inertial motion sensors. It is well known that humans use their otoliths and semicircular canals (SC) to sense linear acceleration and rotational velocity respectively. Otoliths, however, are also sensitive to its rate of change, i.e., jerk [157]. Furthermore, studies described in both [124] and [158] showed that jerk influences the motion perception in motion simulators, making this jerk motion a good choice for a model input.
The semicircular canals on the other hand, only function as a rotational velocity detector during normal head movements. Depending on the motion frequency, semicircular canals function as detectors of rotational velocity ($0.1 - 1Hz$), rotational acceleration ($< 0.1Hz$) or rotational angle ($> 1Hz$) [37]. Additionally, when the rotational velocity is above the perceptual threshold, pitch and roll angles can be derived from a combination of rotational velocity and linear acceleration.
The importance of pitch and yaw angles for the cueing quality was also shown in [92], where it was concluded that subjects relied strongly on mismatches in attitude to determine the cueing quality, at least during the tested longitudinal acceleration manoeuvre. The yaw angle, not generating a change in gravitational acceleration, cannot be derived in this way.
Finally, even though linear velocity cannot directly be sensed by the vestibular system, the longitudinal velocity could be derived from the road rumble frequency: a higher longitudinal velocity results in higher road rumble frequency. For the data used in this study, no road rumble was implemented in the simulator, making the longitudinal velocity physical motion cue in the simulator equal to zero.
In Table 5.1 an overview of all considered motion inputs is provided. As shown in this table, three of the eighteen motion channels are not included as inputs because no difference between actual and simulated vehicle motion could be perceived in these channels. Although the precision and accuracy of the motion perception in the remaining channels vary significantly, they cannot directly be excluded and are therefore used as possible model inputs.

5.2.2. OUTPUT

The model should map the inputs onto a measure of PMI. The model described in this study aims to predict the *average* PMI over a random set of participants, rather than, for example, predicting participant-specific ratings. The measure most related to this time-varying average PMI is the average Motion Incongruence Rating (MIR) measured with the method described in [136].
Such a dataset should include ratings of multiple participants that each rated the same vehicle simulation multiple times. The simplest way of obtaining an average MIR would be to take the mean at each time step over all ratings in the dataset, as was done for the initial model presented in [136]. After further inspection of the dataset, however, the distribution of these ratings at each time step proved to be strongly non-normal, making the mean a poor descriptor for the average MIR. Appendix A in this thesis

Table 5.1: Sensory signals considered as model inputs.

Motion Channel		Symbol	Gravito-inertial Motion Sensors	Included as input
Linear	X	v_x	Indirectly perceived via road rumble	Yes
Velocity	Y	v_y	Not perceived	No
	Z	v_z	Not perceived	No
Linear	X	a_x	Otoliths	Yes
Acceleration	Y	a_y	Otoliths	Yes
	Z	a_z	Otoliths	Yes
Linear	X	j_x	Otoliths	Yes
Jerk	Y	j_y	Otoliths	Yes
	Z	j_z	Otoliths	Yes
Rotational	Roll	θ_r	SC & otoliths combined	Yes
Angle	Pitch	θ_p	SC & otoliths combined	Yes
	Yaw	θ_y	Not perceived	No
Rotational	Roll	ω_r	SC	Yes
Velocity	Pitch	ω_p	SC	Yes
	Yaw	ω_y	SC	Yes
Rotational	Roll	α_r	SC	Yes
Acceleration	Pitch	α_p	SC	Yes
	Yaw	α_y	SC	Yes

provides a full overview of the averaging techniques that have been investigated.
In this study the median, a simple and often applied alternative averaging technique, is used. In Figure 5.1 the mean and median over all ratings are compared to the mode of the estimated distribution (using the Kernel Density Estimation (KDE) method) at each time step. The latter averaging technique is complex, but can be assumed to give the most accurate representation of the average MIR as it estimates the measurement distribution separately for each time step. Figure 5.1 shows that averaging via the median instead of the mean, results in an average MIR much closer to the average MIR obtained with the KDE method. In this study, therefore, the model output will be compared to the median, rather than the mean, of the measured MIR at each time step.

5.2.3. MODEL OVERVIEW

Now that the inputs and output have been selected, a model, representing the input-output mapping, can be defined. Keeping in mind that the output represents the MIR rather than the actual PMI, the model should not only account for perceptual processes, but also include a simplified model of the rating process itself. In [136], therefore, an initial model structure is proposed that includes separate perceptual and response systems.
The perceptual system is a model describing the translation from simulator and vehicle motion to perceived motion incongruence, while the response system models how this perceived incongruence results in the participants' rating. The latter should, for example, account for delays and smoothing that has been shown to occur when rating a variable continuously [111, 125]. The main difference between the two systems is that

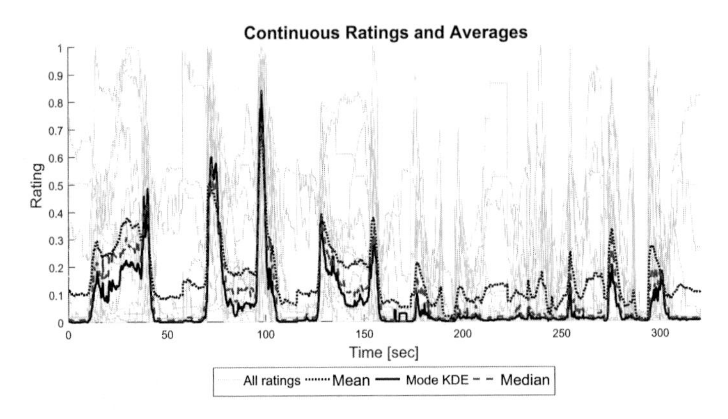

Figure 5.1: Results for different averaging techniques for continuous rating data.

parameters in the response system apply to one combined signal, assumed to be the perceived motion incongruence, while in the perceptual system parameters apply to specific motion channels.

The perceptual system described here is not intended to accurately portray what happens in the human brain during a motion simulation, but rather defines a relation between simulator and vehicle motion differences in all motion channels and the resulting PMI over time. This system is split in three parts, in part one different cueing errors are calculated, part two filters and weights each cueing error and in part three all filtered cueing errors are added up to form one measure of PMI over time. The filtering and weighing of different cueing errors should account for, among others, the filtering behavior that occurs in the human sensory systems and the relative weighing of different cueing errors in the human cognitive system.

As the model describes a human perception process, it is assumed to be subject to a considerable amount of noise, i.e., unpredictable stochastic components, which could enter the system at different places. In the current application and model structure it is likely that most of the noise enters the model before the response system, e.g., in the perceptual system. The exact location of the main noise source entrance is difficult to judge, but for simplicity the current model assumes this to be at the end of the perceptual system. The noise source is an important aspect of the model and its characteristics will be discussed in Section 5.3.3.

In Figure 5.2 the proposed model structure, including these descriptions of the perceptual and response systems, is shown. In the following subsections the non-linear and linear part of the model are described in more detail.

Non-Linear Part

The non-linear part consists of the cueing error calculations. To purpose of this system is to detect which cueing errors require separate filters and weights before being combined into one measure of PMI. Basically, any non-linear transformation of the simulator and vehicle motion should occur in this part.

As can be seen in the non-linear model overview in Figure 5.3, four main non-linear transformations are currently implemented in the general model. These transforma-

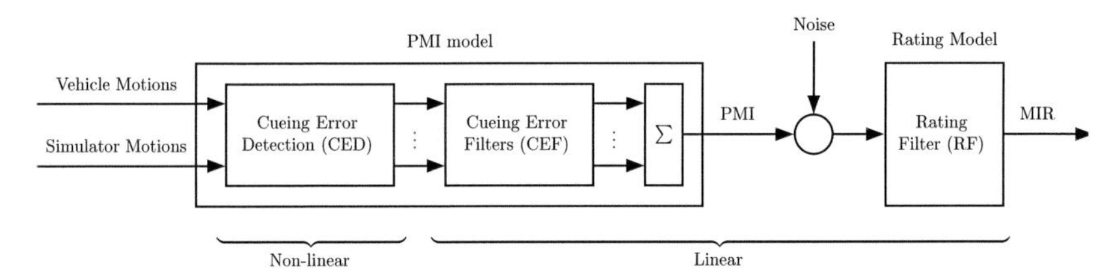

Figure 5.2: General structure of a MIR model consisting of a perceptual system (the PMI model) and response system.

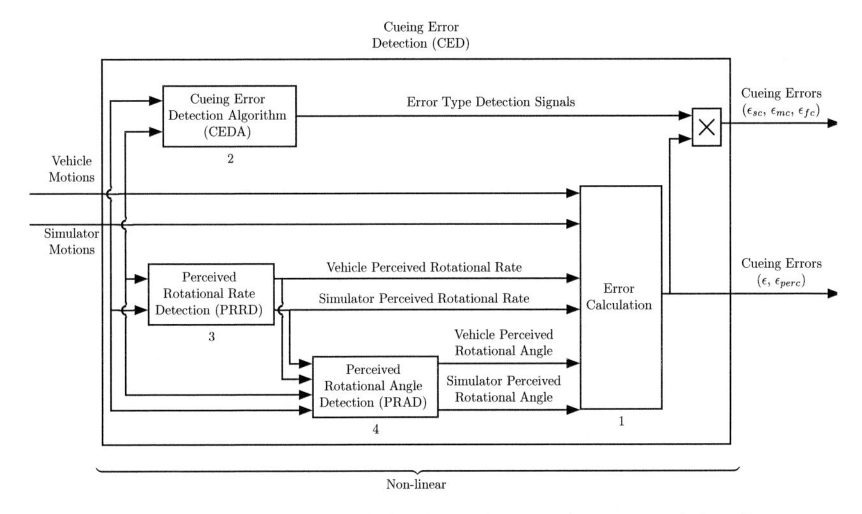

Figure 5.3: Non-linear part of the MIR model including four non-linear transformations to calculate relevant cueing errors.

tions result in three sets of cueing errors: general cueing errors (ϵ), cueing errors split in different types assumed to influence PMI differently (scaled ϵ_{sc}, missing ϵ_{mc} and false ϵ_{fc} cues) and perceived rotational rate and angle errors (ϵ_{perc}) that take into account perceptual motion thresholds.

The general cueing errors ϵ are directly derived from the difference between simulator and vehicle motion signals, e.g., the motion error. In Figure 5.3 this is done in Error Calculation block (1), where a range of different simple transformations can be applied to the motion errors. For example, one can choose to take the absolute or the squared value of these errors. For computational simplicity, the optimization-based MCAs generally use the squared value of the motion errors. When taking into account human perception, however, this might not be the most optimal choice, as it assumes that large errors affect PMI more than small errors. Here therefore the absolute value of the motion errors is used instead.

Within one motion channel, cueing errors of different types, indicated with subscripts ϵ_{sc}, ϵ_{mc} and ϵ_{fc}, can occur. Block (2) in Figure 5.3 represents the Cueing Error Detection

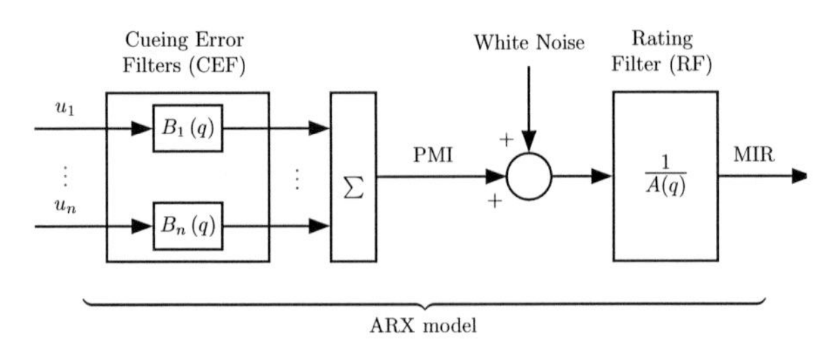

Figure 5.4: Linear part of the MIR model modeled with an ARX structure, consisting of the cueing error and response filters.

Algorithm (CEDA) described in Chapter 4. The algorithm provides three binary time signals that have a value of one when either of the three cueing error types, scaled, missing or false cues, occurs. These output signals are then multiplied with the value of the general cueing error at this time to obtain the specific cueing error.
Another assumption that is often used by MCA tuning experts, is that rotations with a rate below the perceptual thresholds are not perceived by the participants and thus do not have to be minimized further. To account for this fact, subsystems PRRD (3) and PRAD (4), which map actual rotational rates and angles onto perceived rates and angles, respectively, could be implemented. Examples of such subsystems can be found in Appendix B, but, as they have not been experimentally validated, these are not used in the remainder of this study.
Because of the general nature of the model, many different models can be derived depending on the choice of which motion channels and cueing error types to include and which transformations to apply. To make informed choices, motion perception literature can be consulted, as well as comparing the fit of different models based on different sets of choices. This process is further explained in Section 5.3.

Linear Part

The linear part of the model shown in Figure 5.2 is a multiple-input-single-output (MISO) system, with cueing errors as inputs and a MIR as output. Due to its structure, a logical choice is to model this system as an ARX model, for which estimating the optimal parameters has an analytic solution, if the noise can be considered white zero-mean Gaussian noise. Using such a model structure significantly reduces the calculation time compared to more elaborate structures such as Box-Jenkins models. The linear part of the MIR model in the form of an ARX model is shown in Figure 5.4. The cueing error filters are described by n (= the number of cueing errors) polynomials $B_n(q)$ of orders n_b subject to input delays n_k, where n_b and n_k are both vectors of length n. The response system is described with the polynomial $A(q)$ of order n_a.
If an ARX model structure turns out not to suffice in describing the measured MIR, one option could be to use a more complex model structure. Due to the possibly very large increase in computation time, however, it is advised to consider other improvements to the model first, such as the implementation of methods to detect additional cueing error types in the non-linear part of the MIR model.

5.2.4. Model Parameters and Choices

Because of the limited knowledge on what is important for the prediction of the PMI, the model described here is of a general nature. This approach allows for implementation or removal of subsystems, depending on the model fit to experimental data. Due to the limited dataset used for the analysis in this study, not all subsystems presented here were used. However, because these subsystems might be useful for larger datasets, all current options are still discussed.

In Table 5.2 different choices that can be made to obtain sub-models of the general MIR model are shown. Additionally, some other 'free' parameters of the general MIR model are listed in Table 5.3, together with their values used in this chapter.

Table 5.2: MIR model choices

Subject	Choice	Influence on model	Choice in this study
Input set	Different subsets of the total input set can be used	Less inputs will decrease model complexity, but might reduce the model fit.	Several input subsets are introduced and analyzed in Sections 5.4.1 and 5.5, respectively
Error Calculation	Different transformations such as absolute value and even power functions can be used	Determines the general relation between a motion errors and PMI.	Absolute value function
CEDA	The CEDA can be included or not	Removing the CEDA will decrease model complexity, but might reduce the model fit.	Models with and without CEDA are analyzed in Section 5.5
PRRD	The PRRD can be included or not	Removing the PRRD will decrease model complexity, but might reduce the model fit.	Not used
PRAD	The PRAD can be included or not	Removing the PRRD will decrease model complexity, but might reduce the model fit. Currently a non-smooth version of this algorithm is implemented, making it unsuitable for optimization purposes.	Not used
Linear Model Structure	Any model structure derived from the Box-Jenkins model can be used	More complex models than the ARX model are not expected to improve the model fit much as compared to including or removing inputs, but instead will restrict the possibility to obtain quick analytic solutions.	ARX model structure
Output Type	Depending on the experiment data and the goal of the model, the model output can be compared to a MIR obtained via a range of averaging techniques	Non-smooth averaging techniques could complicate the model identification. Averaging over all participants can be used when a general measure for PMI is requested, while averaging over specific subgroups, such as participants with different levels of sensitivity to rotation, can be used to estimate models specific for such a subgroup.	Median

Table 5.3: MIR model parameters

Parameter	Description	Influence on model	Value in this study
n_a	Order of the response system polynomial A	Increased order increases model complexity and most likely initially improve model fit [a]	Maximum 4
n_b	Orders of the cueing error polynomials B_i	Increased order increases model complexity and most likely initially improve model fit [a].	Maximum 4
n_k	Cueing error delays	Incorrect estimation will decrease model fit.	Should be estimated based on experimental data
CEDA	for CEDA parameter descriptions, see Section 4.2.7	Adjusting these parameters influences the detection of cueing error types scaled, missing and false cues.	The parameters from Table 4.2.7 are used
th_{det}	Detection threshold for rotational velocity	Increasing this value reduces the amount of rotational velocity that is assumed to be perceived. For a logical model this parameter should be based on perception literature.	-
th_{discr}	Discrimination threshold for rotational velocity	Increasing this value reduces the amount of rotational velocity that is assumed to be perceived. For a logical model this parameter should be based on perception literature.	-

[a] It is clear that increasing model complexity can result in a better fit, but also a worse prediction. The latter happens when the model is fit to not only the actual PMI but also the system noise.

5.3. System Identification Process

To actually predict the PMI from vehicle and simulator motions, a parametric model needs to be derived from the general MIR model described in the previous section. In this section a system identification process is proposed to select relevant inputs, minimal model orders and parameter values. To simplify the process, only the parameters of the linear part of the model are estimated, while the parameters of the non-linear part are derived either from literature, from Chapter 4 or manually tuned.
The inputs to the linear part of the model are the cueing errors, calculated by the non-linear part of the model, and the output is the MIR. There are several characteristics of the input/output set for the linear part of the MIR model that affect the system identification (SI) process significantly. The characteristics summarized below should be taken into account throughout the SI process.

- **Unknown inputs**: Not all inputs described in Section 5.2.1 have a significant influence on the MIR and a selection thus needs to be made.
  - To make the model less complex and more robust, an iterative approach needs to be adopted to determine which inputs are relevant for the model. In Section 5.3.2 such an input selection strategy is described and tested.
- **Correlated inputs**: Due to the physical relations between the motion channels, the inputs will be correlated. This correlation greatly influences the accuracy of the parameters that will be estimated, as these will be correlated as well.
  - When analyzing the model, accuracy of the input-output system, rather than the accuracy of individual parameters, should therefore be inspected [159].
- **Limited input power**: The vehicle motions cannot be freely designed to have equal power in a predefined frequency range, because they have to concur with (in our case) actual car driving on realistic roads. For example, the often used periodic signals with which specific frequencies are excited in mechanical systems, would, though very informative, not be a representative input for the model. Instead the distribution of signal power can only be designed indirectly and within narrow limits by using different road profiles, cueing algorithms and driving behavior.
  - The limited input power in certain frequencies results in poor estimation of the corresponding parameters with large uncertainties. These large uncertainties, in turn, can result in poor prediction power of the model. It should be kept in mind that for accurate estimation of parameters in higher order models, a large dataset with well distributed input power is needed.
- **Non-negative output**: Cueing errors are reasonably assumed to always reduce, rather than also increase, cueing quality. As the MIR ranges from zero to one and the inputs to the linear-part of the system are always positive, the contribution of these inputs to the modeled MIR should thus also be positive.
  - While the cueing errors are all non-negative, phase shifts can still cause the contribution of a cueing error to the MIR to be negative. During the identification process, attention should be paid to avoid such inconsistent results.

Based on the goal of the model, prediction of PMI from simulator and vehicle motions, and the input/output characteristics, a system identification (SI) process was developed. As mentioned in [159] if the purpose of the model is to simulate the system or to predict the future outputs of the system, then time-domain methods are the most appropriate and these are therefore used here as well. In Figure 5.5 an overview of the developed SI process is given.
As indicated in this figure, before starting the SI process, the inputs and subsystems used in the non-linear part of the MIR model need to be chosen. These choices influence the cueing errors that are used as the input set to the linear part of the MIR model. Additionally, an averaging technique for the measured output needs to be chosen. These choices combined lead to a model definition, which can then be parametrized using the SI process. The SI process itself has some additional parameters that also need to be chosen beforehand and are further explained in Section 5.3.2. The SI process itself can be divided in five steps:

1. Initial parameters selection,
2. Input selection,
3. Order reduction,
4. Parameter estimation, and
5. Model explanatory analysis.

In Step 1 of the SI process the initial model orders and input delay estimates need to be determined. Using this information, the most relevant inputs for describing the MIR output are selected in an iterative process in Step 2. After the relevant input set is selected, a similar iterative process is used in Step 3 to decrease the model order, to reduce model complexity and improve the prediction power of the model. With the inputs, model orders and estimated delays selected, the model parameters can be estimated in Step 4. Next, in Step 5, a model explanatory analysis is performed and a decision is made on whether or not to change initial parameters model n_a, n_b and n_k, and restart the process.

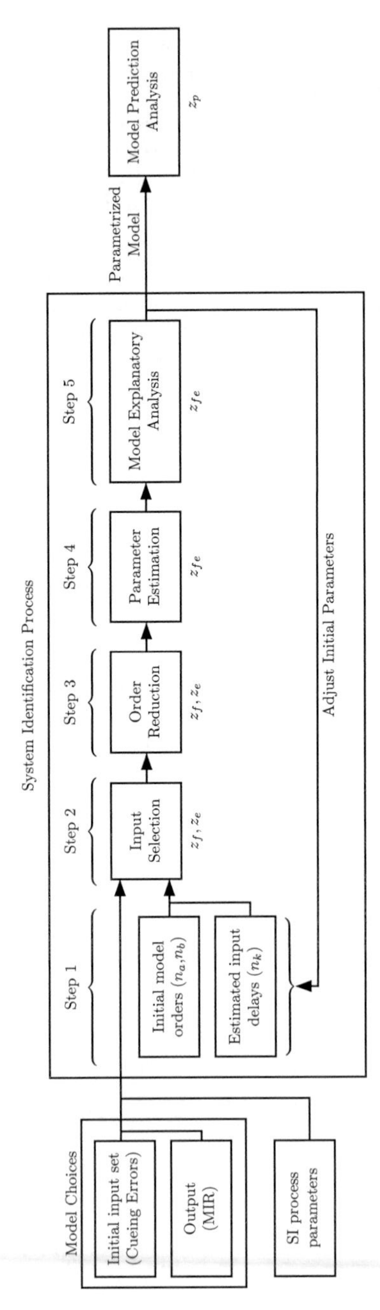

Figure 5.5: Identification process for linear part of the model.

When these initial parameters do not require changes, the prediction power of the parametrized model can be analysed. The SI process can be repeated for several models with different model choices and different SI system parameters, after which the 'best' model can be chosen, based on both the explanatory and prediction analysis results. In Subsections 5.3.1 till 5.3.4 the SI process steps are explained in more detail.
Throughout the SI process different datasets z are used to avoid over-fitting. As the process consists of multiple steps that include model fitting, the data are split in four different datasets.

- Fit dataset z_f: used in Steps 2 and 3 to estimate model parameters.
- Evaluation dataset z_e: used in Steps 2 and 3 to evaluate the cost function and determine the relevance of an input or model order.
- Parameter estimation dataset z_{fe}: used in Steps 4 and 5 to estimate the actual model parameters and analyze the model fit, respectively.
- Prediction dataset z_p: used after the SI process is completed to analyze the model prediction power.

Datasets z_f and z_e should at least include different measurement noise realizations, to avoid over-fitting. Datasets z_{fe} and z_p should contain different input signals as well, such that the prediction capability of the model can be analysed.

5.3.1. Step 1: Initial Parameter Selection

When a specific model is chosen and the SI process parameters are set, the initial model parameters n_a, n_b and n_k can be chosen. It is advisable to choose model orders that are somewhat higher than what one expects them to be, as they will anyway be reduced later during the model order reduction, Step 3 of the SI process. Setting the initial model orders instead too low, can result in the undesired removal of certain inputs during the input selection step.
The initial model delay can be estimated using the cross-correlation between the cueing errors and the measured output. The lag at which the highest correlation coefficient is found should be used as the initial model delay. To avoid unrealistic delays, only a limited range of lags, here taken as 0-3 seconds, should be taken into account. If no clear maximum is found, that particular cueing error is most likely not highly relevant for the MIR, and finding the correct delay is therefore less important. As a time delay is equivalent to a special case of phase shifts, e.g., linear phase shifts, they can be partially accounted for by the phase shifts generated by the polynomials A and B. In cases where no clear time delay is found, it is therefore best to choose a minimal delay.

5.3.2. Steps 2 and 3: Input Selection and Order Reduction

The input selection and order reduction steps of the SI process both deal with decreasing the complexity of the model. Due to the large number of possible inputs and corresponding model orders, a stepwise regression method needs to be adopted for both input selection and order reduction.
With the stepwise regression method, a model is fitted at each iteration to determine whether a variable should be removed or added. As mentioned in the previous section, the model used for input selection should have a relatively high model order to avoid unwanted removal of inputs. For example, assume that one uses only static gains for the input selection and for one input only its high frequency content contributes to the

MIR. Fitting the static gain would probably result in close to zero gain for this input, just because the low frequency content was not filtered out. This input, with valuable information in its higher frequencies, would thus not be selected if the filter order is too low.
The two main approaches for stepwise regression are forward selection and backward elimination. The former approach starts with fitting a model with only one input to the measured output. With the help of a predefined selection criterion (SC), the unique contribution of a specific input (uic_i) can be calculated. The input with the largest unique contribution is then selected and the process is repeated until none of the potential inputs result in a unique contribution higher than a preselected threshold (th_{uic}). Because of the before-mentioned input characteristics and the use of filters, the models that are fitted with few inputs can give very unrealistic results. An input that has power at only a limited time range, for example, can be filtered to (poorly) fit to the MIR over a longer period by low pass filtering it at a very low frequency. Using forward selection therefore gives very unrealistic results and is thus not applicable here.
Backward elimination, instead, starts by fitting a model that uses all inputs and then calculates the unique contribution of each input, uic_i, by removing them one at the time. In this case, more realistic results are expected to be obtained at each iteration step, as all inputs together do have power over the whole time range. This approach, similar to the approach described in [160], is therefore much better suited for use in this study.
The unique contribution of an input needs to be recalculated each time after removing one of the inputs. If two inputs are highly correlated, their unique contribution is very small when both inputs are still present in the dataset. However, when one of the inputs is removed, the unique contribution of the correlated input can increase significantly. With this approach, an input that correlates strongly with the MIR can be removed very early in the elimination process, if it also correlates strongly with another input. This should be taken into account when defining the selection criterion for the input elimination.
After the relevant inputs are selected, the model order is determined. As the model still has relatively many inputs, which can also be partially correlated, simply trying out many different combinations of model orders would take far too much computation time. Therefore a stepwise regression method is also applied here. Instead of removing inputs, at each step one of the polynomial orders is reduced. The selection criterion for this order reduction is then calculated and the unique order contribution (uoc_i) is determined. The order reduction resulting in the lowest unique contribution is applied and the process is repeated until none of the order reductions result in a unique contribution below a chosen threshold (th_{uoc}).
Using backward elimination, instead of forward selection, for order reduction, ensures that the final polynomial orders do not exceed the orders used for input selection. If the model analysis shows that a higher polynomial order is preferred, rather than increasing a single polynomial order, the order of the polynomials for all initial inputs should be increased and the input selection should be redone to ensure a fair comparison of the input contributions.

Selection Criteria

To calculate the unique contribution of a variable, e.g., input or polynomial order, two datasets, z_f and z_e, are used. The model parameters are fitted on dataset z_f by minimizing the prediction error. The fitted model is then evaluated by applying it to dataset z_e. If the prediction error is minimized by fitting the model parameters to noise in

dataset z_f, evaluation of the model on a different dataset makes sure that this does not inadvertently improve the selection criterion.
The selection criterion SC is chosen to compute the normalized weighted root mean square error indicated as a percentage:

$$SC = 100 \cdot \left(1 - \frac{\sqrt{(\mathrm{MIR}_{meas} - \mathrm{MIR}_{mod})^T W (\mathrm{MIR}_{meas} - \mathrm{MIR}_{mod})}}{\sqrt{\left(\mathrm{MIR}_{meas} - \overline{\mathrm{MIR}}_{meas}\right)^T W \left(\mathrm{MIR}_{meas} - \overline{\mathrm{MIR}}_{meas}\right)}} \right), \tag{5.1}$$

where MIR_{meas} is a vector with the measured MIR, MIR_{mod} is a vector with the modeled MIR and $\overline{\mathrm{MIR}}_{meas}$ the mean measured MIR over all time steps. Each time step can also be given a weight via the diagonal matrix W, with the weights (with mean weight of one) per time step on its diagonal. The criterion can have a value between minus infinity, indicating a very poor fit, and hundred, indicating no errors and thus a perfect fit. A criterion value of zero, indicates that the model is not better than a straight line at a value of $\overline{\mathrm{MIR}}_{meas}$.
This criterion is first calculated for all inputs considered in a specific step, which is indicated with SC_{inp_0}. Next the criterion is calculated when one of the considered inputs is removed, which is indicated with SC_{inp_i}, where i indicates the input that is removed. The unique contribution of input i, is then calculated with:

$$uic_i = SC_{inp_0} - SC_{inp_i} \tag{5.2}$$

The unique input contribution thus indicates which percentage of the model fit can be uniquely attributed to a specific input. This unique contribution is compared to a pre-determined threshold th_{uicp}, which can be chosen based on the model usage. If model simplicity is the most important, the threshold should be set higher than when model accuracy is the most important. If all unique contributions are above the threshold th_{uicp}, the process is stopped and no more inputs are removed.
For the order reduction process the unique contribution is calculated with:

$$uoc_p = SC_{ord_0} - SC_{ord_p}, \tag{5.3}$$

where SC_{ord_0} indicates the value of the selection criterion for the model that is considered in the current step, and SC_{ord_p} is the selection criterion for the same model, but where the order of polynomial p is reduced with one. Polynomial p can indicate any of the polynomials, e.g., the response system polynomial A or one of the input polynomials B_n, that at that point in the process has an order higher than zero. The order reduction resulting in the lowest uoc_i is performed, if the uoc_i is below a predefined threshold th_{uoc}. The order reduction is stopped if none of the unique contributions are below this threshold. Appendix C of this thesis describes how the values for the thresholds th_{uic} and th_{uoc} were chosen, based on simulations using synthetic datasets.
As indicated in Figure 5.5 there are some additional SI process parameters that can be set. These parameters only influence the input selection and order reduction process and are therefore discussed here. The first parameter is the weight vector W used in Eq. (5.1), which in this study is set to one for each time step. If certain time steps should be given more importance than other, because, for example, the agreement between participants is higher for certain time steps then for others, this weight can be varied.

In Appendix D some options for this weight vector are further explained. The second parameter is the boolean P_{ANIC}, which is set to one in this study and further explained in the next paragraph.

Avoid Negative Input Contributions (ANIC) As mentioned in the beginning of Section 5.3, one of the characteristics of the MIR is that it is always non-negative. The inputs to the linear part of the model, the cueing errors, are expected to only increase, rather than decrease, the MIR, as it seems illogical that cueing errors improve the MCA quality. The parameter estimation that occurs in the input selection process, however, can result in optimal parameters that cause large phase shifts, resulting in a cueing error being modelled as decreasing the MIR. To have the input selection result in more logical input choices, e.g., inputs that solely increase the MIR, an SI parameter, P_{ANIC}, can be set to one to adjust the input selection process.
During the input selection process the unique contributions of each input are calculated and compared to the threshold th_{uic}. At each iteration there can be multiple inputs that have a unique contribution below this threshold, but when $P_{ANIC} = 0$ only the input with the lowest unique contribution is removed. Setting $P_{ANIC} = 1$ adjusts the selection process when multiple inputs have a below threshold unique contribution. In this case, the models resulting from the removal of each of these inputs are analysed. For each model, the number of inputs that have a decreasing effect on the MIR (N_{NIC}) is calculated. Instead of removing the input with the lowest unique contribution, when $P_{ANIC} = 1$, the input that has a below-threshold unique contribution, and of which the removal results in a model with the smallest N_{NIC}, is removed.
This process reduces, but does not eliminate all inputs with a decreasing effect on the MIR, as this process only operates when multiple inputs have a below-threshold contribution and the lowest N_{NIC} can still be larger than zero. In Appendix D the influence of setting $P_{ANIC} = 1$ on the model fit is shown.

5.3.3. STEP 4: PARAMETER ESTIMATION

The previous steps in the SI process should result in an ARX model definition with a minimum number of inputs and minimum polynomial orders n_a and n_b, as well as estimated delays n_k. The next step in defining the model is the parameter estimation.
For the parameter estimation in this study, the Matlab implementation of the prediction error method using a least squares estimator is used. This estimator equals the maximum likelihood estimator when the system noise is independent and identically distributed (IID) Gaussian noise. For more information on the relation between maximum likelihood and prediction error estimation, see Appendix E.
The measurements in this study, however, indicate that the system noise cannot be described as IID Gaussian noise. To investigate the effect this has on the parameter estimation, a noise model was developed for the data in this study and used to generate noise samples for synthetic datasets. The comparison between back-estimation of random ARX model parameters when using IID Gaussian noise, and noise from the developed noise model shows no significant difference in estimated parameters. Therefore, the parameter estimates using the prediction error method with least squares estimation are still assumed to be the maximum likelihood estimates. More information about the noise modelling and back-estimation results can be found in Appendix E.

5.3.4. Step 5: Model Explanatory Analysis

The previous steps in the SI process resulted in a parametrized model that maps the vehicle and simulator motions onto the motion incongruence rating. To determine whether this model can be used to explain the rating observed in dataset z_{fe} the model explanatory analysis is done. This analysis includes the goodness-of-fit, residual and uncertainty analyses. After the explanatory analysis, a decision can be made to change the initial model orders and/or input delays and restart Steps 2-4 of the SI process. When the analysis shows that the model can explain the measurements sufficiently, the SI process is done and a prediction analysis can be performed using a new dataset, z_p. This analysis, which is not used to alter the model and therefore only discussed in Section 5.4, focuses on the main goal of the model: how well can the model predict the MIR from solely the vehicle and simulator motions?
In the following paragraphs the goodness-of-fit, residual and uncertainty analysis are further explained.

Goodness of fit After a parametrized model is obtained, a goodness-of-fit can be calculated by comparing the measured MIR to the simulated MIR obtained with the parametrized model and the measured inputs. In this study the variance accounted for (VAF) is used to quantify the goodness of fit of each model:

$$VAF = 100 \cdot \left(1 - \frac{\Sigma((\mathrm{MIR}_{meas} - \mathrm{MIR}_{mod})^2}{\Sigma\left(\mathrm{MIR}_{meas} - \overline{\mathrm{MIR}}_{meas}\right)^2} \right) \tag{5.4}$$

Similar to the selection criterion, the VAF can have a value between minus infinity, indicating a very poor fit, and hundred, indicating no errors and thus a perfect fit. A criterion value of zero, again indicates that the model is not better than a straight line at a value of $\overline{\mathrm{MIR}}_{meas}$.

Negative input contributions The influence of each input on the modelled MIR is analysed in the time domain, by analysing the input contributions over time, which result from running each cueing error through their respective $\frac{B_i}{A}$ polynomial pair system. This analysis focuses on the possible negative contributions of a cueing error to the model output. If a large negative input contribution is detected, this is an indication that the model is not a realistic representation of the underlying system.

Residuals If the model is indeed representing the underlying system correctly, the prediction errors, also referred to as the residuals, should show the same characteristics as the assumed system noise. In general the system noise of ARX models estimated with least squares estimators should be independent, homoscedastic and normally distributed [161]. However, as explained in Section 5.3.3, the actual system noise is probably heteroscedastic and non-normal. Appendix E showed that violations of these noise assumptions do not influence the parameter estimation significantly. Instead, therefore, the residuals will only be tested for their independence of inputs and outputs.
If the residuals depend on their own past values or on input values, this means that the information from the inputs and past outputs is not correctly captured by the model. To correct for this, a first step can be to increase the polynomial orders, where high autocorrelations usually relate to polynomial A and high cross-correlations to polynomial

B or incorrectly estimated delays. If very high orders are needed to reduce the correlations sufficiently, it is possible that other model structures than the ARX structure should be used. For example, if high orders for polynomial A are needed, it is possible that an ARMAX structure, which includes a moving average term, would describe the system better.

Uncertainty Analysis If the residuals have been shown to be independent of the input and output, the uncertainty analysis can be performed. In this analysis the covariance matrix of the parameters is used to investigate relations between parameters as well as the confidence intervals of each parameter. High correlation coefficients between parameters of the same B polynomial, can give an indication of wrongly estimated model orders, while high correlation between parameters of different B polynomials relates to input correlations. Therefore, in case high correlations are found, the model order should probably be adjusted. When the 95% confidence intervals of the parameters are large, this can, for example, indicate that the model order is chosen too high, or that the frequency of the signal related to this parameter has low power.
From the uncertainties of the model parameters, also confidence intervals of each $\frac{B_i}{A}$ polynomial pair system are investigated. Large confidence intervals in certain frequency ranges can be an indication of low input power at these frequencies. The model is therefore not expected to have a high prediction power for inputs with high power in these frequency ranges.

5.3.5. SI PROCESS PARAMETERS

In Table 5.4 the SI process parameters discussed in the previous sections are shown, together with their influence on the SI process and the values used in this study. The values for th_{uic} and th_{uoc} are chosen based on the synthetic data results discussed in Appendix C. Parameter vector W is set to provide equal weights for all time steps, e.g., $W = W_1$, and P_{ANIC} is set to one to avoid the selection of inputs that provide a negative contribution to the modelled output, see Appendix D. To demonstrate the influence of these parameters on the model design process, Section 5.5 will discuss results for the basic model with different values for parameters W and P_{ANIC}.

5.4. ANALYSIS

The model design procedure described in the previous sections aims to structure the process of designing a simple model to describe the, for motion cueing, most important features of the highly non-linear process of motion perception in vehicle simulation and the corresponding rating behaviour. With this process, many different models can be designed, depending on the model and SI process parameters shown in Tables 5.2, 5.3 and 5.4. To analyse the SI process, three initial parameter sets are chosen and the resulting three models are analysed using the data from the experiment described in Chapter 2. In the following sections the three parameter sets leading to three different models, the datasets and the analysis methods are further explained.

5.4.1. MODELS

The three parameter sets, named Basic, Additional Input (AI) and Cueing Error Detection Algorithm (CEDA), were chosen to validate the SI process described in the previous sections. These models are chosen as a baseline, to show the impact of including derivatives and integrals of standard motion channels and also to show the impact of

Table 5.4: System identification process parameters.

Symbol	Description	Influence on model	Value in this study
th_{uic}	Unique contribution threshold for input selection	Lower value results in more inputs being selected	0.2
th_{uoc}	Unique contribution threshold for polynomial order reduction	Lower value results in less order reduction	0.2
W	Time step weights used for input selection and order reduction	Can be used to give more importance to the fit at certain time steps such as there where participant ratings are in agreement	W_1
P_{ANIC}	Switch parameter to turn on negative input contribution avoidance in input selection	Set to one this will result in less inputs that give an illogical negative contribution to the output	1

adding the cueing detection algorithm in the non-linear part of the model, respectively. Each model used the SI process described in Section 5.3 to select the inputs, model order and parameter values of the linear part of the model. All models use the ARX structure for the linear part of the model. The only difference between the linear part of the three models is the initial set of cueing errors, resulting from choices made in the non-linear part of the model, to start the SI process with.
For each model the measured output is calculated as the median of the MIR over all participants, the initial polynomial orders n_a and n_b are set to four, and the input delays are initially calculated via cross correlation between each input and the model output for dataset z_{fe}. The SI process parameter W is set to W_1 such that each time step is equally important in the fit criterion for input and order selection. Parameter P_{ANIC} is set to one such that inputs with a negative input contribution are less likely to be selected during the input selection process.

Basic The basic model is the simplest model derived from the general model described in Section 5.2.3 and serves as a baseline. The initial input set includes the cueing errors in the six standard motion channels: a_x, a_y, a_z, ω_r, ω_p and ω_y. The model has a non-linear part that only includes the absolute value calculation of the difference between these vehicle and simulator motion inputs.
This model is most similar to the cost calculations, e.g., weighted sums of these cueing errors, that are currently used in many MCA optimization algorithms. The model differs from these cost functions in that it also contains dynamics. A static weight measure of the relative importance between cueing errors derived from this basic model could, for example, be the low-frequency (DC) gain for each input-output pair.

Additional Inputs The model with additional inputs (AI) increases the model complexity by simply adding integrals and derivatives of the cueing errors to the initial input set of the basic model. The initial input set includes the cueing errors in fifteen motion channels: a_x, a_y, a_z, ω_r, ω_p, ω_y, v_x, θ_r, θ_p, j_x, j_y, j_z, α_r, α_p and α_y. Similar to the basic model, this model has a non-linear part that only includes the absolute value calculation of the difference between these vehicle and simulator motion inputs.
Using cueing error weights derived from this model in a cost function for MCA optimization, would require additional motion channels to be analyzed by the cost function, but no additional non-linear calculations.

Including CEDA The third model includes not only the derivatives and integrals of the standard cueing errors, but also uses the CEDA described in Chapter 4 to distinguish between scaled (*sc*), missing (*mc*) and false cues (*fc*) in the a_x and a_y motion channels. The initial input set includes nineteen cueing errors: $a_{x_{sc}}$, $a_{x_{fc}}$, $a_{x_{mc}}$, $a_{y_{sc}}$, $a_{y_{fc}}$, $a_{y_{mc}}$, a_z, ω_r, ω_p, ω_y, v_x, θ_r, θ_p, j_x, j_y, j_z, α_r, α_p and α_y. This model has a non-linear part that includes the CEDA and includes the absolute value calculation of the difference between these vehicle and simulator motion inputs. This model is highly non-linear due to inclusion of the CEDA, and cannot easily be implemented as part of the cost function for MCA optimization.

5.4.2. Datasets

As mentioned in Section 5.3, the SI process makes use of four different datasets: z_f, z_e, z_{fe} and z_p. Datasets z_f and z_e are used for input selection and order reduction and should contain different measurement noise, to avoid over-fitting. Dataset z_{fe} is used for parameter estimation and is a combination of the previous mentioned datasets. Dataset z_p should contain different input signals, and is used to analyse the prediction capability of the model.
In this study, the total dataset z refers to the data obtained in the experiment presented in Chapter 2. Dataset z_{fe} contains measurements from manoeuvres 'Curve Driving' (CD) and 'Braking/Accelerating' (BA), each repeated with three different MCAs: MCA_{scal}, MCA_{TRL} and MCA_{NL}. These three MCAs cause global motion scaling with a gain of 0.6, tilt-rate limiting to 1 $[deg/sec]$ and large tilt-rates as no limiting is applied, respectively. Dataset z_p instead consists of data from the manoeuvre 'Braking, Curve Driving, Accelerating' (BCDA) repeated with all three MCAs, such that both z_{fe} and z_p contain power in the same cueing errors. Hence, we expect the model fitted on z_{fe} to predict the output in dataset z_p.
Dataset z_f contains MIR measurements from trials one and two, while dataset z_e contains measurements from trial three. The difference between these two datasets should thus only be due to noise within participants. Another option could have been to use different participants groups per dataset, rather than different trials. However, the differences between participants are not expected to be solely due to rating noise, but rather represent actual biological- or preference-related differences. As the goal of the model in this study is to predict a MIR that provides insight in the *average* participant, each dataset should include as many participants as possible.
In Figure 5.6 the resulting measured median MIR, together with six cueing errors (absolute differences between vehicle and simulator motion in channels a_x, a_y, a_z, ω_r, ω_p and ω_y) are shown for datasets z_f, z_e and z_{fe}. In Figure 5.7 the same information is shown for dataset z_p.

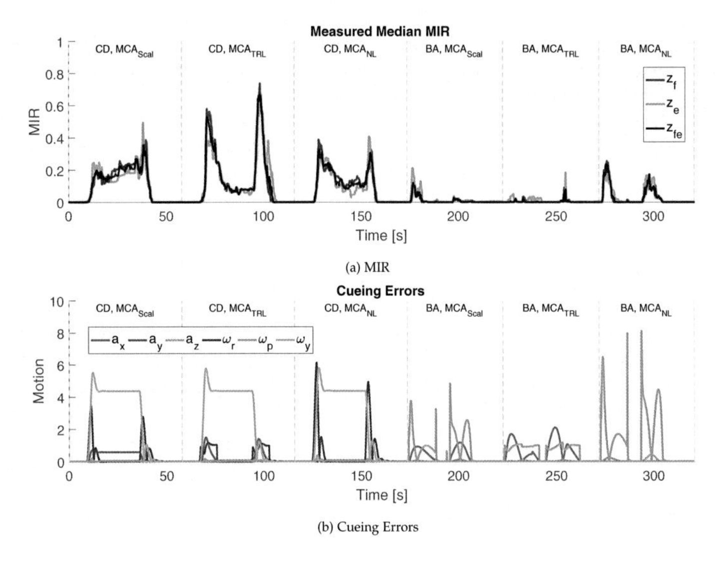

(a) MIR

(b) Cueing Errors

Figure 5.6: Measured median MIR and six possible cueing errors (absolute differences between vehicle and simulator motion in channels a_x, a_y, a_z, ω_r, ω_p and ω_y) for datasets z_f, z_e and z_{fe}.

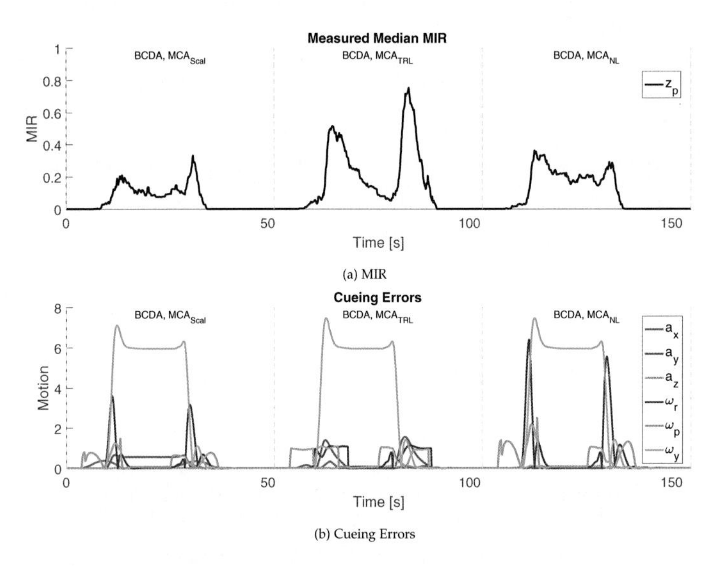

(a) MIR

(b) Cueing Errors

Figure 5.7: Measured median MIR and six possible cueing errors (absolute differences between vehicle and simulator motion in channels a_x, a_y, a_z, ω_r, ω_p and ω_y) for dataset z_p.

5.4.3. METHODS

To investigate the validity of the models, three analyses are performed: the model structure, the explanatory and the prediction analysis. The first analysis considers the SI process outcome in terms of which inputs, model orders and delays are chosen. The second analysis shows how well the model can describe the data that were used for fitting the model, while the latter shows how well the model can predict the average rating of a new dataset that was not used to design the model.

Model Structure Analysis In the model structure analysis the models that resulted from the SI process when using the different initial SI parameter sets described in Section 5.4.1 are introduced and the logic of the choices for input, order and delays are discussed.
To give an indication of how important each input is for the model, two additional parameters per model input are calculated for each of the models: the low-frequency (DC) gain (G_{DC}) and the output contribution percentage (OCP). The former could replace weights in a cost function for MCA optimization and shows the relative importance of the cueing errors with the same unit. The latter, instead, shows the relative importance of each cueing error for this specific rating, irrespective of the unit, and is calculated with:

$$OCP = 100 \cdot \frac{\sum_{t=0}^{T} |IC|}{\sum_{t=0}^{T} |y_{mod}|}, \tag{5.5}$$

where IC indicates the input contribution for one cueing error, calculated by multiplying the cueing errors with their respective A/B polynomial pair, t indicates the time step and T the total time duration of the rating in this dataset.

Model Explanatory Analysis For the explanatory analysis, dataset z_{fe} is used. This analysis is similar to the analysis described in Section 5.3.4 and is split up as follows:

- Goodness of fit
  - The results of the model output fits to the average rating is shown.
  - The goodness of fit is quantified with the Variance Accounted For.
- Negative input contributions
  - The input contributions for each cueing error are shown per model. The input contribution represents the cueing error after transformation by its respective input-output polynomial pair.
  - A visual inspection of the input contributions is done to determine if the model results in illogical input contributions.
- Residual analysis
  - The auto-correlation of the residuals and the cross-correlations the residuals with the inputs is shown for a lag/lead of three seconds.
  - For the model to be valid, the correlations should not significantly exceed the 95% confidence interval.
- Uncertainty analysis

- The covariance matrix of the model parameters is shown to determine if the correlation between parameters related to the same input channel are sufficiently low.
- The parameter values and their 95% confidence intervals are shown to indicate which parameter estimate should be trusted.
- Bode plots for each input-output polynomial pair, together with the 95% confidence intervals, are shown to indicate over which frequency ranges the model is expected to be valid.

Model Prediction Analysis For the prediction analysis, dataset z_p is used. As this dataset was not used to fit the model, only the goodness of fit, negative input contributions and residual analysis can be performed.
As the goal of the model is to predict the average rating over time from different cueing errors, the most important analysis is the goodness of fit analysis.

5.5. RESULTS

The initial model parameters introduced in Section 5.4.1 were used in the SI process and analysed using the methods described in Section 5.4.3. In this section, the results from these analysis methods are shown for all three resulting models.

5.5.1. MODEL STRUCTURE ANALYSIS

In Figure 5.8 the block diagrams for each of the three models resulting from the initial SI parameter sets described in Section 5.4.1 are shown.
The structure of the Basic and AI models are the same, but they have different inputs. The CEDA model also has a different non-linear part, where the cueing error detection algorithm is used to split the linear accelerations in x and y direction in three different types of cueing errors (scaled, missing and false cues).
In Table 5.5 the estimated polynomial orders, input delays, input/output DC gains and the output contribution percentage (OCP) for each of these models are shown.
For the Basic model, the input selection resulted in the removal of longitudinal vehicle motion related inputs a_x and ω_p, leaving the model with two sets of each four inputs in total. The input selection for the AI model resulted in additional cueing errors in j_y and j_z compared to the basic model, resulting in a model with two sets of each six inputs. Finally the CEDA model, when compared to the Basic model, also includes a cueing error in ω_p, replaces ω_y with θ_r and uses all cueing error types of a_y ($a_{y_{sc}}$, $a_{y_{fc}}$, $a_{y_{mc}}$) instead of a_y itself.
The optimal delays are similar between models for the same inputs and are mostly below one second, with exceptions for $a_{y_{sc}}$, $a_{y_{fc}}$, j_y and θ_r. The cueing error types for a_y do show a clear difference between type and also with respect to the total cueing error a_y used by the Basic and AI models. The scaled and missing cues have a delay twice as large as the false cue and the total cueing error a_y in the other models.
The DC gains for the rotational velocities are between a factor five and ten lower than those for linear motions, at least partly due to the differences in unit. Another interesting finding is the differences in DC gain between the different cueing error types in a_y. The missing cue has a gain of one and a half times the gain of the scaled cue and the false cue a gain of almost two times the gain of the scaled cue. This shows that for equal amplitudes, a false cue is twice as detrimental to PMI than a scaled cue.

The importance of the different cueing errors for the rating in this dataset specifically is calculated with the output contribution percentage (OCP), which shows that for all models a_y gives a large contribution to the modelled output, as well as either ω_y or θ_r.

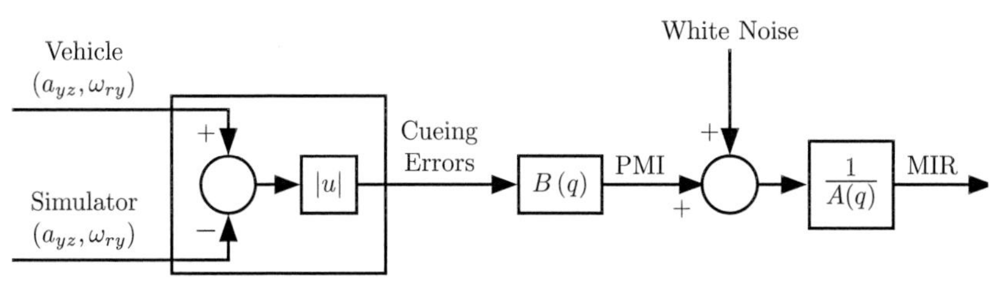

(a) Basic model derived from general MIR model.

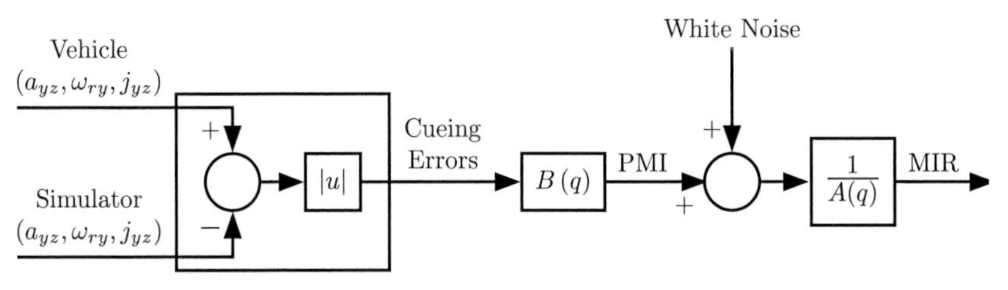

(b) AI model derived from general MIR model.

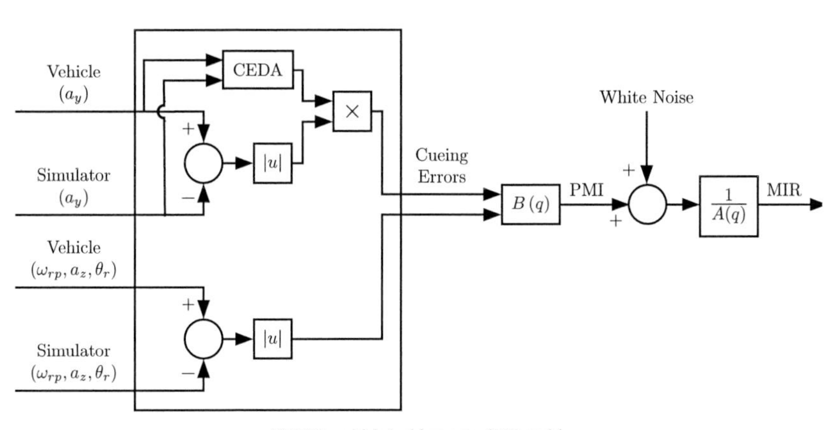

(c) CEDA model derived from general MIR model.

Figure 5.8: Model Structures.

Table 5.5: Estimated Model Parameters

Model/Signal	a_y	$a_{y_{sc}}$	$a_{y_{fc}}$	$a_{y_{mc}}$	a_z	j_y	j_z	θ_r	ω_r	ω_p	ω_y	MIR	**nr. param**
Polynomial orders A (MIR) and B (cueing errors)													
Basic	3				2				1		2	4	12
AI	2				2	1	1		1		2	4	13
CEDA		1	2	2	2			2	1	1		2	13
Input delays per cueing errors													
Basic	0.7				0.4				0.7		0.7		
AI	0.8				0.4	1.8	0.1		0.7		0.7		
CEDA		1.3	1.5	0.7	0.4			2	0.7	0.5			
DC Gain per cueing errors													
Basic	0.27				0.27				0.08		0.01		
AI	0.18				0.05	0.35	0.90		0.06		0.02		
CEDA		0.20	0.31	0.38	0.07			0.01	0.07	0.02			
OCP per cueing errors													
Basic	35				15				23		27		
AI	23				9	11	11		14		32		
CEDA		15	8	9	9			30	19	11			

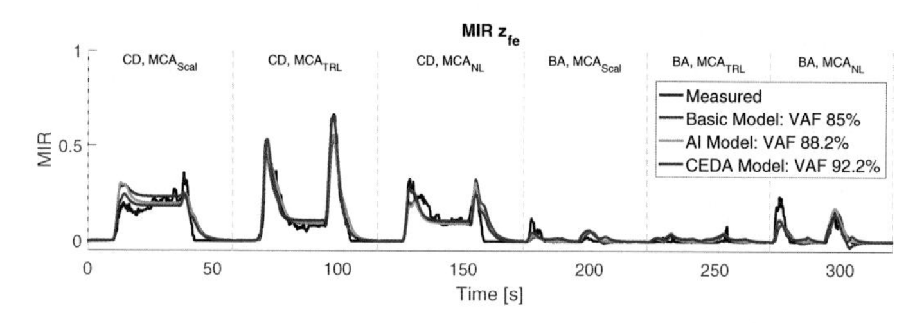

Figure 5.9: Basic, AI and CEDA model fits to dataset z_{fe}.

5.5.2. MODEL EXPLANATORY ANALYSIS

The explanatory analysis consists of the goodness of fit, the negative input contribution, residual and uncertainty analysis. In Figure 5.9 the goodness of fit of all models is shown. The goodness of fit is highest for the CEDA and lowest for the Basic model, with a difference in VAF of 7.2% at the cost of one additional parameters. The AI model only yields an improvement in the fit of 3.2% compared to the Basic model, at the cost of one additional model parameter. All models show a better fit for the CD maneuvers than for the BA maneuvers. The increased VAF for the CEDA model is mainly due to a better fit for the CD maneuvers.

To determine whether the input contributions make sense, i.e., cueing errors should not decrease the PMI and thus increase the cueing quality, the input contributions are shown in Figure 5.10. This figure shows that the Basic and AI models use ω_y to decrease the modelled rating at the beginning of each CD manoeuvre, while the CEDA model does not show this (physically: flawed) relation between cueing error and rating during these manoeuvres. To different degrees, all models also show some negative contribution with a_z, mainly during BA with MCA_{NL}.

Figure 5.10 also shows that the increased rating during CD manoeuvres is captured by introducing ω_y in the Basic and AI model, but with θ_r in the CEDA model. The rating during BA manoeuvres is modelled with the cueing errors in a_z by all models, and additionally by j_z and ω_p for the AI and CEDA models, respectively.

In Figure 5.11 the auto-correlations of the residuals and cross-correlations of the residuals with the inputs for each model are shown, in the left and remaining subplots, respectively. None of the models result in residual correlations significantly higher than the 95% confidence interval (CI). The cross-correlation between residuals and input ω_r, however, slightly exceeds the 95% CI around the three seconds lag for all models, indicating that information in this input is possibly not modelled completely.

In Figure 5.12 the results of the parameter uncertainty analysis are shown. In the top plots for each sub-figure the parameters and their 95% CI are shown and in the bottom plots the correlation matrix is shown. In Figure 5.13 the bode plots for each $\frac{B_i}{A}$ polynomial pair are shown for each model together with their 95% CI resulting from the parameter uncertainty. The uncertainty analysis results show that the parameters related to the same input at consecutive time steps always have a high negative correlation. This can be explained by the high correlation between the neighbouring time steps themselves and is inherent for ARX model structures. More interesting is that parameters related to different inputs do not show high correlation. This indicates that the input selection process indeed removed correlated inputs sufficiently.

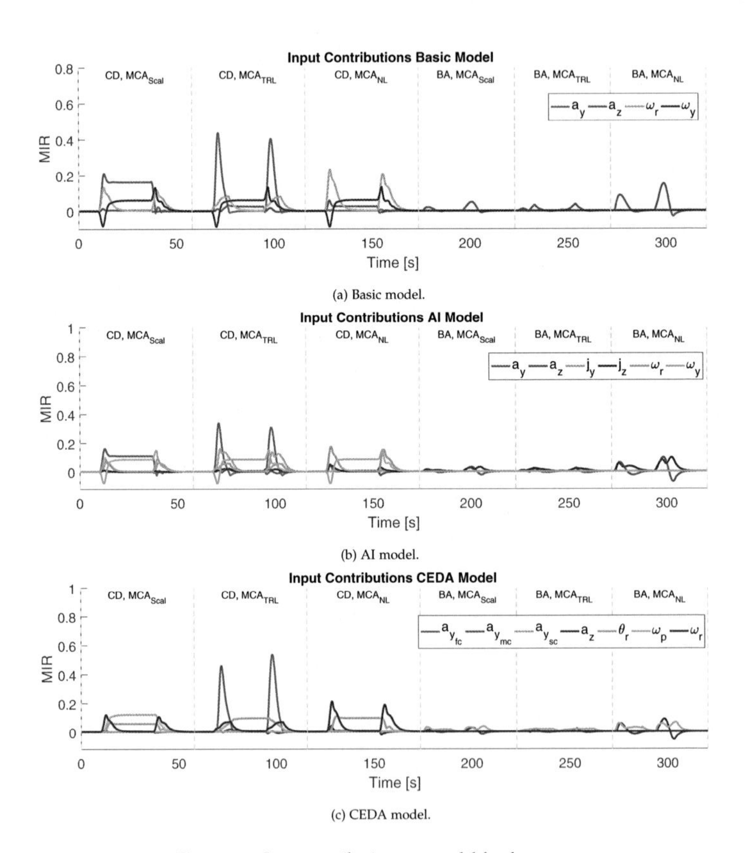

(a) Basic model.

(b) AI model.

(c) CEDA model.

Figure 5.10: Input contributions per model for dataset z_{fe}.

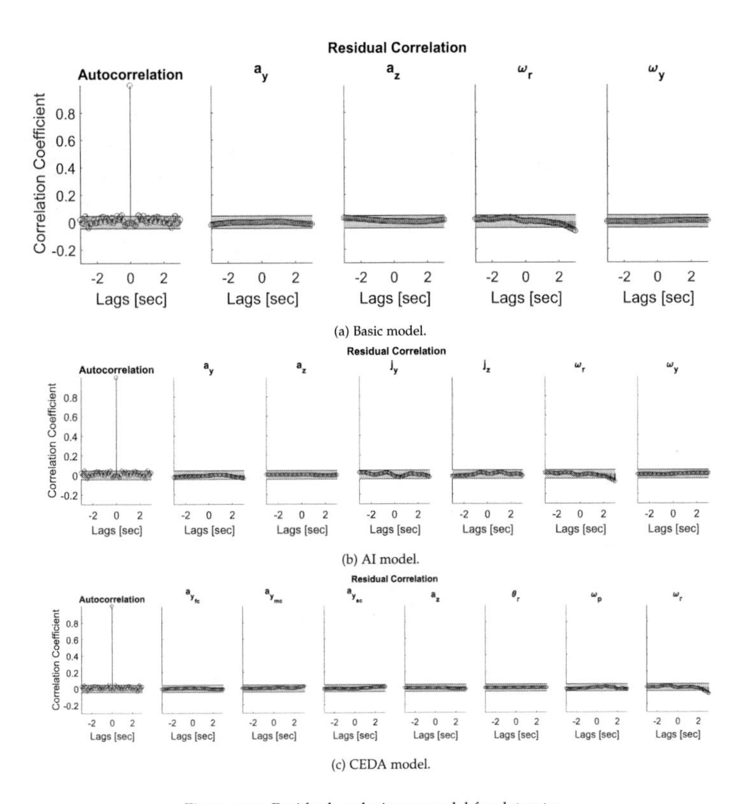

(a) Basic model.

(b) AI model.

(c) CEDA model.

Figure 5.11: Residual analysis per model for dataset z_{fe}.

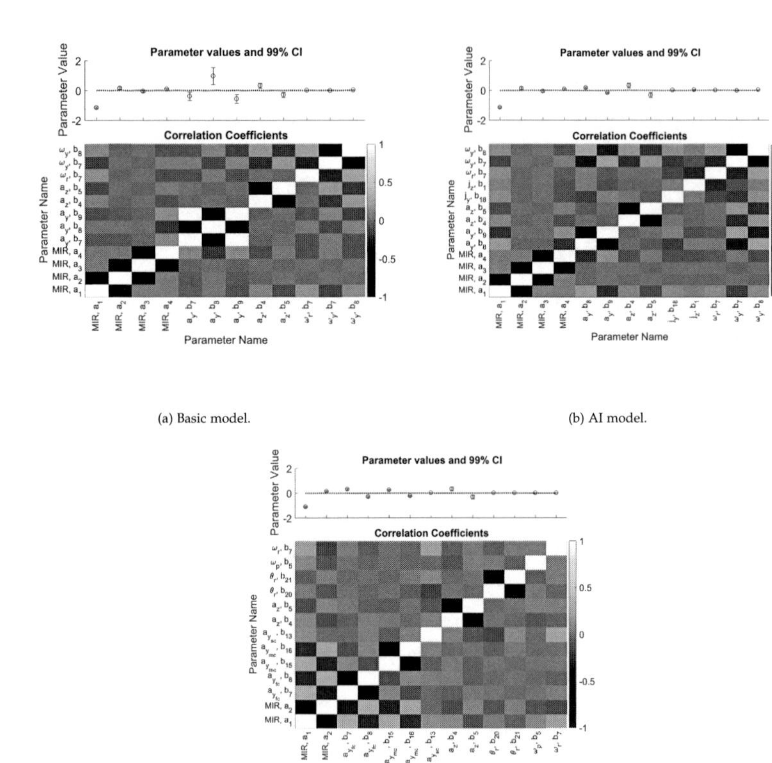

(a) Basic model.

(b) AI model.

(c) CEDA model.

Figure 5.12: Uncertainty analysis for model parameters.

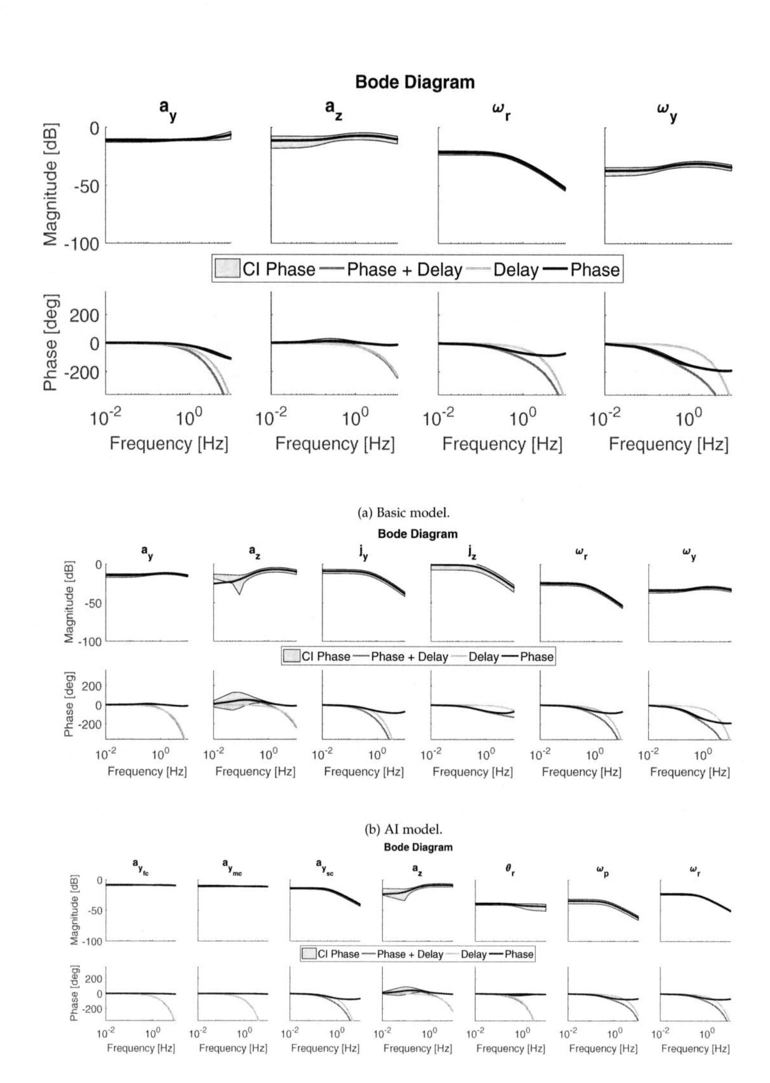

(a) Basic model.

(b) AI model.

(c) CEDA model.

Figure 5.13: Bode plots for each $\frac{B_i}{A}$ polynomial pair per model.

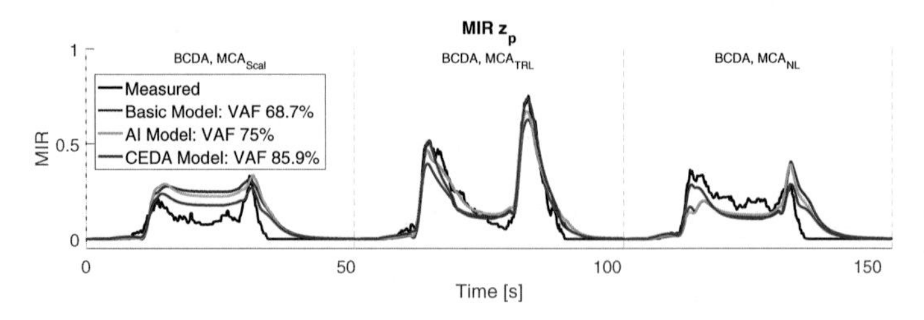

Figure 5.14: Basic,AI and CEDA model prediction for dataset z_p.

Figure 5.13 indicates that cueing errors in a_z mainly lack low-frequency content, while the polynomials for θ_r are most uncertain in the high frequency range. The phase plots for a_z also show the phase delays which cause the negative contribution for this cueing error.

5.5.3. Model Prediction Analysis

The prediction analysis consists of the goodness of fit, the negative input contribution, and the residual analysis. In Figure 5.14 the goodness of fit for the prediction of the rating in dataset z_p is shown for each model. The prediction power of the CEDA model is highest, the Basic model the lowest, with a difference in VAF of 17.2%. The AI model also has an improved prediction power with a difference in VAF of 6.3% compared to the Basic model. The improvement of the CEDA model is visible in all sections of the dataset, with MCA_{TRL} showing the best prediction.
The input contributions for dataset z_p are shown in Figure 5.15. The input contributions for dataset z_p show similar effects as for dataset z_{fe}, with mainly cueing errors in ω_y causing negative contributions with the Basic and AI models.
In Figure 5.16 the auto-correlations of the residuals and the cross-correlations of the residuals with the inputs are shown for dataset z_p. The autocorrelations of the residuals show that not all information from previous outputs is captured by any of the models. The large negative correlations visible for the AI and CEDA models for inputs a_y, ω_r and ω_y are due to incorrectly estimated delays, which could also partly explain the observed autocorrelation of the residuals.

5.6. Discussion

In this chapter a generalized model design process for models that aim to predict motion incongruence ratings was introduced. The design process consists of five steps that should be iteratively followed. The process includes several parameters and choices, such as initial input set, that result in different models. Three representative models were constructed and the results for the model structure, explanatory and prediction analysis were shown.

5.6.1. SI Process

The system identification (SI) process described in this chapter was set up to deal with the inherent challenges of modelling PMI, caused by the characteristics of the system that needs to be modelled and its corresponding inputs and outputs. To deal with the selection of the many possible, often highly correlated, inputs, an iterative input selec-

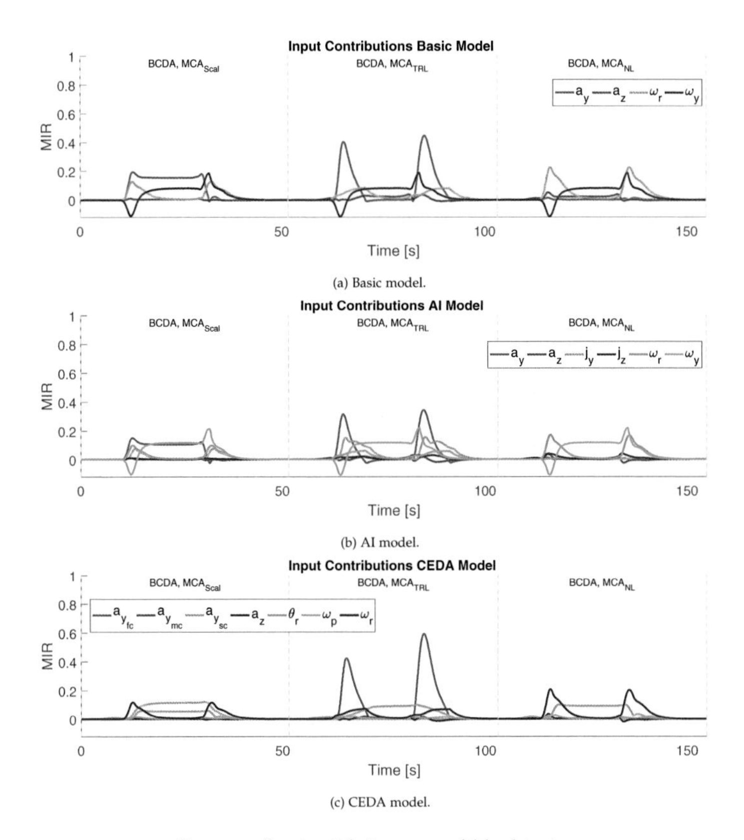

(a) Basic model.

(b) AI model.

(c) CEDA model.

Figure 5.15: Input contributions per model for dataset z_p.

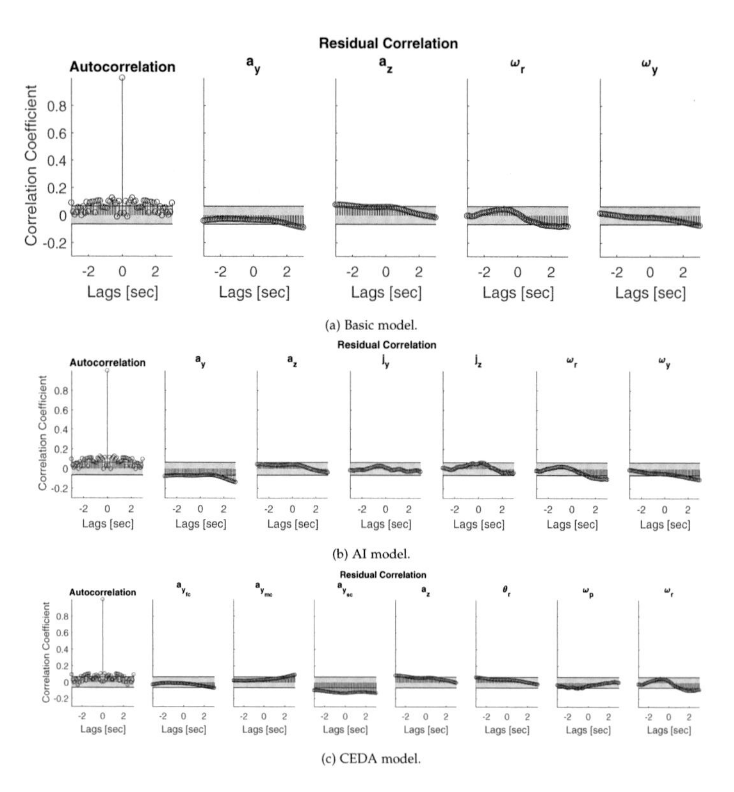

(a) Basic model.

(b) AI model.

(c) CEDA model.

Figure 5.16: Residual analysis per model for dataset z_p.

tion algorithm, based on backward elimination, was introduced.
The criterion thresholds used for eliminating an input were chosen based on simulations, and a threshold resulting in a minimal amount of extra inputs, e.g., noise fitting, was chosen. The uncertainty analysis of the model resulting from this SI process showed that indeed no highly correlated inputs were selected for the final models, indicating that the process works as expected. From the threshold simulation analysis, however, it was shown that it is still possible that incorrect inputs are found, and that the number of incorrect inputs increases with increasing number of actual system inputs. This effect can partly be explained by the low input power in a wide range of frequencies, which is inherent to the experiment set up where real vehicle motions are used. Increasing the size of the dataset and its cueing error variability, however, is assumed to reduce the incorrect selection of inputs, and make the process more robust.
For the input selection also the ANIC parameter was introduced, which provides an additional cost for inputs that contribute negatively to the modelled output when set to one. This input-output relation is undesirable as it infers that a cueing error can *improve* cueing quality. In Appendix D the effect of setting the parameter ANIC to zero is shown by the selection of an input that only contributes negatively to the modelled output. Even though some partially negative input contributions were found in the final models shown in Section 5.5, none of the inputs provided a major negative contribution to the output, providing another indicator that the models could indeed be representing the underlying system.
The low input power mentioned before, also influences the model order selection. The generally used AIC method for selecting the polynomial orders was shown to result in the over-estimation of the model orders. Instead a model order reduction algorithm, also based on backward elimination, was presented that resulted in a much stricter order selection, causing less over-estimation of the model orders. The criterion threshold was again chosen based on simulations, which also showed the leniency of the AIC selection method. During the model analysis step in the SI process, any order reductions that were too strict were corrected, resulting in final models that showed no significant residual/input cross-correlations.
When a model structure is chosen, the parameters in the linear part of the model can be estimated. The ARX structure was chosen to represent the linear part of the model, due to its simplicity and resemblance to the author's basic idea of processing cueing errors into PMI. Due to the many parameter estimation iterations used in the SI process, the low calculation time of the linear least squares estimation method is desirable. However, the system that is modelled here violates the noise assumptions that would make this estimator a maximum likelihood estimator. Via simulations, using a developed noise model that imitates the real system noise, it was shown that the violations do not result in significant departures from the maximum likelihood estimation. However, the parameter estimation through least squares is therefore reasonably assumed to be still close to the maximum likelihood estimate. It should be noted, however, that some small differences between the noise model and the measured noise were visible. For future work it is recommended to improve the noise model and confirm that the noise does not significantly influence the parameter estimation.

5.6.2. MODELS

The goal of this study was to develop a general method to obtain a model that can predict the Motion Incongruence Rating (MIR) from the vehicle and simulator motions. The developed general model structure and system identification process were used to develop three models. The first model, the Basic model, resembles current cost functions, in that it only maps the six linear acceleration and rotational velocity motion channels onto the MIR. The second model, the Additional Inputs (AI) model, differs from the Basic model by also allowing the SI process to select cueing errors in the derivatives and integrals of these six motion channels as the model input. Finally, the CEDA model includes the algorithm to split cueing error in a specific motion channel into scaled, missing and false cues, described in Chapter 4, implemented in the non-linear part of the model.

Each of the models were fitted to a first dataset and used to predict the MIR of a new dataset from the same experiment. To different degrees, each of the models could explain and predict main features of the MIR.

The Basic model resulted in the lowest goodness of fit and prediction, with VAF values of 85% and 68.7% respectively, and the CEDA model in the highest goodness of fit and prediction, with VAF values of 92.2% and 85.9% respectively. The AI model resulted in a only slightly higher VAF of 88.2% for the goodness of fit, while the prediction was considerably higher than that of the Basic model with a VAF of 75%.

These results show that simply including additional motion cues as inputs to the model, such as linear jerk, already improves the prediction power of the model. For further improvements, however, a more sophisticated model is needed, i.e., one that takes into account the variable influence that cueing errors of different types have on the MIR, such as done with the CEDA model.

The residual analysis showed that all models capture the information in the selected inputs and previous output for the fit dataset adequately. Even though the model predictions showed that the main features in the prediction dataset were captured by all models, the residual analysis for this dataset shows that not all information in the inputs and previous output is used accurately. As the uncertainty analysis showed high uncertainties for some of the input-output pairs in certain frequency ranges, it is likely that the prediction will improve when fitting the data to a larger and more variable dataset. Additionally, the prediction dataset included both longitudinal and lateral motions at the same time, this interaction was not present in the fit dataset and thus not modelled. Including interaction effects in the model would probably require additional algorithms in the non-linear part of the model and would be interesting for further research.

Even though the size of the initial input sets differed largely, after the SI process each of the three final models have a similar number of parameters, with the Basic model having twelve and the AI and CEDA models each having thirteen parameters. The prediction power improvement of the AI and CEDA versus the Basic model of 6.3% and 17.2% respectively, was thus obtained with only one additional parameter.

The input selection resulted in lateral acceleration and roll rate being used by each model to simulate a large part of the MIR during the curve driving (CD) manoeuvres. For the Basic and AI models the constant increase in MIR throughout the curve is modelled with the yaw rate, while the CEDA model instead uses the roll angle. The former is partly contributing negatively to the output, making the roll angle the more logical choice. Additionally, the actual perceived yaw rate cueing error during constant turns likely differs from the yaw rate cueing error used here, as human perception of constant yaw rate decays over time [162, 163]. From a physical point of view, therefore, it is not likely that the yaw rate cueing error during manoeuvre CD as modelled here influences the PMI.

The CEDA model shows the best fit during the CD manoeuvres, mainly because it dis-

tinguishes between the scaled, missing and false cues with the first two MCAs. The scaled cue in lateral acceleration was estimated to have a DC gain of half that of the false cue, indicating that for the same amplitude, a false cue is perceived as twice as detrimental to the cueing quality than a scaled cue. The AI model tries to simulate this difference instead with the higher lateral jerk obtained with the second MCA. As this fit is clearly worse than the fit of the CEDA model, it is more likely that the difference in cueing error type is the cause of the difference in MIR.
Surprisingly, the errors for the braking and accelerating (BA) manoeuvres are not modelled with cueing errors in longitudinal acceleration, but instead with errors in *vertical* acceleration, which are mainly caused by tilt-coordination. The AI and CEDA models additionally explain the MIR during BA manoeuvres with the vertical jerk and the pitch rate, respectively.
In general, it thus seems that during BA manoeuvres the rotation was causing the increase in MIR rather than errors in longitudinal acceleration. This can be explained by the low accuracy with which humans can extract longitudinal acceleration from visuals [96] and the resulting large cueing errors in longitudinal acceleration that are perceived as congruent. The rotation, however, is possibly compared to the experience, rather than the direct visuals, which dictates that any large perceived rotations are not realistic for car driving.
The fit during the BA manoeuvre is worse than during the CD manoeuvre with all models, indicating that some improvements can be made here. On the one hand, additional cueing errors can be calculated that take the human perception during these kind of manoeuvre more explicitly into account. On the other hand, the cueing errors in the BA manoeuvre were much less pronounced, and a dataset with larger cueing errors in the longitudinal motion channels, and thus a lower signal to noise ratio, would possibly result in a better fit and prediction.
As one of the goals of predicting the MIR is to use this measure as a cost in an optimization-based MCA, the models were chosen and analysed with this in mind. The Basic model could be used to tune current cost functions by replacing the motion channel weights with the DC gain for each of the cueing errors in this model. Alternatively, the full model could be used to replace the cost function entirely. It should be noted that the main interest for MCA optimization is the PMI, and not the MIR which is modelled by the models. The rating system, represented in this model by polynomial A, does therefore not have to be included in this cost function.
For the AI model a similar implementation is possible, although it requires the optimization algorithm to calculate errors in linear jerk as well as in linear acceleration. Due to the complexity of the CEDA it is more difficult, but not impossible, to implement the CEDA model as a cost function for optimization-based MCAs. Instead, it is easier to use the model to estimate the quality of an MCA off-line, and aid in manual tuning of the algorithms. Optionally, the algorithm could also be used to optimize specific MCA parameters off-line, by using the model in a non-linear optimization algorithm.

5.6.3. Future work and recommendations

Due to the bottom-up approach of this method for developing PMI prediction models, there are many possible representations for the black box system that is the human perception of motion cueing quality in vehicle simulation. Moving towards a grey box model by implementing knowledge on the human perception system could reduce the number of possibilities and possibly also improve the model prediction. It is therefore recommended to develop additional and more accurate algorithms that can be implemented in the non-linear part of the model.

The current models were fitted on a small dataset with limited power in the different cueing error inputs. It is therefore encouraged to apply the SI process to data from a much larger and more variable dataset. As the continuous rating scale currently is linked to the experiment, where the maximum rating is anchored to the maximum incongruence perceived during the whole experiment, it is somewhat challenging to combine data from different experiments to gain larger datasets. Chapter 6 therefore presents a way to combine data obtained from different experiments, such that more accurate models can be developed.

Currently the linear part of the model uses an ARX structure. The author has briefly tried applying different model structures, such as the ARMAX structure where a moving average is included, but did not find any significant improvements in the model fit and prediction or reductions of the number of parameters. As this was only briefly investigated, and the choice for an ARX structure is mainly based on an initial guess and for simplicity, it could be interesting to investigate the effects of different model structures more thoroughly, and possibly improve the model prediction further.

For simplicity, the parameter estimation was done using a least squares estimation method, but it was also shown the system noise violated the assumptions of white Gaussian noise that make this estimator a maximum likelihood estimator. Even though for the current dataset the system noise was shown to not have a significant influence on the parameter estimation, it is advised to check this for any future datasets as well and use a more suitable estimator if the noise assumption violates significantly influence the parameter estimation.

5.7. Conclusion

In this study a method was developed to design a model that can predict the motion incongruence rating (MIR) from the vehicle and simulator motions. This method was applied to continuous rating data from a human-in-the-loop experiment and is shown to produce models that were able to predict the main features of the motion incongruence rating for similar manoeuvres in the same experiment.

The developed models could be used to tune cost functions for optimization-based MCAs or to evaluate the MCA quality off-line, without human-in-the-loop experiments. However, it is recommended to apply and validate the proposed model design process to larger and more variable datasets for more accurate models.

6

Model Transfer Between Experiments

This chapter introduces the Model Transfer Parameter (MTP) which can be used to make PMI data obtained with different experiments comparable. This parameter can be used for the between-experiment prediction analysis of Perceived Motion Incongruence (PMI) models, or to aggregate data obtained with different experiments into larger datasets for the development Perceived Motion Incongruences (PMIs) models. First, a validation of the MTP estimation process is presented and used to estimate the MTP between data obtained with experiments described in Chapters 2 and 3. This MTP is then used to analyse the between-experiment prediction capabilities of different PMI models. The results show that these capabilities strongly depend on the richness of the dataset to which the models were fitted. For rich datasets, these models have good between-experiment prediction capabilities.

6.1. Introduction

Predicting the cueing quality used for vehicle motion simulators can aid in improving the cueing algorithms and reduce simulator motion sickness. A model that predicts a measure related to the cueing quality, the Perceived Motion Incongruences (PMIs) between vehicle and simulator motions, was introduced in Chapter 5. As for any model, its predictive power strongly depends on the size and variability of the dataset to which the model parameters are fitted. Being able to combine data from different experiments to fit the models is therefore desired.

While the average PMI resulting from a matched set of simulator and vehicle motions is expected to be largely comparable between experiments, the rating scale used for continuous rating of motion incongruence, by design, is not. Data obtained from different experiments can thus in general not directly be compared. In this chapter, a method is introduced to compensate for these rating scale differences. The "model transfer parameter" that is estimated with this method can be used to evaluate a model's prediction power between datasets from different experiments. Moreover, it allows for combination of data from different experiments, which could lead to the development of more accurate and widely applicable motion incongruence prediction models.

Subjective assessment of cueing algorithms for flight simulation is often done using categorical scales, such as the Cooper-Harper Handling Quality Rating Scale [127], the Motion Fidelity Rating Scale [50], and the Simulator Fidelity Rating Scale [164]. These scales can provide a good measure of the absolute simulator fidelity when assessed by experienced evaluation pilots and can be used for simulator classification. As with these calibrated rating scales an absolute measure of simulator fidelity is obtained, the results from different experiments are directly comparable. When not using experienced evaluators, simpler questionnaires combined with, for example, Likert scales [38, 165], visual analogue scales [36], or qualitative scales [61, 166] are often used. Not many studies exist where, like in this chapter, the results are compared *between* experiments, but if done so, as in [165], scale linking or aligning [167] can be used to make the corresponding scales better comparable.

One type of scale linking is referred to as scale anchoring. In this procedure the scales used in two different experiments are mapped onto a common scale by equating the rating for anchor items that occur in both tests. This procedure can be used if two experiments contain the same anchor items (e.g., reference conditions), or if a third experiment is performed that contains items or conditions of both experiments. The former is, for example, applied to compare results from three simulator studies described in [165]. The latter is, for example, done in [168], where the rating scales used in several different video quality experiments are mapped onto the common scale of an additional anchor test.

With continuous rating of PMI, the rating scale minimum and maximum are per definition anchored to the minimum and maximum PMI obtained throughout an entire experiment. While most experiments will contain a similar minimum PMI of zero, i.e., where both the visual and the physical motion cues are zero or match perfectly, such as during a full stop, the maximum PMI generally differs, resulting in different scales for each experiment.

As the experiments used for this chapter, described in Chapters 2 and 3, do not contain similar items, e.g., similar cueing errors, direct linking using anchors is not possible. An option would be to perform a third experiment where cueing errors similar to segments of both experiments are rated, but this would require a lot of time, as it can be complex to combine these different sections in a realistic experiment. Additionally, the rating of

PMI is a dynamic process, where past measurements influence future measurements, which makes it difficult to isolate and compare specific items.
Another recently introduced procedure for scale linking was proposed in [169], where the pain intensity scales of two different questionnaires were mapped onto a common scale. This mapping was done by performing two estimations simultaneously: an estimation of the parameters of a model that relates the item responses to the latent trait (pain intensity) and an estimation of the linking transformation between scales. The linking here is possible because the two questionnaires both contain some of the same items. In [170] a similar process, referred to as "calibrated projection", is introduced. Here, instead of common items between questionnaires, some participants responded to items of both questionnaires.
In [168] the scale linking procedure using an anchor test is compared to a procedure similar to calibrated projection. To achieve linked scales, two least squares problems are solved iteratively: the first problem includes fitting the parameters of a model relating video quality variables to the video quality rating, and the second involves fitting the scale transformation parameters to map the scales from different experiments to one common scale. In such a case, the link between different experiments is thus not made through equal items or equal participants, but rather through similar model inputs, e.g., video quality variables.
In this chapter a process similar to the calibrated projected approach of [168] is used to estimate the constant scaling factor, the Model Transfer Parameter (MTP), to map PMI ratings (Motion Incongruence Rating (MIR)) from a first experiment to the rating scale of a second experiment. The resulting MTP estimate is then used to evaluate the across-dataset *prediction power* of models that were fitted to the experiments described in Chapters 2 and 3. Additionally, the MTP estimate is used to *combine* data from both experiments for the development of a more generally applicable MIR model.
In Section 6.2 first the datasets from the two experiments used in this study are briefly summarized. The model transfer parameter estimation method is explained in Section 6.3. In Section 6.4 first the results of the MTP estimation for the experiments in this study are shown, after which the use of the MTP for analysing the prediction power of a model between experiments, and the use of the MTP when combining datasets from different experiments to obtain more accurate models, is shown. A discussion on the results is provided in Section 6.5, followed by a conclusion in Section 6.6.

6.2. EXPERIMENTS

In this study the data from two previously described experiments are used: an experiment performed at the CyberMotion Simulator (CMS) at the Max Planck Institute, described in Chapter 2, and an experiment performed in the Daimler driving simulator, described in Chapter 3.

6.2.1. CMS EXPERIMENT

The dataset obtained in the CMS experiment contains ratings from sixteen participants, who all rated the PMI of a nine-minute vehicle motion simulation three times. The simulation consisted of nine segments that each included a different combination of one out of three manoeuvres and one out of three tested motion cueing algorithms (MCA). The manoeuvres were as follows:

- CD: Curve Driving at 70 km/h, on a curve with a 257 meter radius and a 120 degrees deflection angle

- BA: Braking from 70 km/h to full stop and again Accelerating to 70 km/h on a straight road
- BCDA: Braking from 70 km/h to 50 km/h while entering the curve, Curve Driving at 50 km/h and Accelerating again to 70 km/h when exiting the curve, on a curve with a 131 meter radius and a 120 degrees deflection angle

The MCAs were defined as follows:

- *Scal*: Scaling
  - Motion scaling (gain=0.6), which leads to scaling and small rotational rate errors (<4 deg/sec)
- *TRL*: Tilt-Rate Limiting
  - Rotation rate limiting to 1 deg/sec, which leads to missing or false cues, and very small rotational rate errors
- *NL*: No Limiting
  - Neither tilt rate limiting nor scaling is applied, which leads to large rotational rate errors (>8 deg/sec)

In Figure 6.1 the motion incongruence ratings and the introduced cueing errors in linear acceleration and rotational velocity throughout the CMS experiment are shown. The nine different segments are separated with vertical dotted lines and the corresponding segment names, here shown as the horizontal axis labels, are a combination of the manoeuvre (first line) and MCA (second line) abbreviations.

6.2.2. DAIMLER EXPERIMENT

The Daimler dataset also contains ratings from sixteen participants, who each rated the PMI of a nine-minute vehicle motion simulation three times. The simulation consisted of a single simulated test track that included several different manoeuvres, which was repeated with two different MCAs. The MCAs were a classical washout-based MCA developed and tuned by Daimler, here referred to the Daimler MCA, and an optimization-based MCA developed by the MPI for Biological Cybernetics, here referred to as the MPI MCA.
The manoeuvres included in the track were as follows:

- Rural Curves (RC): drive over a rural road consisting of a large radius left, right and left curve, during which a constant speed was maintained.
- Overtake (OT): double lane change manoeuvre at constant speed to avoid a car parked on the right-hand side of the road.
- Slow Down (SD): upon entering an urban area the speed is initially reduced from 100 to 70 km/h and then from 70 to 50 km/h.
- Traffic Light Deceleration (TLD): driving through a gentle curve and decelerating to a full stop in front of a red traffic light.
- Traffic Light Wait (TLW): standing still in front of the red traffic light for 6 seconds.

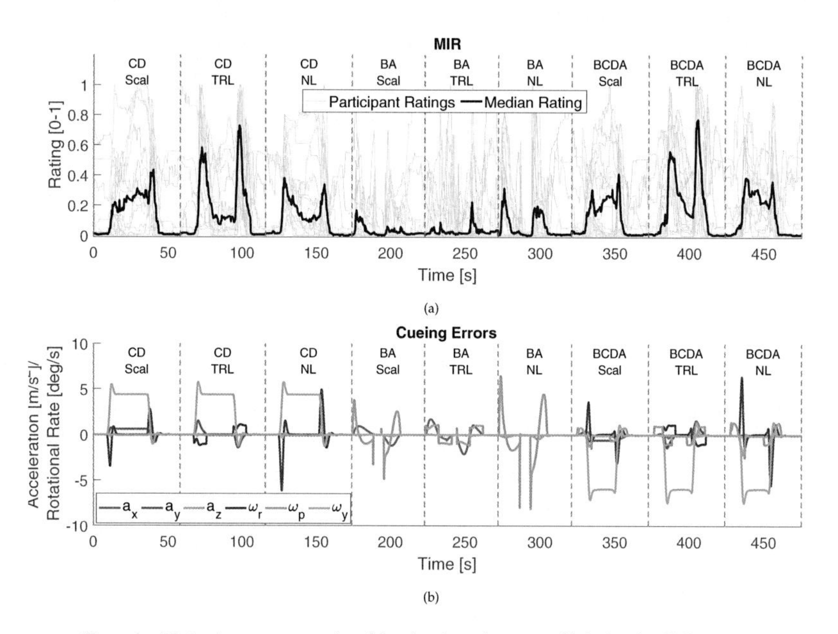

Figure 6.1: Motion incongruence ratings (a) and main cueing errors (b) during the CMS experiment.

- Traffic Light Acceleration (TLA): accelerating from the full stop to 50 km/h after the traffic light switched to green
- City 1 (Ci1): multiple gentle curves through the city at a constant speed of 50 km/h.
- Roundabout (Ro): decelerating to 20 km/h, driving through a four-exit roundabout, exiting at the second exit and accelerating back to 50 km/h.
- City 2 (Ci2): multiple curves through the city at a constant speed of 50 km/h.
- Turn Left (TL): decelerating to 20 km/h, driving through a 90-degrees left turn and accelerating back to 50 km/h.
- City 3 (Ci3): multiple gentle curves through the city at a constant speed of 50 km/h.

In Figure 6.2 the motion incongruence ratings and cueing errors in linear acceleration and rotational velocity throughout the Daimler experiment are shown. The different segments, each containing a different combination of manoeuvre and MCA, are again indicated with vertical dotted lines. Each segment is indicated with the manoeuvre name abbreviation (first line). The MCA name (second line) is shown once for all manoeuvres. Due to the limited space on the horizontal axis, the names of the traffic light manoeuvres are combined into TrL, representing TLD/TLW/TLA.

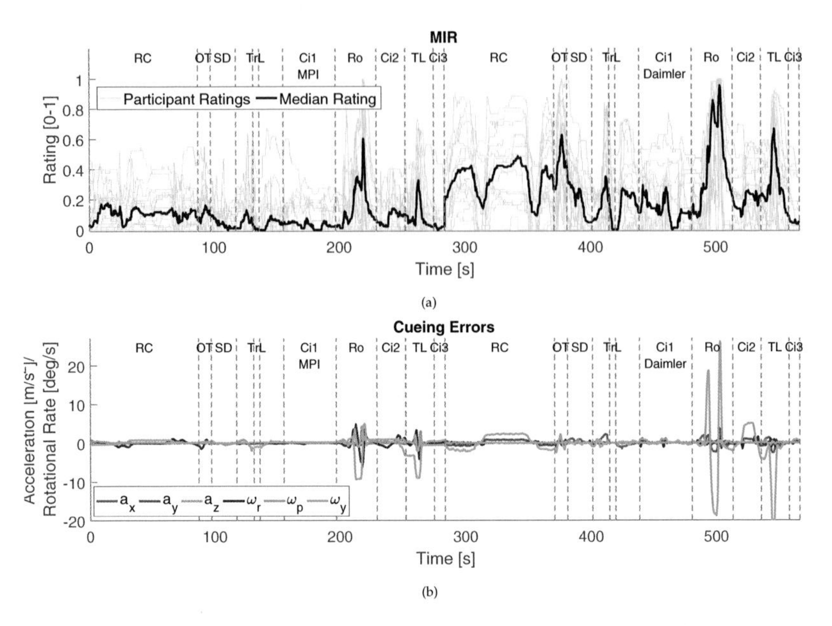

Figure 6.2: Motion incongruence ratings (a) and main cueing errors (b) during the Daimler experiment.

6.3. Model Transfer Parameter

In this section the need for the model transfer parameter, as well as the method used for its estimation, are explained in detail. First a short description of the models and their identification process, initially introduced in Chapter 5, is given, after which the need for the MTP due to differences in rating scales between experiments is further explained. Finally, the estimation of the MTP using an iterative process similar to those introduced in [165], [169] and [170], is explained. As a sanity check, this MTP estimation method is validated in Appendix F, by applying it to a set of data all obtained within one experiment, where the MTP should, per definition, be unity. Additionally, to explain why this relatively complex parallel method is used, the parallel method is compared to a simpler serial estimation method.

6.3.1. MIR Model

As introduced in Chapter 2, the considered general Motion Incongruence Rating (MIR) model is a combination of a PMI model and a rating model as shown in Figure 6.3.
The PMI model consists of the calculation, filtering, weighing and summation of different cueing errors. The rating system outputs the modelled MIR by passing the PMI through a low-pass rating filter (RF) after adding noise. The system identification process explained in Chapter 5 is used to select the inputs and filter orders and estimate the model parameters.
Different models can be derived by adjusting the model choices and the SI process parameters. The initial parameters and choices of the three models, Basic, Additional Inputs (AI) and Cueing Error Detection Algorithm (CEDA), all introduced in Chapter 5,

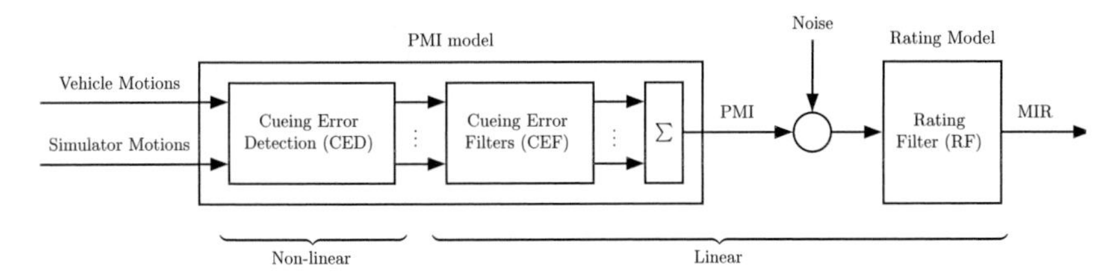

Figure 6.3: General structure of a MIR model consisting of a perceptual system (the PMI model) and a response system (the Rating model).

are also used in this chapter. However, as the SI process is applied to different datasets in this chapter, the resulting final models will differ from those reported in Chapter 5. The three sets of initial model choices and parameters used in this chapter are as follows:

- Basic:
  - Motion channels: linear acceleration (a_x, a_y, a_z) and rotational velocity (ω_r, ω_p and ω_y).
  - Cueing error calculation: absolute difference between vehicle and simulator motions.
- Additional Inputs (AI):
  - Motion channels: linear acceleration (a_x, a_y, a_z), rotational velocity (ω_r, ω_p, ω_y), longitudinal velocity (v_x), roll and pitch angle (θ_r, θ_p), linear jerk (j_x, j_y, j_z) and rotational acceleration $()\alpha_r, \alpha_p, \alpha_y)$.
  - Cueing error calculation: absolute difference between vehicle and simulator motions.
- Cueing Error Detection Algorithm (CEDA):
  - Motion channels: scaled, missing and false cues in longitudinal and lateral acceleration ($a_{x_{sc/mc/fc}}$, $a_{y_{sc/mc/fc}}$), vertical acceleration (a_z), rotational velocity (ω_r, ω_p and ω_y), longitudinal velocity (v_x), roll and pitch angle (θ_r, θ_p), linear jerk (j_x, j_y, j_z) and rotational acceleration ($\alpha_r, \alpha_p, \alpha_y$).
  - Cueing error calculation: absolute difference between vehicle and simulator motions and an additional split in false, missing and scaled cues in lateral and longitudinal acceleration, calculated with the CEDA described in Chapter 4.

6.3.2. Rating scales

The rating model for the experiments in this study, shown in Figure 6.3, should describe the rating system adopted with the continuous rating method. The rating system can, for example, be influenced by the rating device, e.g., rating devices with lower input resolution will result in a less accurate and less smooth ratings over time.

A more pronounced difference between rating systems of different experiments, however, is the rating *scale*. The rating scale for continuous rating is anchored to the

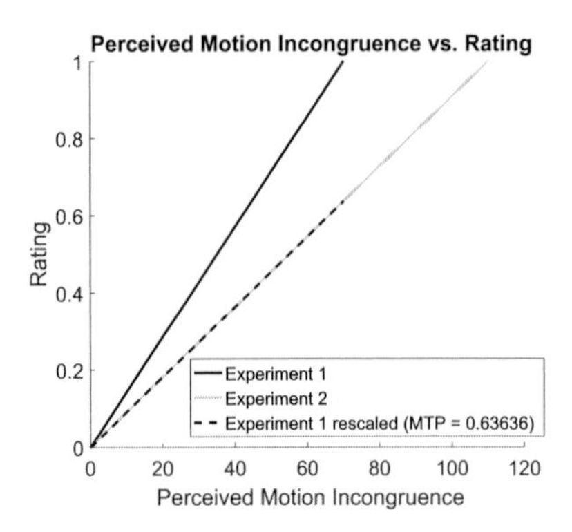

Figure 6.4: Example of applying an MTP to map ratings from two experiments to the same PMI scale.

minimum and maximum motion incongruence perceived during the experiment. This means that the maximum rating in one experiment, does not necessarily have the same meaning, or correspondence to the same PMI value, as a maximum rating in another experiment. Comparing rating data from two experiments therefore requires at least a rescaling of one of the ratings.
Before explaining this scale adjustment, first several assumptions on the rating system and scale are made:

- The minimum rating of zero is anchored to parts of the simulation where no simulator and no vehicle motion is present and thus by definition no discrepancies between vehicle and simulator motion are present. As this occurs in basically any realistic driving simulation (for example, at the start and the end of the simulation) this anchor can be assumed to be equal between experiments.

- The rating scale linearly increases between the minimum and maximum rating with increasing perceived motion incongruence.

- The rating dynamics are similar between the experiments used in this study. For example, no significant differences in rating dynamics due to, for example, different rating devices are assumed.

As the final MIR ranges between zero and unity, in case the maximum PMI differs between experiments, the same MIR value will correspond to a different PMI value. In Figure 6.4 the mathematical relation between MIR and PMI is shown with the continuous lines.
To account for this difference in rating scale, the Model Transfer Parameter (MTP) is introduced:

$$\mathrm{MTP} = \frac{max\left(\mathrm{PMI}_1(t)\right)}{max\left(\mathrm{PMI}_2(t)\right)} \tag{6.1}$$

where PMI_1 and PMI_2 indicate the PMI present during a first and a second experiment. For the example of Figure 6.4, during Experiment 1 a maximum PMI of about 70 and during Experiment 2 a maximum of 110 fictional units was present. By scaling the rating from Experiment 1 with the MTP, the scales for both experiments are mapped onto

a common scale, i.e., the scale from Experiment 2, as indicated with the striped line in Figure 6.4.
Prediction of the MIR based on a model derived from the data of Experiment 1 should thus be rescaled with the MTP in order to be compared to the MIR obtained in Experiment 2. Note that the choice for the scale of Experiment 2 to be the common scale is arbitrary, i.e., any scale can be used as a common scale to which the ratings of both experiments will be mapped. Using one of the existing scales as a common scale, however, requires only one MTP for rescaling.
The MTP should thus account for differences in rating scales, but an MTP different from unity should not increase or decrease the system noise. In the model block diagram in Figure 6.3 the MTP should therefore apply just before the noise enters the system. In this chapter, however, the multiplication is applied to the model input directly, for practical reasons. In the results shown in Section 6.4, the MIR data are therefore presented as it was measured during the experiments, i.e., unscaled, with values between zero and one. To test the predictive power of an estimated model, this model is first fitted to a first dataset and the estimated MTP is used to apply the model to a second dataset. When combining two datasets from different experiments, the inputs of the second experiment are multiplied with the estimated MTP before fitting a model to the combined dataset.

6.3.3. MTP Estimation Process

The model transfer parameter is estimated using an iterative process. First, using the SI process described in Chapter 5, a MIR model is fitted to data from two experiments, where the input data from the second experiment are first multiplied with an MTP. Second, the MTP is adjusted and a new model is found by repeating the SI process with this new MTP. An optimal MTP is found using a gradient-descent method that minimizes the sum of the prediction error of the model fit.
Ideally, the MTP estimation process can be applied once to the complete dataset, i.e., including all segments of both experiments, using a global search algorithm to find the best MTP estimate. Practically, however, this is not the most effective method as the following needs to be considered:

- Robustness
    - To avoid over-fitting and to ensure that the MTP estimate is not biased by differences between experiments other than the rating scale, the estimation should be repeated using different SI process inputs and outputs and an average MTP should be used as the final estimate.
        - ◇ Repeat the MTP estimation with different models, e.g., initial model choices and parameters, for the estimation process.
        - ◇ Repeat the MTP estimation with different subsets of segments from both experiments.
    - The adopted search algorithm cannot distinguish between global and local minima. To reduce the effects of finding local minima on the average MTP estimate, the initial MTP should be randomized for each estimation.
- Computation time
    - A global search method would require months of computation as during

each iteration the whole SI process is repeated. Instead, for this study, a simplified gradient-descent search method is adopted.

- As the PMI can only contain positive values, the MTP does as well. To decrease the search space, a constrained optimization, allowing only MTPs above zero, is therefore used.
- To further decrease computation time, the initial MTP used by the search algorithm should be close to an initial estimate of the MTP. This initial estimate can for example be obtained by fitting a simple model to the first dataset and searching for the MTP resulting in the lowest prediction error when applying the model and MTP to the second dataset.

Therefore, instead of estimating the MTP only once, the MTP estimation process is repeated for different combinations of the following inputs.

- Set of initial MIR model parameters and choices, described in Chapter 5,
- Data subset pair, where each part of the pair includes a subset of the input/output data in one experiment, and
- Randomized initial MTP that is in the order of the initial estimate of the MTP.

The final output of the MTP estimation method is the median over all MTP estimates resulting from the different combinations of these inputs.
The reliability of the MTP estimation strongly depends on how well the MIR model fits to the data, which is why only reasonable model choices and SI process parameters should be used. In this study the three sets of initial model choices and parameters, Basic, AI and CEDA, described in Chapter 5, are used. The different subset pairs were chosen to each contain input power similar to the complete dataset. In total 20 different subset pairs were used, each containing at least 29 of the 32 segments present in the two experiments,
An initial MTP estimate using the two complete datasets was found to be 0.8. To minimize the computation time the randomized initial MTP was kept close to this estimate by bounding the randomization between 0.1 and 2.
The MTP estimation method was applied to each of the three models with 20 different subset pairs using one randomized initial MTP estimate, resulting in 60 MTP estimations. The median of these 60 MTP estimations is then taken as the resulting MTP estimate between the two experiments.
In Appendix F a detailed validation of this method is presented. Additionally, the parallel estimation method described in this chapter is compared to a simpler serial estimation method, which shows to be less accurate but has a much faster computation time.

6.4. Results

In this section, first the results of the MTP estimation are presented, after which two different uses of the model transfer parameter are demonstrated. The use of the MTP for analyzing the prediction power of a model between two experiments is shown first. Secondly, the MTP estimate is used to combine the datasets from both experiments to obtain a more accurate and widely applicable MIR model.
In this section, the Variance Accounted For (VAF) values of each model on the fit and prediction datasets are shown, together with the input contributions to the modeled

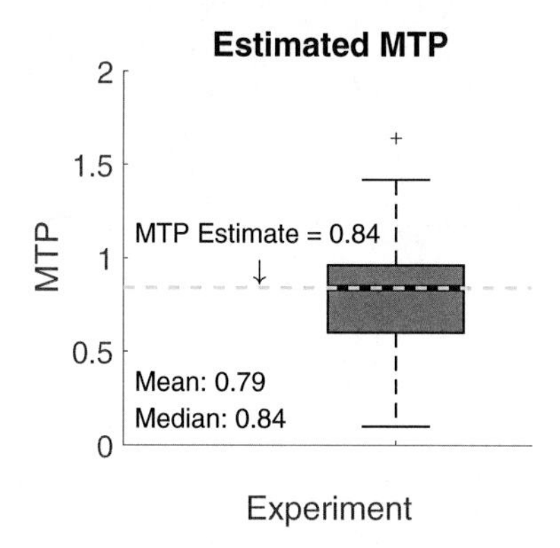

Figure 6.5: Boxplot showing the median and interquartile ranges for MTP estimation between the CMS and Daimler experiments.

output of each of the different cueing errors accounted for in the model. Additionally, in Appendix G the model structure, residual and uncertainty analyses for all three fitted models are presented.

6.4.1. MTP ESTIMATION

Using the MTP estimation process described in Section 6.3.3, the MTP mapping the CMS experiment rating scale onto the rating scale of the Daimler experiment was estimated. In Figure 6.5 the results for the MTP estimation are shown in a boxplot, which shows the 60 MTP estimates discussed in Section 6.3.3.

The MTP estimate, the median MTP over all different estimates, is 0.84. Given that the MTP is equal to the maximum PMI ratio between two experiments, this MTP estimate indicates that the maximum PMI was higher during the Daimler experiment than during the CMS experiment.

6.4.2. BETWEEN-EXPERIMENT PREDICTION

Predicting the PMI of a second experiment with a model fitted to a first experiment can, naturally, be done without the use of the MTP. However, if verification of this prediction is required, and given that during an experiment only the MIR can be measured, the MTP is needed to map the MIR to the same scale as the one used for the model prediction.

In this section, three models, the Basic, AI and CEDA models, are fitted to both the CMS or the Daimler experiment dataset, after which the predictive power of the resulting models is analysed using the dataset of the other experiment. As the determined MTP of 0.84 maps the CMS rating scale onto the Daimler rating scale, a model made using the CMS dataset should be multiplied with the MTP before it can be used to describe the Daimler dataset. The inputs of the Daimler prediction dataset are therefore first multiplied with the MTP of 0.84, while the inputs of the CMS prediction dataset are instead multiplied with its inverse, before they are put through the MIR models.

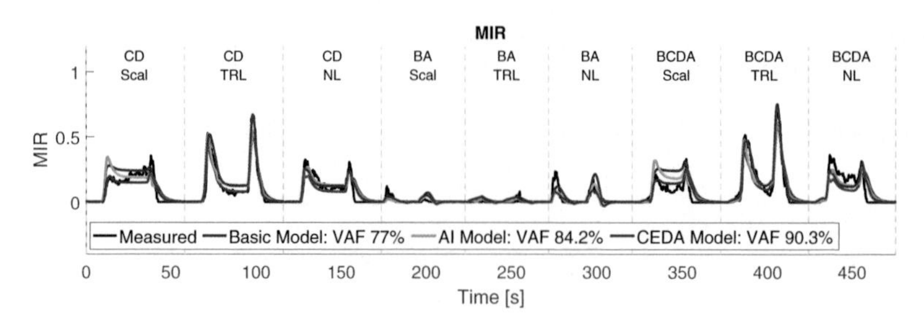

Figure 6.6: Outputs of Basic, AI and CEDA models, fitted and applied to the CMS experiment dataset.

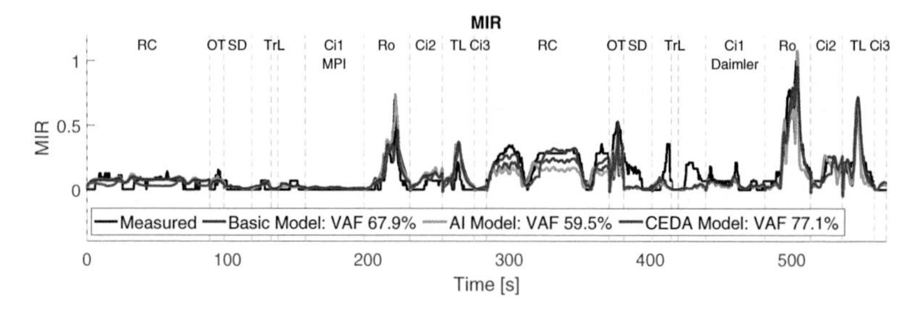

Figure 6.7: Predictions of the Basic, AI and CEDA models, fitted to the CMS experiment dataset, and applied the Daimler experiment dataset.

CMS Data Results

In this section the results of models fitted to the dataset of the CMS experiment and their prediction for the data from the Daimler experiment are shown. Additionally, in Appendix G.1 the model structure, the residual and the uncertainty analyses are presented. Figure 6.6 shows the output of the models fitted to the CMS experiment dataset and Figure 6.7 shows the predictions obtained with these models for the Daimler experiment dataset.

Figure 6.6 shows that, consistent with Chapter 5, the fit improves with complexity of the model, where the simplest model (the Basic model) has a VAF of 77%, the AI model improves the VAF with 7.2% and the most complex model, the CEDA model, further improves the VAF with 13.2%. The largest difference between the CEDA model and the Basic and AI models is seen during the CD and BCDA maneuvers for MCA *Scal*. Additionally, peaks during the same maneuvers for MCA *TRL* and MCA *NL* are slightly less underpredicted with increasing model complexity.

Figure 6.7 shows that these models fitted to the CMS experiment dataset can predict most features of the ratings given during the Daimler experiment as well. However, the VAF values for the prediction of the Daimler data are 9.1%, 24.7% and 13.1% lower for the Basic, AI and CEDA models, respectively. While the model fit improves with increasing model complexity, the VAF of the predictions only improves for the CEDA structure (9.1% increase compared to Basic model), while the prediction becomes less accurate when only the additional cueing errors are included (AI model, 8.4% decrease). The main differences between the Basic and CEDA models can be observed

in the prediction of the roundabout (Ro) maneuver for the MPI MCA in Figure 6.7, where the Basic model predicts a large overshoot, and during the rural road curves (RC) for both MCAs, where the Basic model underpredicts the ratings. The AI model does not underpredict this maneuver with the MPI MCA, but does so for the Daimler MCA. None of the models correctly predict the rating during the deceleration and acceleration at the traffic light (TrL). During the left turn (TL) maneuver all models show an overshoot for the MPI MCA.
In Figures 6.8 and 6.9 the contributions of all of the inputs, i.e., the different cueing errors, to the total output for each model are shown, when applied to the CMS and the Daimler experiment datasets, respectively.
Figures 6.8(a) and 6.9(a) show that the Basic model contains some negative input contributions due to the yaw rate cueing error. Negative input contributions would imply that a cueing error in fact *improves* the cueing quality, which is unlikely and thus a modelling artefact. For the CMS experiment dataset to which these models were fitted, the yaw cueing error has a shape that resembles the roll angle cueing error. The roll angle error, only a possible input in the AI and CEDA models, seems to be a more likely input, as Figures 6.8(b), 6.8(c),6.9(b) and 6.9(c) show that including this error does not result in any negative input contributions. Figures 6.8(b) and 6.9(b) show that the AI model does have small negative input contributions for the lateral acceleration. Splitting the lateral acceleration cueing error in three different cueing errors, the main difference between the AI and CEDA models, reduces these negative input contributions of this cueing error.
While for the CMS dataset the PMI during longitudinal accelerations was modelled with the cueing error in vertical acceleration (Basic and AI models) or pitch angle (CEDA model), Figure 6.9 shows that these cueing errors have close to no power in the Daimler experiment dataset during manoeuvres with longitudinal acceleration, such as the "slow down" (SD) and "traffic light" (TrL) manoeuvres. Figure 6.2 shows that, instead, these manoeuvres with the Daimler MCA have power in the longitudinal acceleration cueing error, which is not included in any of the models fitted to the CMS experiment dataset. This largely explains why these manoeuvres with the Daimler MCA are underpredicted.
While many different cueing errors contribute to the model outputs for the CMS dataset, Figure 6.9 shows that most of the rating data for the Daimler MCA are in fact modelled with the lateral acceleration cueing error. For the CEDA model, the dominant contribution is seen to be the missing cue in lateral acceleration, while the corresponding false and scaled cues have much smaller contributions. Other cueing errors that have noticeable contributions are those in roll and yaw rate for the Basic model, and the errors in roll rate and lateral jerk for the AI model, particularly during the roundabout and left turn manoeuvres, see Figure 6.9. The CEDA model uses cueing errors in roll rate to model these manoeuvres when the MPI MCA is used.
Figure 6.9 shows that the AI and CEDA models both describe the ratings during the Rural Curves (RC) manoeuvre using the MPI MCA, see Figure 6.9 around 50 sec, with the roll angle. The underprediction of the rating during this segment by the Basic model, is a direct result of the lack of the roll angle error in this model.

Daimler Data Results

In this section the results of models fitted to the complete dataset of the Daimler experiment and their prediction for the data from the CMS experiment are shown. Additionally, in Appendix G.2 the full details of the corresponding model structure, residual and

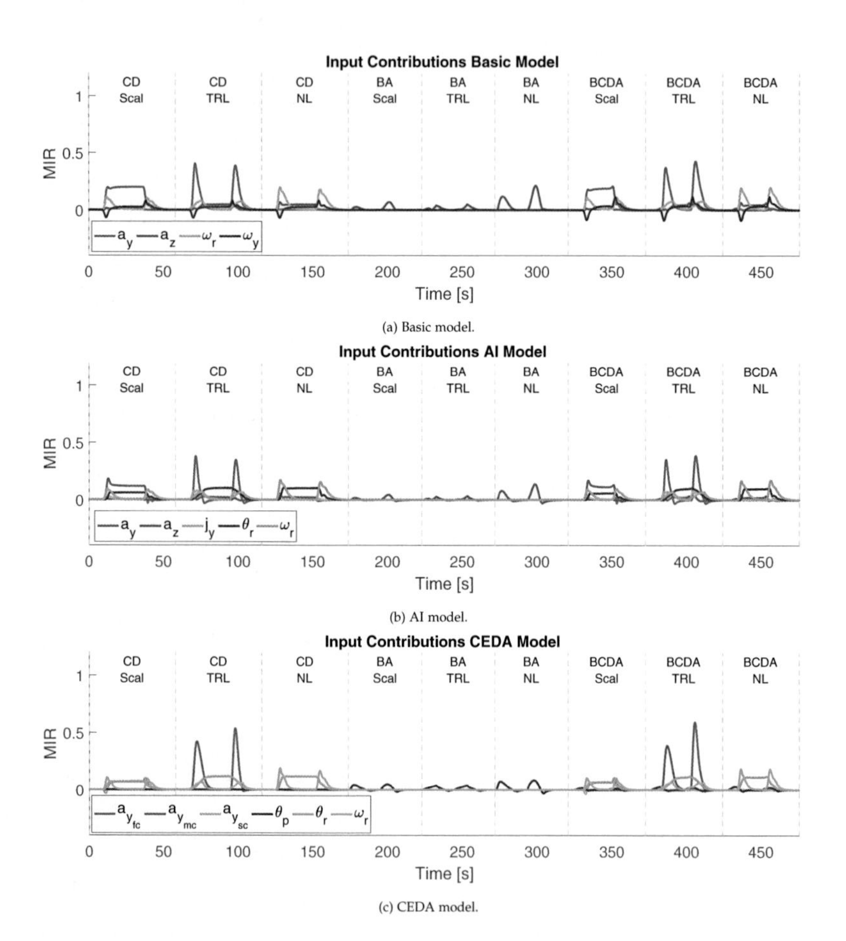

(a) Basic model.

(b) AI model.

(c) CEDA model.

Figure 6.8: Input contributions of the Basic, AI and CEDA models, each fitted and applied to the CMS experiment dataset.

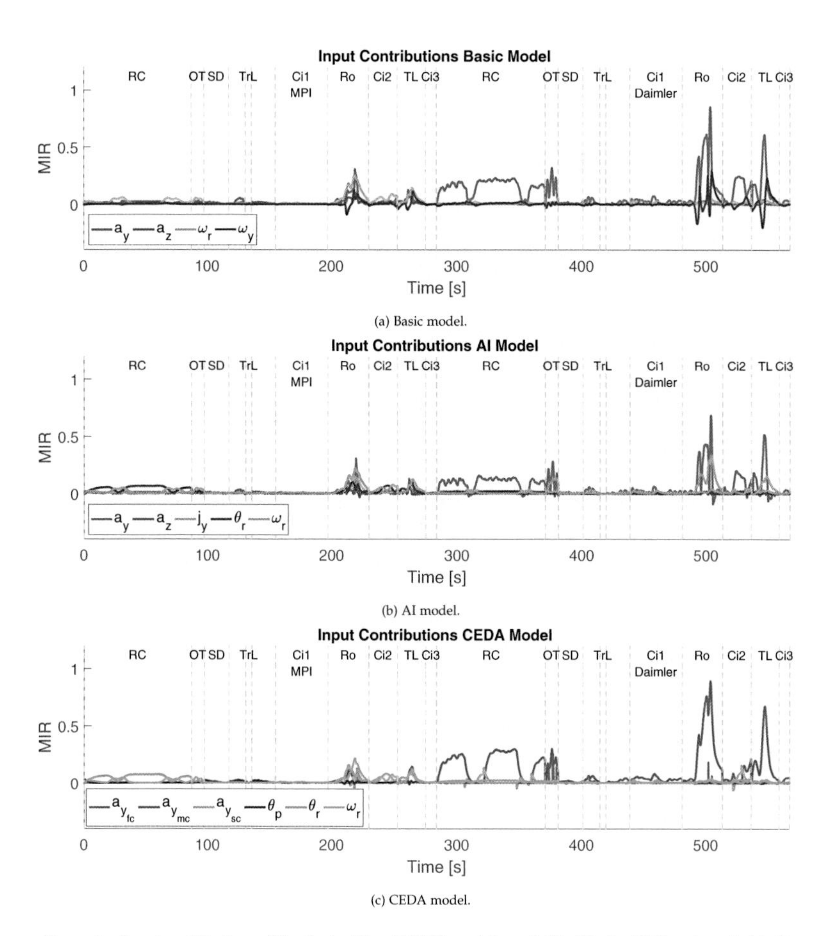

(a) Basic model.

(b) AI model.

(c) CEDA model.

Figure 6.9: Input contributions of the Basic, AI and CEDA models, each fitted to the CMS and applied to the Daimler experiment dataset.

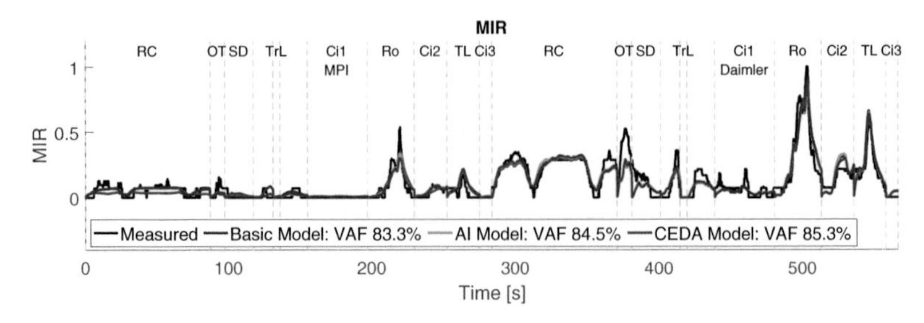

Figure 6.10: Outputs of Basic, AI and CEDA models, fitted and applied to the Daimler experiment dataset.

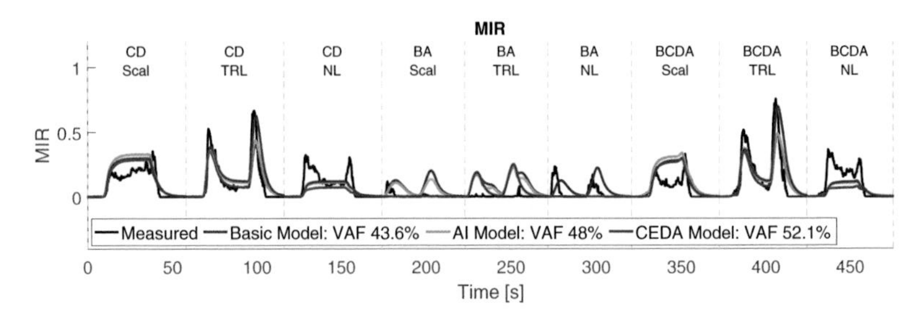

Figure 6.11: Predictions of the Basic, AI and CEDA models, fitted to the Daimler experiment dataset, and applied the CMS experiment dataset.

uncertainty analyses are presented. The overall model fits to the Daimler experiment data are shown in Figure 6.10 and the model predictions for the CMS experiment data are shown in Figure 6.11. While Figure 6.10 shows that also for the models fitted to the Daimler experiment dataset the VAF increases with increasing model complexity, the differences here are marginal, with increases in VAF of only 1.2% and 2% for the AI and CEDA models, respectively, as compared to the Basic model. The main difference between model fits can be seen in the modelling of the Rural Curves (RC) manoeuvre when using the MPI MCA, where only the Basic model again underpredicts the rating. The models can all explain a large part of the rating, but underestimate the rating during the Over Take (OT), Slow Down (SD) and Traffic Light (TrL) manoeuvres when using the Daimler MCA.
Figure 6.11 shows that the CMS experiment rating is poorly predicted by the models fitted to the Daimler experiment data, with VAF values of 43.6%, 48% and 52.1% for the Basic, AI and CEDA models, respectively. While hardly any difference in fit was found between the models, the prediction power does differ considerably, with a VAF increase of almost 9% between the Basic and CEDA models. The predicted rating during the false cues of the CD and BCDA manoeuvres is around 1.5 times higher with the CEDA than with the Basic and AI models. During the BA manoeuvres all models predict very different ratings. While the Basic model predicts an increased rating for all MCAs, the AI model predicts only an increase when using MCA *Scal* or MCA *TRL*, and the CEDA model only predicts an increase for MCA *TRL*. All models, however, predict the measured ratings poorly, as the highest rating for this manoeuvre occurs

when using MCA *NL*.
In Figures 6.12 and 6.13 the input contributions for each model are shown when applied to the Daimler and CMS experiment dataset, respectively. None of the models use negative input contributions and all models use the cueing error in lateral acceleration to model most of the rating, especially with the Daimler MCA. The MIR during Slow Down (SD) and Traffic Light (TrL) manoeuvres, when using this MCA, are modelled with the cueing error in longitudinal acceleration. Figure 6.13, however, shows that contributions of these cueing errors for the CMS experiment dataset result in a poor prediction, especially during the BA manoeuvres. For the Basic model, the vertical acceleration also contributes to these large prediction errors. The CEDA model has a smaller prediction error than the Basic model during these manoeuvres, as only the missing cue in longitudinal acceleration contributes.

6.4.3. COMBINED EXPERIMENT FITTING

With the combined dataset, including the original data from the CMS experiment and the data from the Daimler experiment that was rescaled with the MTP of 0.84, a new, overall more accurate, model can be fitted. Here the system identification process presented in Chapter 5 with the initial model choices and parameters of the three models, Basic, AI and CEDA, are again applied. As all currently available continuous rating data are used for the system identification, only the model fit and not the between-experiment prediction power can be shown. A prediction analysis can then only be performed when more data are available. The model fits are shown in Figures 6.14 and 6.15 for the CMS and Daimler experiment data, respectively. Figure 6.14 shows that a higher complexity of the combined models again results in increasing VAF values when applying them to the CMS experiment dataset alone, with the CEDA model having a VAF value that is 16.7% higher than the VAF value obtained for the Basic model. The Basic and AI models overpredict the rating during the CD and BCDA manoeuvres with MCA *Scal* much more than the CEDA manoeuvre, likely because the scaled cues in lateral acceleration, mainly present in these segments, are weighed differently from the missing and false cues when using the CEDA model. The BA manoeuvre with MCA *TRL* is overpredicted by the Basic and CEDA models.
The longitudinal acceleration maneuvers in the Daimler experiment dataset, shown in Figure 6.15 are instead underpredicted by all three models. This figure also shows that for the Daimler experiment dataset the VAF values do not show a noticeable increase with added model complexity: the differences between the models are much smaller (max: 4.7%) than for the CMS experiment dataset.
In Figures 6.16 and 6.17 the input contributions for each of the models fitted to the combined dataset are shown for the datasets from the CMS and Daimler experiment, respectively. None of the cueing errors provide negative contributions to the modelled rating for any of the models. The cueing errors in lateral acceleration clearly represent the dominant contribution to the modelled rating for all models. The addition of the cueing error in roll angle for the AI and CEDA models also contributes to the fit, mainly during rural road curves when using the MPI MCA, and reduces the weight of the cueing errors in vertical acceleration by replacing part of its contribution during the CD and BCDA manoeuvres in the CMS experiment.
To compare the results of the combined models with the models fitted to the individual CMS and Daimler datasets, Table 6.1 shows the VAF values of all models applied to either their fit or their prediction dataset. The VAF values of the combined models are all higher than the VAF values of the prediction of the other models, but lower than the

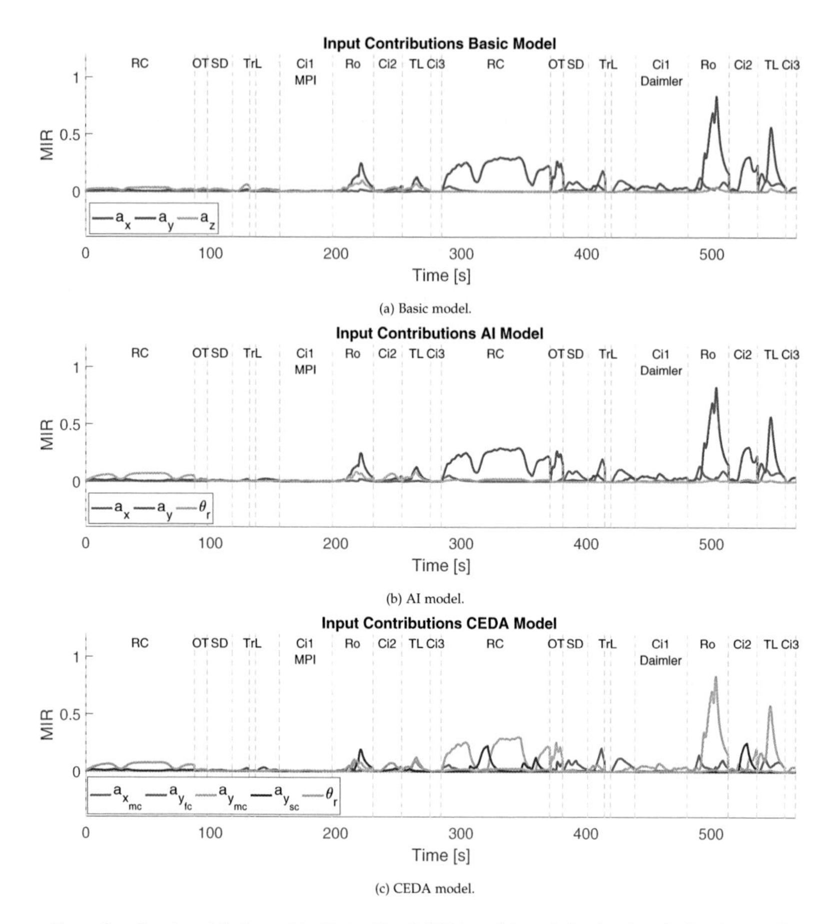

(a) Basic model.

(b) AI model.

(c) CEDA model.

Figure 6.12: Input contributions of the Basic, AI and CEDA models, each fitted and applied to the Daimler experiment dataset.

Table 6.1: VAF values per model when applied to either the fit or the prediction dataset. C stands for the CMS and D for the Daimler experiment dataset.

	Fit Dataset					
	CMS		Daimler		CMS-Daimler	
Model	Fit (C)	Pred. (D)	Fit (D)	Pred. (C)	Fit (C)	Fit (D)
Basic	77.02%	67.91%	83.28%	43.62%	63.21%	79.13%
AI	84.20%	59.47%	84.51%	47.97%	68.27%	78.60%
CEDA	90.26%	77.13%	85.30%	52.05%	79.95%	83.32%

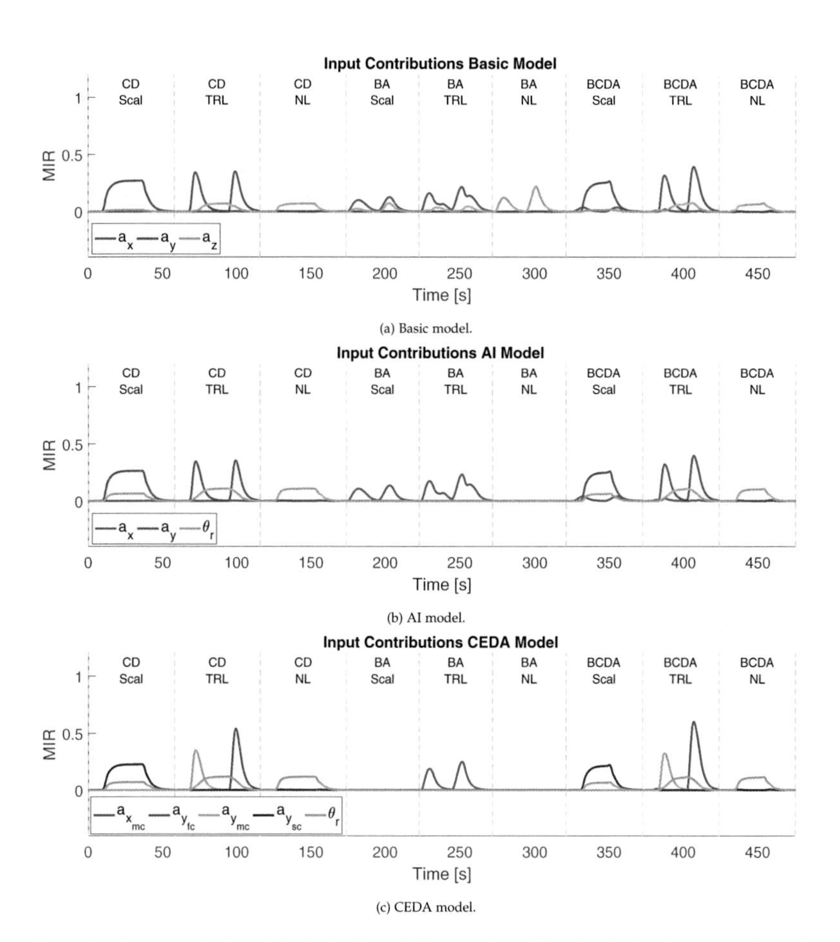

(a) Basic model.

(b) AI model.

(c) CEDA model.

Figure 6.13: Input contributions of the Basic, AI and CEDA models, each fitted to the Daimler and applied to the CMS experiment dataset.

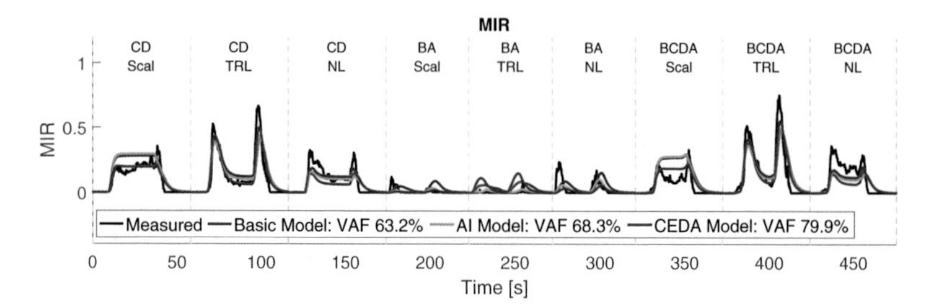

Figure 6.14: Outputs of Basic, AI and CEDA models, fitted to the combined CMS-Daimler dataset and applied to the CMS dataset.

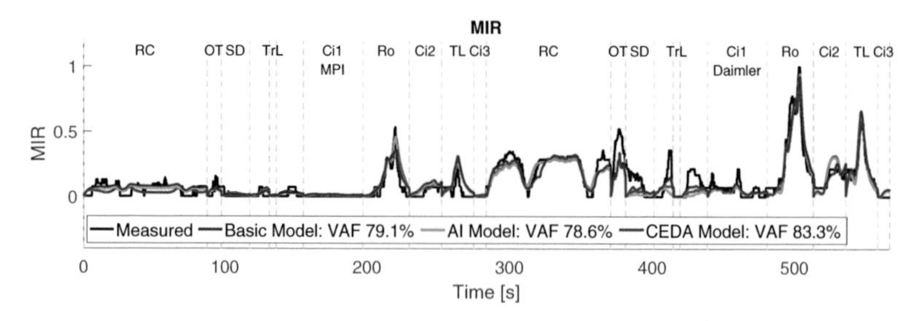

Figure 6.15: Outputs of Basic, AI and CEDA models, fitted to the combined CMS-Daimler dataset and applied to the Daimler dataset.

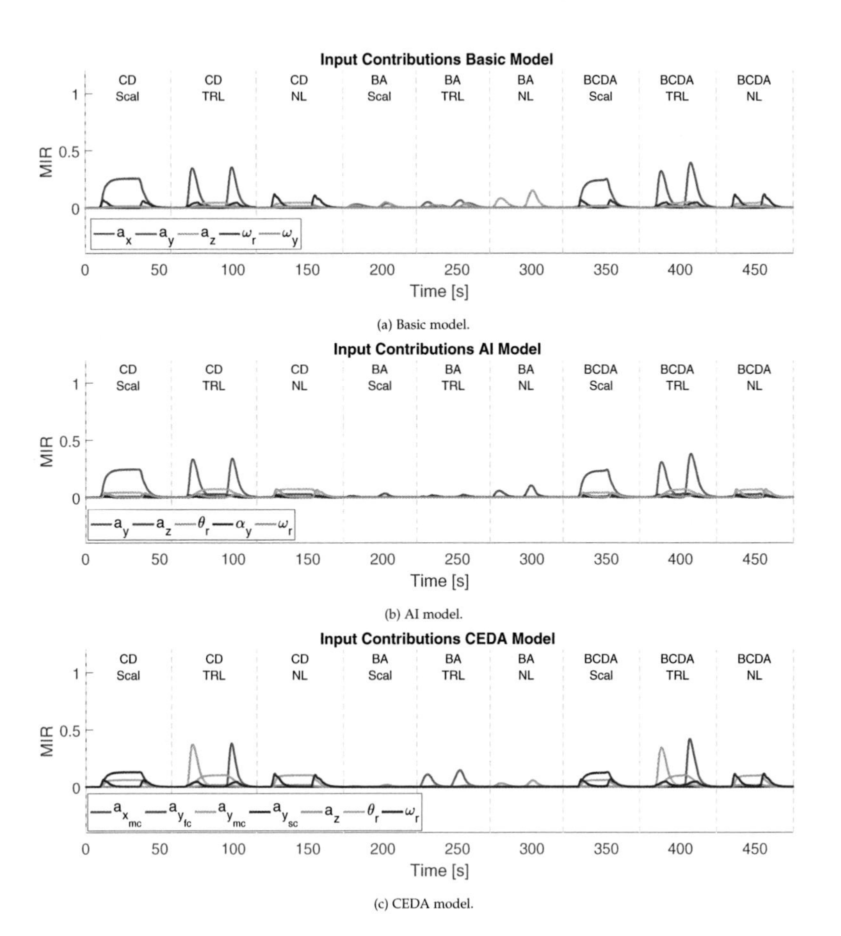

(a) Basic model.

(b) AI model.

(c) CEDA model.

Figure 6.16: Input contributions of the Basic, AI and CEDA models, fitted to the combined CMS-Daimler dataset and applied to the CMS dataset.

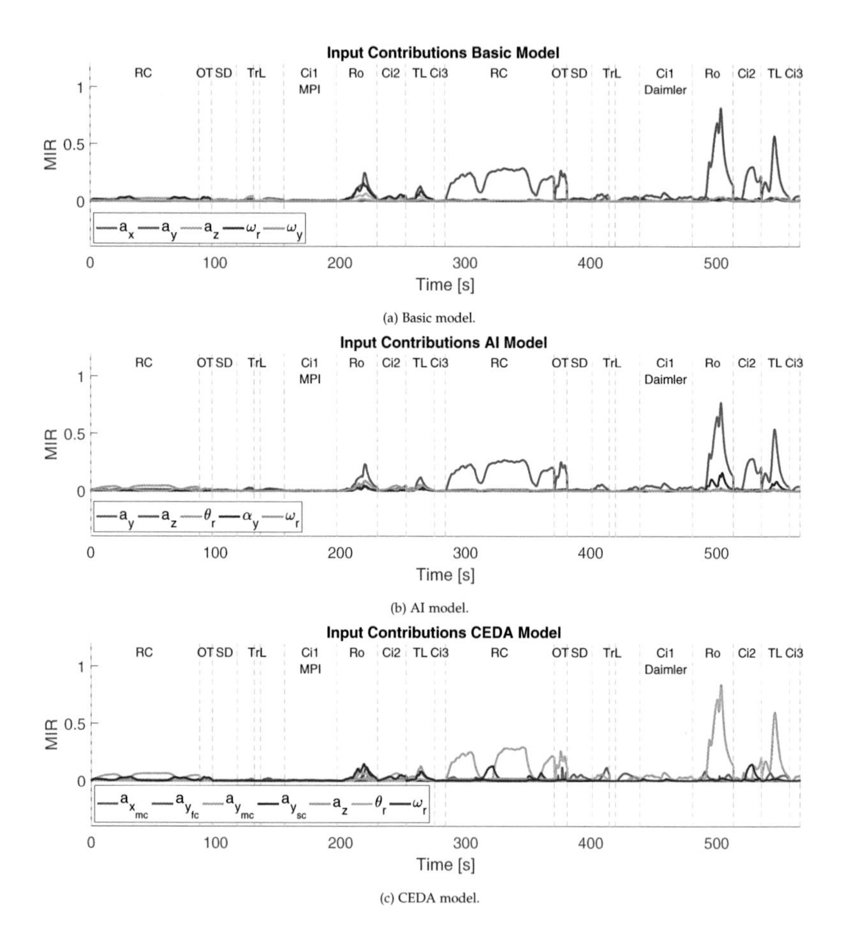

(a) Basic model.

(b) AI model.

(c) CEDA model.

Figure 6.17: Input contributions of the Basic, AI and CEDA models, fitted to the combined CMS-Daimler dataset and applied to the Daimler dataset.

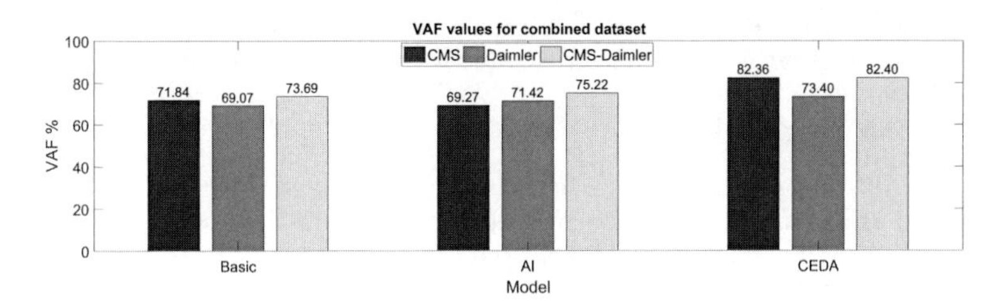

Figure 6.18: VAF values per model, fitted to the CMS (blue), Daimler (green) or combined CMS-Daimler (yellow) dataset, when applied to combined CMS-Daimler dataset

Table 6.2: Number of parameters per model

	Fit Dataset		
Model	CMS	Daimler	CMS-Daimler
Basic	11	7	9
AI	13	6	12
CEDA	13	8	14

VAF values of the fits of the models fitted to one dataset only. The differences between fit and prediction VAF values for the latter models are on average 15.65% and 36.48% for models fitted to the CMS and Daimler experiment dataset, respectively. This differences between VAF values of the CMS and Daimler datasets is reduced to on average 9.88% when fitting the models to the combined dataset. In Figure 6.18 the VAF values of each model when applied to the combined dataset are shown, while the corresponding number of parameters are listed in Table 6.2.

Figure 6.18 shows that the VAF values of all models increase when fitting to the combined dataset as compared CMS experiment dataset (VAF increases of 1.85%, 5.94% and 0.04% for the Basic, AI and CEDA models, respectively) and as compared to the Daimler experiment dataset (VAF increases of 4.62%, 3.80% and 9.00% for the Basic, AI and CEDA models, respectively).

Table 6.2 shows, however, that the VAF increases with respect to fitting to the Daimler dataset come at a cost of a larger number of model parameters, i.e., respectively, 2 and 6 parameters more for the Basic and the AI and CEDA models. The increases in VAF compared to fitting to the CMS dataset for the Basic and AI models, instead are obtained with, respectively, 2 and 1 parameters less. The very small ($< 1\%$) increase in VAF for the CEDA model comes at a cost of 1 model parameter.

The additional parameters are mainly due to more cueing errors being included in the models when fitting to the combined dataset. This indicates that especially the Daimler dataset lacks power in certain cueing errors, which results in simpler, but less widely applicable models.

6.5. DISCUSSION

A method for comparing continuous ratings of Perceived Motion Incongruence (PMI) obtained from different experiments was introduced. As the considered MIR rating

scale is anchored to the PMI range in a specific experiment, the main challenge is to transfer the PMI ratings from these different experiments onto a common scale. For this scale transfer, a method was developed to estimate the Model Transfer Parameter (MTP), which can be used to map the ratings from one experiment onto the scale of a second experiment.
Using this MTP, the prediction power of different models, fitted to two different experiments and using three different initial model parameter settings, Basic, AI and CEDA, was analysed. Additionally, the MTP was used to combine data from two experiments into a large dataset to which also three different models were fitted. For all analyses, the datasets collected in the CMS and the Daimler experiments, described in Chapters 2 and 3, respectively, were used.

6.5.1. MTP estimation

The MTP estimation method optimizes the model fit to a dataset containing data from two different experiments, via iteratively adjusting the model parameters and the MTP. The estimation method results in a set of MTP estimations by fitting multiple models, i.e., the Basic, AI and CEDA models, to different sections of the total dataset, i.e., combinations of segments with similar input power. By repeating the MTP estimation under different conditions and averaging over these estimations, the risk of over-fitting, and thus obtaining an MTP estimate that accounts for more than just the scale differences, is reduced.
The MTP that can be used to map the ratings of the CMS to the ratings of the Daimler experiment was estimated to be 0.84. This indicates that the maximum PMI during the Daimler experiment was larger than during the CMS experiment and that the motion cueing quality during the former was thus lower. This finding is supported by the fact that two out of 18 participants had to quit the Daimler experiment due to motion sickness, while none of the participants became sick during the CMS experiment.

6.5.2. Between Experiment Prediction

Using the MTP estimated with the MTP estimation method, the predictive power of models fitted to both the CMS and Daimler datasets was analysed. To each dataset, three models (the Basic, AI and CEDA models) were fitted. The model fits improved with increased model complexity, with VAF values between 77% and 90.3% and values between 83.3% and 85.3% for models fitted to the CMS and Daimler experiment data, respectively. The difference between the models fitted to the Daimler experiment dataset was found to be notably smaller (differences in VAF values of up to 2.0%) than those fitted to the CMS experiment dataset (differences in VAF values of up to 13.2%). This can be explained by the difference in cueing error diversity between these two experiments. The CMS experiment was deliberately set up to include many different types of cueing errors, while the cueing errors in the Daimler experiment were generally very small for the MPI MCA, and were mainly caused by missing cues in lateral acceleration for the Daimler MCA. This explains why the benefits of including additional cueing errors in the more complex models were not reflected by the attained model fits.
Applying the models fitted to the CMS experiment dataset to the Daimler experiment dataset resulted in notably lower VAF values than applying them to the CMS experiment dataset (reductions in VAF of 9.1%, 24.7% and 13.1% for the Basic, AI and CEDA models, respectively), but still resulted in the prediction of the main features of the

Daimler experiment ratings. The Basic model had a higher prediction power than the AI model (VAF 67.9% compared to 59.5%), and the CEDA model showed the best prediction power with a VAF of 77%. The Basic and CEDA models thus show a relatively good (VAF > 60%) prediction of the Daimler ratings, especially given the large differences between the experiments, .e.g., different participants, motion platforms, cueing algorithms and simulated vehicle, and visual field-of-view.
The prediction power results and the large differences between experiments suggest that a general perceived motion incongruence model exists and can be developed with data from multiple experiments. However, the differences between the fit and prediction power for between-experiment prediction, rather than within-experiment prediction, as was shown in Chapter 5, indicate that more accurate and specialized models can be obtained by decreasing the differences between the datasets.
The between-experiment prediction power of the model fitted to the CMS dataset for the longitudinal manoeuvres, i.e., those involving deceleration and acceleration, was relatively poor. One explanation for this poor prediction is the large difference in visual motion cues between the two experiments, i.e., 360 (Daimler) compared to 140 (CMS) degrees field-of-view combined with higher quality of the visuals and more visual details and increased optical flow in the Daimler simulator.
While the expected lateral acceleration can be partially derived from the shape of the road, i.e., curves, the longitudinal acceleration is mainly derived from the optical flow [155, 171]. The improved visuals in the Daimler simulator would have made is easier to accurately derive the longitudinal acceleration than possible in the CMS. More accuracy in the visual acceleration cue reduces the spread of physical motion cues that can be perceived as congruent. This more accurate perception of differences between visual and physical motion cues in turn results in higher PMI ratings for the same manoeuvre.
For future research, it would be interesting to investigate which transformations can be included in the non-linear part of the PMI model that account for differences in visual quality between simulators. A simple example of such a transformation could be to increase the weight for longitudinal acceleration errors with increasing visual quality, as cueing errors are more easily detected when the visual quality is high.
More complex transformations could describe cueing errors as the likelihood that the physical and visual motion cues are coming from the same source, as also used in causal inference models of multi-sensory integration [67], and have the precision of the visual longitudinal acceleration estimate be partially dependent on factors influencing visual quality, such as field-of-view, update rate and resolution.
The between-experiment prediction of the model fitted to the Daimler dataset was considerably poorer (VAF values between 43.6% and 52.1%). The low prediction power was mainly visible in the longitudinal acceleration manoeuvres. The difference in weights between scaled, missing and false cues as possible in the CEDA model, was the main reason for the increased prediction power of this model.
One reason for the poor prediction power of the models fitted to the Daimler dataset, is that this dataset has much less diversity in its cueing errors than the CMS experiment dataset. This was also shown by the lower number of cueing errors selected by each of the models as compared to the models fitted to the CMS dataset (3, 3 and 5 cueing errors vs. 4, 5 and 6 cueing errors for the Basic, AI and CEDA models).
The prediction power of MIR models thus strongly depends on the richness of the dataset. When developing a MIR model, it should be taken into account that not only the input power is well spread over the desired frequency range, but also that the input power is well spread over cueing errors in different motion channels and different

cueing error types such as scaled, missing and false cues.
The off-line prediction of the PMI can aid in the optimization of cueing algorithms, without the need for human-in-the-loop experiments. Additionally, the development and improvement of a PMI prediction model can aid in better understanding the challenges in motion cueing, as bad cueing can be identified over time and correlated to specific differences between visual and physical motion cues. Depending on the use of the PMI prediction models, the requirements for the accuracy of the prediction differ.
In future research (Chapter 7), it will be tested whether the current accuracy of the models is high enough to optimize motion cueing algorithms.

6.5.3. COMBINED EXPERIMENT FITTING

Using the estimated MTP, the datasets from the CMS and Daimler experiment were combined and the three considered MIR models (Basic, AI and CEDA) were fitted to the aggregated dataset. As no further data were available to analyse the prediction power of these models, only an explanatory analysis was performed.
The model fits had VAF values ranging from 73.7% to 82.4%, again increasing with increasing model complexity. The main increase in VAF was caused by the splitting of different cueing error types in longitudinal and lateral acceleration, as included in the CEDA model. The VAF values of these models when applying them to the combined dataset are higher than those of the models fitted to either individual dataset. Only a marginal VAF increase (0.04%) was found between the CEDA model when fitted to the combined dataset or when fitted to the CMS dataset. This could indicate that the input power in the CMS dataset does not differ largely from the input power in the combined dataset.
The input power related to longitudinal acceleration seemed to cause the main difference between these models: fitting to the CMS dataset results in including the pitch angle cueing error in the model, while fitting to the combined dataset instead results in including vertical acceleration cueing error and the longitudinal acceleration missing cues.
The models fitted to the combined dataset also showed less difference in VAF values between the two datasets than when fitting the models to one of the datasets only. For the CEDA model that was fitted to the CMS experiment dataset, the VAF value of the fit to this dataset was 90% while the VAF value of the prediction of the Daimler experiment dataset was 77%. For the model that was fitted to the combined dataset these VAF values were 80% and 83%, respectively. This continuity of performance increases the overall faith in these models. When more data is available, a prediction analysis of these models should determine which of these models, both showing very comparable VAF values for their fit to the combined dataset, describes the underlying system better.
The three models fitted to the combined dataset all used significant contributions of cueing errors in lateral and vertical acceleration, as well as roll rate cueing errors. As these models were fitted to a large dataset, this provides a good indication of the importance of cueing errors in these motion channels for motion cueing quality in driving scenarios and it is thus advised to focus on the reduction of cueing errors in these channels.
Notable improvements in model fit were observed when adding the roll angle as well. While roll angle is not unambiguously perceivable by the human perceptual system when visual information is present, it is strongly related to the perceived linear acceleration and a rotational rate above threshold, and has also been related to cueing quality

in previous research [92]. Reducing the roll angle cueing error in vehicle motion simulation is therefore also advised.
Finally, the residual analyses discussed in Appendix G show that all models had relatively high output polynomial orders of four and still showed some residual autocorrelations above the 95% confidence interval, indicating that still not all output information was correctly captured by the models. As mentioned in [172], if high output polynomial orders for ARX models are needed this might indicate that other model structures such as the ARMAX should be used. For future work it would be interesting to investigate the benefits and drawbacks of other, more complex, model structures.

6.6. CONCLUSIONS

In this chapter a method for comparing continuous rating data for Perceived Motion Incongruence (PMI) from different experiments was introduced. Applying the method to known scale transformations showed that accurate Model Transfer Parameters (MTPs), used to map the continuous ratings of different experiments onto a common scale, can be obtained with this method.
The use of the MTP estimation method for analysing the prediction power between experiments, as well as combining data from different experiments for more accurate and generally applicable models was demonstrated. For datasets with a sufficiently rich input set, models fitted to data from one experiment can be used to predict the perceived incongruence rating of a different experiment, where the simulators, visuals and participant groups differ significantly. This result suggests that it may be feasible to derive a general model for perceived motion incongruence with this approach, which can then be used as a general off-line predictor of PMI. In future research an attempt will be made to use these models for the optimization of a motion cueing algorithm.

III

Minimizing Perceived Motion Incongruence

7 Optimizing Motion Cueing with a MIR Model

In this chapter the potential of Motion Incongruence Rating (MIR) models for Motion Cueing Algorithm (MCA) optimization is investigated. In a human-in-the-loop experiment two optimization-based MCAs are compared on a medium-stroke hexapod simulator. The first MCA uses standard cueing error weights in its cost function, while for the second MCA these weights were based on a fitted MIR model from Chapter 6. The results show that such models provide a promising cueing error weight estimation method for optimization-based MCAs, but they also highlight the limitations of these models due to, for example, their dependency on the richness of the datasets to which they are fitted.

7.1. Introduction

One of the challenges in vehicle motion simulation is finding the simulator motions, given vehicle motions and simulator limitations, which result in the highest perceived cueing quality. Lately, more and more optimization-based Motion Cueing Algorithms (MCAs) are being developed which use a cost function based on the difference between simulator and vehicle motions to find an optimal simulator input at each time step [42, 57, 86, 173]. Such MCAs avoid worst case parameter tuning such as is usually done for the classical washout filter-based MCA[174]. The cost function in an optimization-based MCA, however, also contains parameters that need to be tuned.
Currently, as off-line automatic cueing quality assessment is still difficult, this tuning is done by experts [40, 55, 56]. In the previous chapters a method for modelling and predicting the perceived motion mismatch between visual and vestibular motions during a vehicle simulation was introduced. Here this process is used to calculate optimal cost function parameters of an optimization-based MCA developed at the Max Planck Institute (MPI) for Biological Cybernetics [42, 44, 57].
The optimization-based MCA uses Model Predictive Control (MPC) to optimize the simulator control inputs at each time step. This method makes use of a mathematical model of the simulator and a prediction of the desired vehicle motions over a specified prediction horizon. The optimization uses a cost function that includes the weighted error between the reference motion, i.e., the vehicle motion, and the simulator motion over this horizon. By tuning the output error weights, the perceived motion mismatch during the simulation can be minimized.
In the implementation of the MPC-based MCA used here, the output includes linear acceleration and rotational velocity. Currently the algorithm uses output weights based on the amplitude difference in a large set of vehicle motions between linear acceleration and rotational velocity. It was found that amplitudes for linear acceleration are on average around ten times smaller than the amplitudes of the rotational velocity when measured in $[m/s^2]$ and $[rad/s]$, respectively [87]. These output error weights are thus not chosen based on human motion perception information, but purely chosen to account for the differences in units between linear acceleration and rotational velocity.
In an attempt to include knowledge of human perception of motion mismatches, this chapter uses the Motion Incongruence Rating (MIR) modelling process described in the previous chapters to generate perception-based output error weights. During this process a static MIR model, designed as the output related part of the MPC-based MCA cost function, is fitted to continuous rating data from the two previous car driving experiments of Chapters 2 and 3.
The resulting cueing error weights corresponding to the best fit to the continuous rating are then used as optimized output error weights in the MPC-based MCA cost function. To assess if the perception-based output error weights indeed result in improved MCA quality, a human-in-the-loop experiment was performed, where both error weight sets were compared during a car driving simulation.
The experiment has a similar set up as the experiments described in [136] and [137], where a continuous rating method was adopted to obtain a time-varying measure of cueing quality during a motion simulation. Additionally, participants were requested to provide one overall rating of the cueing quality per condition. For each condition, a car driving scenario, including manoeuvres with both lateral and longitudinal accelerations, was presented to the participant, while the MPC-based MCA was either using the perception-based or the standard output error weights. After the experiment, participants were requested to fill out a questionnaire to provide more information on the

factors influencing their rating.
This chapter is organized as follows. In Section 7.2 a brief introduction to the MPC-based MCA is given, with a focus on its cost function. The usage of the MIR modelling process to design perception-based output error weights is also briefly discussed. In Section 7.3 the human-in-the-loop experiment comparing the perception-based output error weights to the current weights is further explained. The results of this experiment are shown in Section 7.4 and discussed in Section 7.5. A conclusion on the usage of the MIR modelling process for the design of perception-based output error weights is given in Section 7.6.

7.2. MPC-BASED MCA

For the experiment explained in this chapter the MPC-based MCA designed at the MPI for Biological Cybernetics [42, 57] is used. This MCA uses a linearised model of the hexapod simulator to compute the *future* simulator inputs over a given prediction horizon for a provided reference motion over this horizon. An advantage of this type of MCA, compared to the classical washout filter, is that the simulator limits are evaluated at each time step [175]. This avoids the need for worst case MCA parameter tuning, often resulting in largely scaled-down motions, as is done with classical washout filters [174]. The MCA has many design options of which the most important, those related to the prediction horizon and the cost function, are further explained below.

7.2.1. PREDICTION

The MPC-based MCA optimizes the simulator inputs, platform linear and rotational accelerations, for a desired simulator output, linear acceleration and rotational velocity in head frame, over a specific prediction horizon. This optimization is repeated each simulation time step and only the first optimized simulator input is actually sent to the simulator. The simulator output for a given input is compared to a reference motion over this prediction horizon.
While ideally this reference motion equals the exact vehicle motions that will occur during this prediction horizon, during active driving these future motions are, of course, unknown. Even though the passive driving experiment described in this chapter does not require active driving, for application purposes and realism, the reference motion and prediction horizon length were chosen to be independent of actual future vehicle motions. The reference motion is therefore assumed to be equal to the current vehicle motion and kept constant over the prediction horizon as also used in [57].
With this type of reference motion the cueing algorithm can anticipate hitting simulator limits in the near future. For long prediction horizons, however, the actual future vehicle motions will differ so much from this reference motion, that unnecessary anticipation occurs, resulting in large reductions in cueing quality. A prediction horizon of two seconds was, through trial and error tuning, found to provide a good trade-off between necessary and unnecessary anticipatory MCA behaviour.

7.2.2. Cost Function

During the optimization a cost (J) is calculated for a certain set of simulator inputs over the prediction horizon with:

$$J = \|\mathbf{x}_N - \hat{\mathbf{x}}_N\|^2_{W_{x_N}} + \sum_{k=0}^{N-1} \left(\|\mathbf{x}_k - \hat{\mathbf{x}}_k\|^2_{W_x} + \|\mathbf{u}_k - \hat{\mathbf{u}}_k\|^2_{W_u} + \|\mathbf{y}(\mathbf{x}_k, \mathbf{u}_k) - \hat{\mathbf{y}}_k\|^2_{W_y} \right), \tag{7.1}$$

where k is the discrete time step, N the prediction horizon length, $\mathbf{x}$ the simulator states (platform position/orientation and linear/rotational velocity), $\mathbf{u}$ the simulator inputs (platform linear/rotational acceleration) and $\mathbf{y}$ the simulator outputs (platform linear acceleration and rotational velocity). The variables $\hat{\mathbf{x}}$, $\hat{\mathbf{u}}$ and $\hat{\mathbf{y}}$ indicate the state, input and output references, respectively. The diagonal matrices $W_{x_N} = diag(\mathbf{w}_{\mathbf{x}N})^2$, $W_x = diag(\mathbf{w}_\mathbf{x})^2$, $W_u = diag(\mathbf{w}_\mathbf{u})^2$ and $W_y = diag(\mathbf{w}_\mathbf{y})^2$ indicate the terminal state, state, input and output error weighting matrices, respectively.
The input and terminal state related costs are used for stability and insure convexity of the optimization problem [57]. Their references ($\hat{\mathbf{u}}$ and $\hat{\mathbf{x}}_N$) are here set to zero. The weights $\mathbf{w}_\mathbf{u}$ and $\mathbf{w}_{\mathbf{x}N}$ were tuned for stability of the output and set to 1 and 2.5, respectively. The state-related cost can be used for washout of the simulator motion when setting its reference $\hat{x}$ to zero over the full prediction horizon. As only the platform positions needed to be washed out, the position-related weights of $\mathbf{w}_\mathbf{x}$ were set to four and the velocity related states were set to zero. Finally, the output-related cost influences how well a certain output channel is being followed by setting different weights for the different motion channels linear acceleration and rotational velocity in $\mathbf{w}_\mathbf{y}$.
The reference outputs $\hat{\mathbf{y}}$ are here chosen to be equal to the current vehicle motions and remain constant over the prediction horizon. The weights in $\mathbf{w}_\mathbf{y}$ that are standard used for this algorithm are based on the relative variance between linear acceleration and rotational velocity in typical car manoeuvres, which differs by about a factor of ten, resulting in the standard weights of $\mathbf{w}_{\mathbf{y}_\mathbf{s}} = [1, 1, 1, 10, 10, 10]$.

Perception-based Weights

The standard output error weights in $\mathbf{w}_{\mathbf{y}_\mathbf{s}}$ basically only account for the differences in units between linear acceleration and rotational velocity. The relative importance of errors in different motion channels for the perceived cueing quality is thus not explicitly taken into account.
In the experiment described here, output error weights based on perceived motion incongruence measures $\left(\mathbf{w}_{\mathbf{y}_\mathbf{p}}\right)$ are compared to the standard weights in $\left(\mathbf{w}_{\mathbf{y}_\mathbf{s}}\right)$. For $\mathbf{w}_{\mathbf{y}_\mathbf{p}}$, the output error weights per motion channel are chosen such that the corresponding output cost best fits the measured motion incongruence, i.e., the Motion Incongruence Rating (MIR), from the experiments described in Chapters 2 and 3.
To this end, the MIR model design process described in Chapter 5 was applied to the combined datasets presented in Chapter 6, using a static, zero input delay model with linear acceleration and rotational velocity cueing errors as inputs. These model choices result in a static MIR model, hereby named 'CF6', that is comparable to the output error related part of the cost function described in Section 7.2.2. In the top plot of Figure 7.1 the fit of this model to the combined dataset is shown ($VAF = 56.5\%$).
The contributions of the different cueing errors to the model output, shown in the

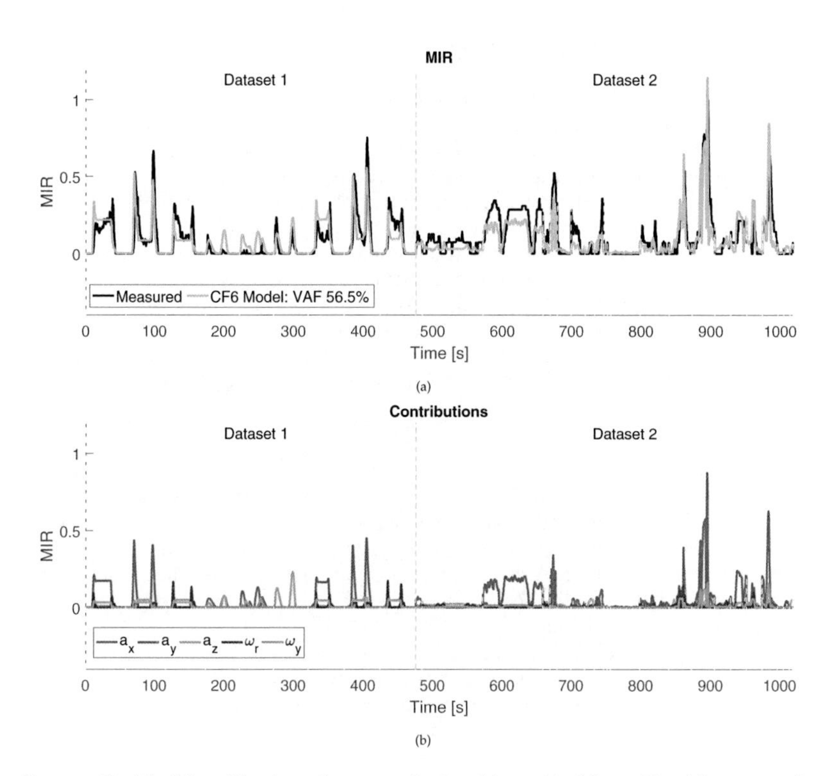

Figure 7.1: Fit of the CF6 model to the median measured rating of the combined dataset (a) and the corresponding contributions of the different inputs (b).

bottom plot of Figure 7.1, indicate that all but the pitch rate error receive a non-zero perception-based weight. The cueing error weights in the CF6 model that best fit the measured MIR are found to be 0.0620 for a_x, 0.2881 for a_y, 0.5347 for a_z, 0.0270 for ω_r, 0.0000 for ω_p and 0.0075 for ω_y. To use these weights in the MPC-based MCA cost function, first the weights for rotational velocity need to be converted from *sec/deg* to *sec/rad*. To obtain an equivalent influence of the perception-based weights on the total cost function as the standard weights have, the perception-based weights should also be scaled accordingly. Here the perception based weights are scaled such that the sum of the perception-based weights equals the sum of the standard weights. These two transformations result in the perception-based weights $\mathbf{w}_{\mathbf{y}_\mathrm{p}} = [0.7147, 3.3222, 6.1652, 17.8658, 0, 4.9321]$.
The perception-based weights thus give about three times more weight to a_y and about six times more weight to a_z, while the weight for a_x is reduced by about one third, compared to the standard weights. Using $\mathbf{w}_{\mathbf{y}_\mathrm{p}}$ in the MPI MCA is thus expected to result in a better following of the lateral and vertical acceleration. The better following of the, in a car, zero vertical acceleration in turn relates to smaller rotation angels.
The perception-based weights also have about one and one third more weight for roll rate and about half the weight for yaw rate compared to the standard weights. Cueing errors in pitch rate are not penalized when using the perception-based weights, while they are when using the standard weights. As the roll and pitch rates are close to zero during car driving, using $\mathbf{w}_{\mathbf{y}_\mathrm{p}}$ is expected to result in lower roll rates but higher pitch rates than when using $\mathbf{w}_{\mathbf{y}_\mathrm{s}}$. The yaw rate, which does have significant power during car driving, is expected to be followed somewhat better when using $\mathbf{w}_{\mathbf{y}_\mathrm{s}}$ then when using $\mathbf{w}_{\mathbf{y}_\mathrm{p}}$.

7.2.3. MCA Output

To investigate the effects of the different output error weights on the motion quality, a short car simulation was designed, including a set of maneuvers with both lateral and longitudinal accelerations. During the simulation the car first accelerates from standstill to 50 km/h and then slows down to 30 km/h to enter a roundabout. While exiting the roundabout at the second exit, the car accelerates back to 50 km/h and finally decelerates to a full stop.
In Figure 7.2 the vehicle linear accelerations and rotational velocities for this simulation are shown. Additionally, the motions calculated with the MPC-based MCA, using either $\mathbf{w}_{\mathbf{y}_\mathrm{s}}$ or $\mathbf{w}_{\mathbf{y}_\mathrm{p}}$ for the output error weights, are shown. Figure 7.2 also shows the corresponding motions as measured with the IMU on the simulator platform. The plots are divided into five sections, where RA stands for "Roundabout".
Figure 7.2 shows that the different weights mainly affect the amount of tilt coordination that is used to follow the vehicle lateral acceleration. When using $\mathbf{w}_{\mathbf{y}_\mathrm{p}}$ the average scaling difference between vehicle and simulator lateral acceleration is around 0.9, while the use of $\mathbf{w}_{\mathbf{y}_\mathrm{s}}$ results in a much lower gain of around 0.6. Additionally, the reduced weight on yaw for $\mathbf{w}_{\mathbf{y}_\mathrm{p}}$ results in a small degradation in reproducing the vehicle yaw rate.

Predicted MIR

To check if using the optimized weights $\mathbf{w}_{\mathbf{y}_\mathrm{p}}$ indeed reduces the predicted MIR, the MCA outputs were fed through the three MIR models described in Chapter 6. For the Basic model only the cueing errors in linear acceleration and rotational velocity are

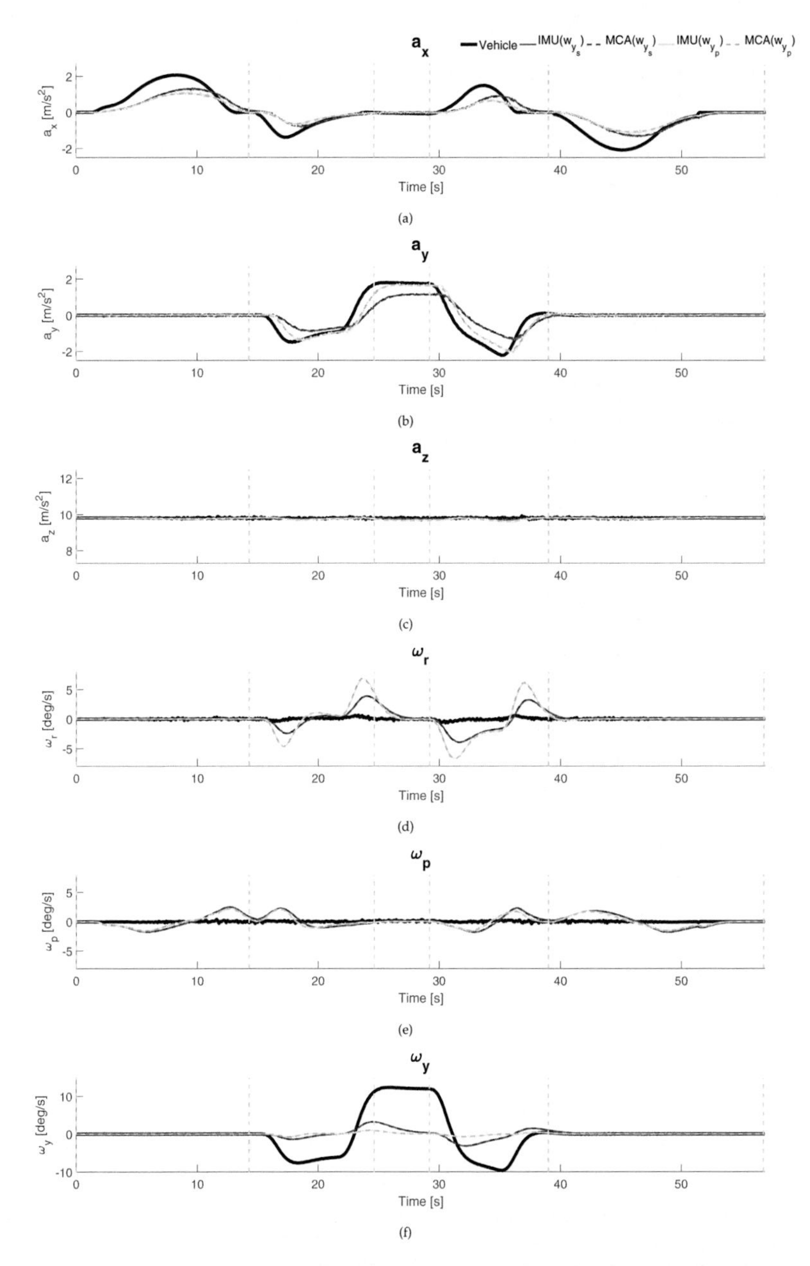

Figure 7.2: Linear acceleration ((a)-(c)) and rotational velocity ((d)-(f)) of the car during the car simulation, together with the motions calculated with the MCA using either $\mathbf{w_{y_s}}$ or $\mathbf{w_{y_p}}$ and the corresponding measured (IMU) motions on the simulator.

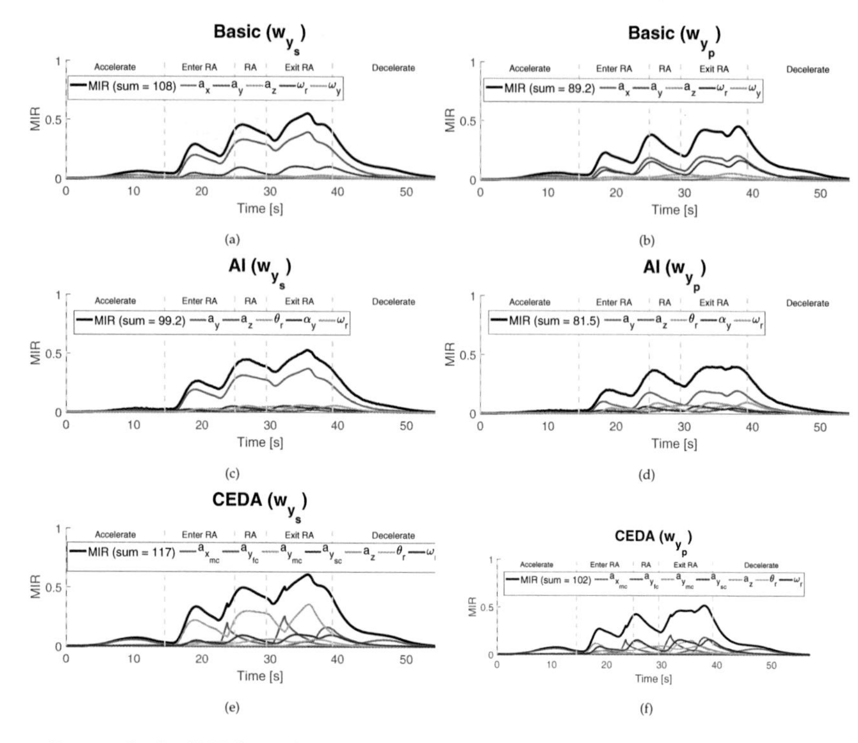

Figure 7.3: Predicted MIR for simulator motions resulting from either $MCA(\mathbf{w}_{\mathbf{y}_\mathbf{s}})$ (left column) or $MCA(\mathbf{w}_{\mathbf{y}_\mathbf{p}})$ (right column), using the Basic, AI and CEDA MIR models.

considered, while for the Additional Input (AI) model also their derivatives (linear jerk and rotational acceleration) as well as the rotational angle and the forward velocity are considered. The third model, the CEDA model, additionally makes use of a cueing error detection algorithm, described in Chapter 4, that splits the cueing errors in linear acceleration into scaled, missing and false cues.

In Figure 7.3 the predicted MIR from the Basic, AI and CEDA models and the contributions of cueing errors in the different motion channels are shown. The sum of the predicted MIR over time for each MCA setting is indicated in the title of each plot and shows that all models predict a higher motion quality, i.e., overall lower MIR, when using $\mathbf{w}_{\mathbf{y}_\mathbf{p}}$ instead of $\mathbf{w}_{\mathbf{y}_\mathbf{s}}$ in the MPC-based MCA. The average MIR decrease over all models is 17% when using $\mathbf{w}_{\mathbf{y}_\mathbf{p}}$ instead of $\mathbf{w}_{\mathbf{y}_\mathbf{s}}$. The largest differences occur during the roundabout, where the difference between MIR for the two settings is on average (over time and all three models) about 0.06, while during the acceleration and deceleration sections this difference is smaller than 0.005.

7.3. Experiment

The experiment goal is to determine whether the perception-based output error weights for the MPC-based MCA improve the overall MCA quality. To this end, MCA outputs were generated for a set of car manoeuvres, using either the standard of the perception-based output error weights. The motion quality was measured using the same subjective rating method as in Ref. [136].

7.3.1. Independent Variables and Dependent Measures

The only independent variable is the output error weight setting of the MPC-based MCA. We compare the baseline ($\mathbf{w}_{y_s}$) weights with those optimized based on previously measured perceived motion incongruence ($\mathbf{w}_{y_p}$).
The dependent measures are the measures for MCA quality, obtained using the continuous subjective rating method from [136], resulting in a Continuous Rating (CR) of the cueing quality throughout the simulation for each participant, as well as with an Overall Rating (OR) after each simulation, resulting in one quality rating per condition. To be able to verify the consistency of the participant ratings, both measurements were repeated three times for both MCA settings.

7.3.2. Hypothesis

As the MIR model analysis predicts a lower MIR when using the optimized $\mathbf{w}_{y_p}$ output error weights instead of the heuristically-tuned weights $\mathbf{w}_{y_s}$, the hypothesis is that the motion cueing quality improves. This improvement should be visible in both the measured overall rating and the continuous rating, especially throughout the roundabout.

7.3.3. Apparatus

The experiment was performed in the CyberPod Simulator at the MPI for Biological Cybernetics, which has a hexapod motion platform (eMotion-1500-6DOF-650-MK1 from Bosch Rexroth). The experiment set up is shown in Figure 7.4. The visuals were projected on a screen about one meter in front of the participant using a VPixx technologies ProPixx beamer with 1920x1080 resolution and a 120 Hz update rate, providing the participants with a 80° horizontal field-of-view (FOV). To avoid perception of the earth fixed environment, all lights apart from the large simulator screen were turned off during the vehicle simulations.
The vehicle motions were generated using CarSim software and the visuals were generated using Unity. The steering wheel in the visuals was animated using the steering wheel angle provided by CarSim. The participants used a custom made rotary knob with visual feedback in the form of a rating bar to provide their rating during the experiment.

7.3.4. Participants

15 participants, of which five female, aged between 20 and 61 years (mean 30 years) performed the experiment. All possessed a valid car driving license, and all but three participants had no prior knowledge or experience with motion cueing algorithms. All participants participated voluntarily in the experiment and those not working at the MPI were compensated for their time (eight euros per hour).

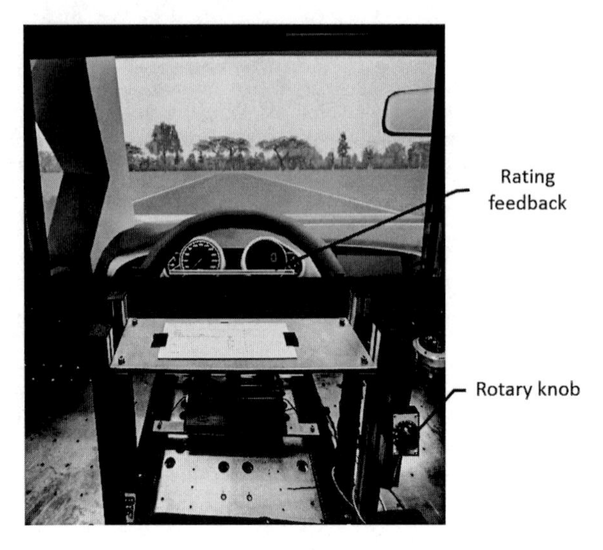

Figure 7.4: Experiment set up with rotary knob and visual rating feedback used to provide the continuous perceived motion incongruence rating during the experiment.

7.3.5. Instructions and Procedures

The task of the participants was to rate the perceived motion incongruence, or mismatch, between the visual and physical motions during a passive car driving simulation, i.e., rate the mismatch between the physical motions in the simulator and the motions you would expect in a real vehicle based on the simulator visuals. After reading the experiment instructions, participants were further briefed about the goal of the experiment and their tasks verbally.

The experiment started with a training phase in which the simulation was repeated twice for each MCA setting, i.e., four simulations in total. Throughout the experiment, the simulations occurred in pairs including both MCA settings, but the order of the setting within one pair was randomized for all participants. During the first repetition of such a simulation pair, participants were asked to observe the mismatches and try to anchor the rating scale to the minimum and maximum mismatch perceived over both simulations. During the second repetition, participants were requested to use the anchored rating scale to provide a rating continuously (CR) throughout the simulations, by moving the rating bar on the screen using a rotary knob. After each simulation they were also requested to provide an overall rating (OR) on the same scale, indicating a summary of their continuous rating, using the rotary knob.

After this training another three repetitions per MCA setting, i.e., in total six simulations, were done in which the participants were asked to provide a continuous and overall rating using their rating scale, and to try being as consistent as possible.

To monitor simulator sickness, also a misery score (MISC) [176] was requested after each simulation pair. At the end of the experiment, participants were requested to fill in a questionnaire, with questions related to their rating and driving experience. In Appendix H the questionnaire is included, together with the definitions of the abbreviations used in the following sections. The total experiment lasted around 45 minutes per participant.

7.4. RESULTS

During the experiment the participants were asked to judge the perceived motion incongruence by giving both a CR and an OR for each MCA setting separately. At the end of the experiment each participant also filled out a questionnaire. To determine if differences between MCA quality exist, in the next sections the rating and relevant questionnaire results are shown.

7.4.1. DATA PROCESSING

For each participant the OR and the CR were collected for three repetitions of a simulation pair, including both MCA settings. Each participant was explicitly asked to anchor the rating scale to the maximum and minimum mismatch present during the two simulation and thus use the whole rating scale for their CR for each simulation pair. To correct for any deviations from this task, the CR was normalized, such that both the maximum and minimum ratings were obtained at least once during each simulation pair. For each participant, the OR or CR per MCA setting is calculated as the mean over all three repetitions.
To determine whether significant differences are present between the MCA settings, statistical tests were performed. For those comparisons that include normal data a two-sample t-test was performed and the t-statistic, degrees of freedom and p-value are reported. For those comparisons that include non-normal data, a Wilcoxon signed-ranks test was used instead and the corresponding W-statistic, or for bigger samples the Z-statistic, and p-value are reported. If significant differences are found with the appropriate test, also the data means are reported.
For very small sample sizes, instead the non-parametric common language effect size A [177], which is robust to non-normally distributed samples [178], is calculated. This effect size measure indicates the probability that a score sampled at random from one distribution will be greater than a score sampled from some other distribution [179].

7.4.2. CONSISTENCY

OVERALL RATING

Although subjective ratings are likely not to be exactly the same for each repetition, it is expected that participants at least consistently prefer the same MCA. To determine if the participant understood and performed the given task correctly, here the consistency of this preference is checked. All but two of the 15 participants indeed showed a consistent preference for one of the MCAs. Figure 7.5 shows that both participants 1 and 11 changed their preference during the second repetition, as compared to the first and last repetition, and thus did not have a consistent preference. Their results are therefore excluded from further OR analysis.

CONTINUOUS RATING

As providing a CR is a different task, the consistency of this rating is checked separately for each of the fifteen participants as well. Following the same approach taken in previous studies [109, 135–137], the consistency is calculated with Cronbach's Alpha (α) [117]. Sets of ratings with an α below 0.7 are considered inconsistent [118] and are excluded from further CR analysis.

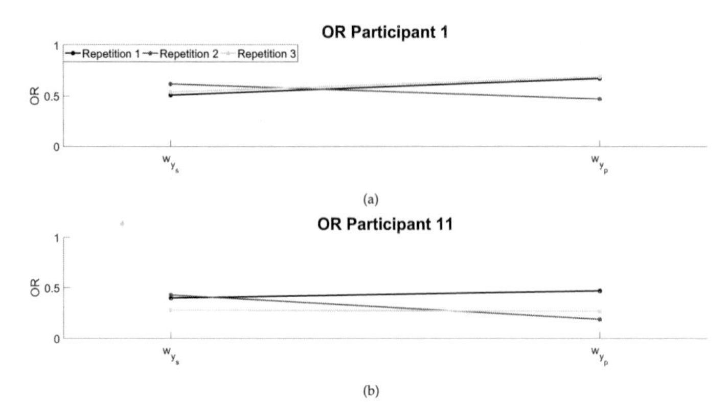

Figure 7.5: Participants with inconsistent overall ratings.

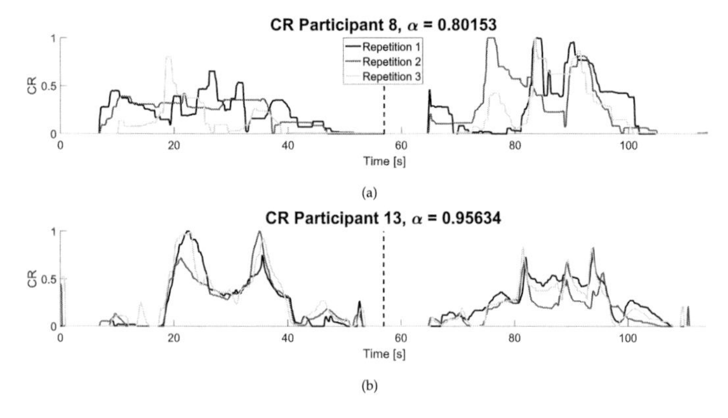

Figure 7.6: Participants with consistent continuous ratings.

In total eleven participants showed consistent ratings (mean: 0.89, std: 0.061) and four participants had to be excluded from further analysis due to inconsistency of their CR data. Figure 7.6 shows the CR of two participants with consistent ratings over the three trials.

The ratings shown in Figure 7.6 result in the maximum and minimum consistent α values of 0.96 and 0.80, respectively, and thus give an indication of all ratings that are here assumed to be sufficiently consistent. In Figure 7.7, instead, the CR of participants with inconsistent ratings for the three repetitions is shown. These ratings result in α values of 0.58, 0.53, 0.55 and 0.51, respectively, and are well below the threshold of 0.7. As participants 1, 3, 9 and 12 gave inconsistent CR, their results are excluded from the CR analysis.

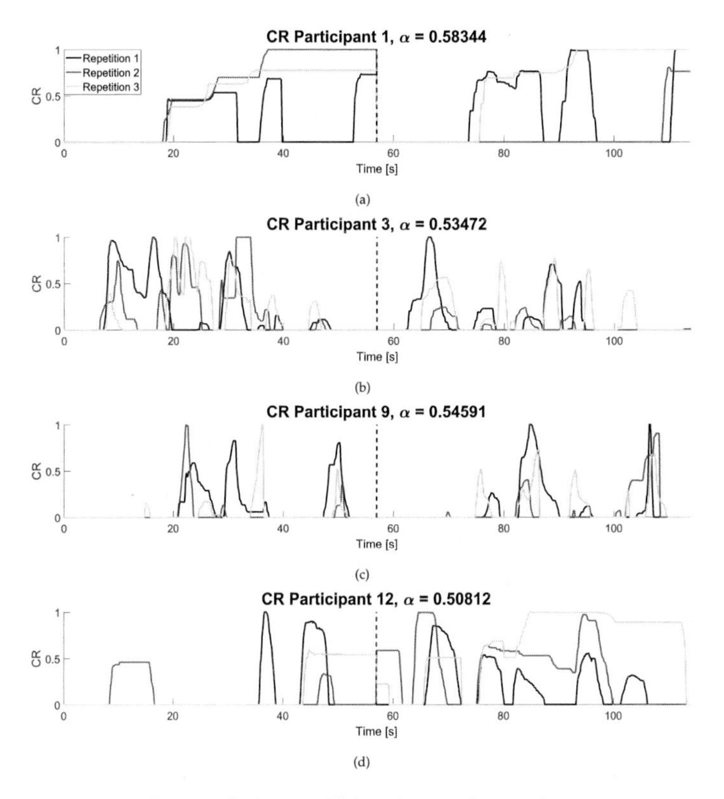

Figure 7.7: Participants with inconsistent continuous ratings.

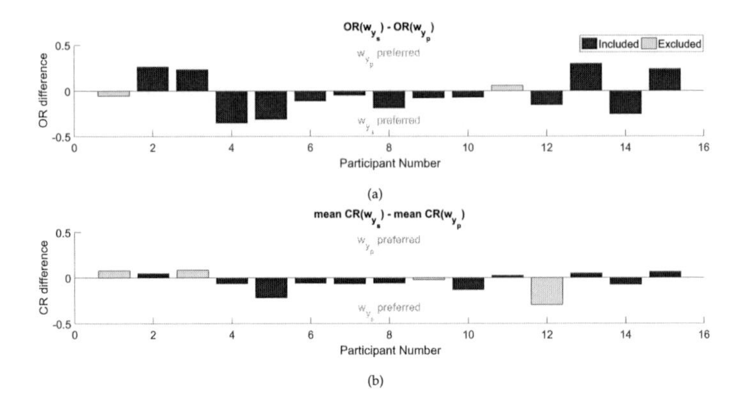

Figure 7.8: Rating differences between both MCA settings per participant. A positive or negative difference indicates a preference for settings $MCA\left(\mathbf{w_{yp}}\right)$ and $MCA\left(\mathbf{w_{ys}}\right)$, respectively.

7.4.3. Participant Groups

The data showed that not all participants had the same MCA preference. Next to presenting the overall (averaged) rating results, the results of two participant groups, namely those preferring $MCA\left(\mathbf{w_{y_s}}\right)$ and those preferring $MCA\left(\mathbf{w_{y_p}}\right)$, will therefore also be shown.

In Figure 7.8 the relative rating differences between MCA settings is shown for both the mean OR and CR per participant. The magenta coloured bars indicate the participants that were excluded from further analysis of either OR or CR.

Figure 7.8 shows that all participants except Participant 1, showed a consistent preference when rating with either OR or CR. Five out of the fourteen participants preferred the $MCA\left(\mathbf{w_{y_p}}\right)$, while nine preferred the $MCA\left(\mathbf{w_{y_s}}\right)$. In the following figures, the results for these two groups are shown separately.

To determine what made these two participant groups rate the MCA settings differently, their questionnaire responses are also compared. In the questionnaire, participants were asked to rate the extent to which certain factors influenced their rating (Question 2 in Appendix H). When taking all participants into account, no clear difference between participant groups for any of the factors was found. However, when only taking into account those participants that gave a consistent OR and CR, some differences between participant groups emerge, as can be seen in Figure 7.9.

These results give an indication on what the preference of these two participant groups was based on, but due to the small sample sizes of each group ($\mathbf{w_{y_s}}: N=7$, $\mathbf{w_{y_p}}: N=3$) these results should be considered with care. The clearest differences between participant groups ($A > 0.8$) are found for the factors "stronger" ($A = 0.80$, $\mathbf{w_{y_s}} > \mathbf{w_{y_p}}$), "weaker" ($A = 0.90$, $\mathbf{w_{y_s}} < \mathbf{w_{y_p}}$) and "direction" ($A = 0.86$, $\mathbf{w_{y_s}} < \mathbf{w_{y_p}}$). All other factors showed common language effect scores smaller than 0.8.

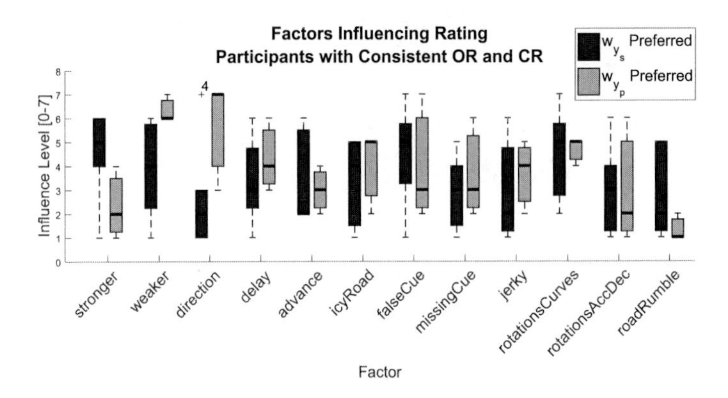

Figure 7.9: Boxplots showing the median and interquartile ranges of the level of influence of different factors on the participant ratings of the group preferring $MCA\left(\mathbf{w_{y_s}}\right)$ (7 participants) and the group preferring $MCA\left(\mathbf{w_{y_p}}\right)$ (4 participants). The results only include participants with consistent CR.

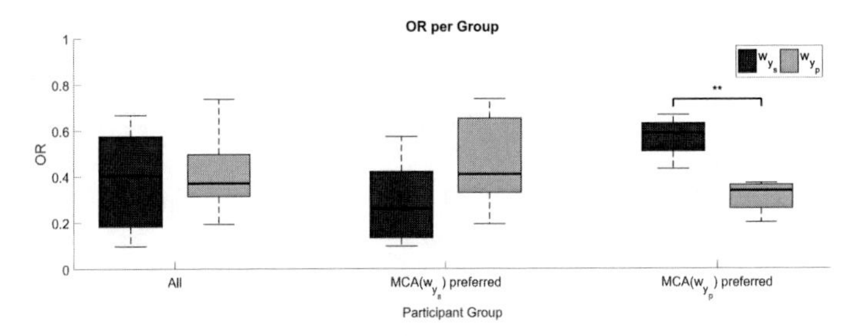

Figure 7.10: Boxplots of the mean OR over three repetitions for participants with a consistent OR. The box shows the interquartile range and the median over all ratings.

7.4.4. RATING RESULTS

For the analysis of the OR, the consistent data of 13 participants are used. For each participant one OR per MCA setting is calculated as the average over three repetitions. In Figure 7.10 the OR data results for all consistently rating participants, as well as for each participant group ($\mathbf{w_{y_s}}$ preferred $N = 9$, $\mathbf{w_{y_p}}$ preferred $N = 4$), are shown in a boxplot.

No significant difference between the two MCA settings was found for all participants combined ($t(24) = 0.58, p \geq 0.05$), nor for the participant group that preferred $MCA\left(\mathbf{w_{y_s}}\right)$ ($t(16) = 2.09, p \geq 0.05$). For the participants preferring $MCA\left(\mathbf{w_{y_p}}\right)$ the difference in OR between the MCA settings (mean $MCA\left(\mathbf{w_{y_s}}\right)$ = 0.57, mean $MCA\left(\mathbf{w_{y_p}}\right)$ = 0.31) was found to be significant ($t(6) = 4.15, p < 0.05$).

For the CR analysis, the consistent data of 11 participants are used. The CR per participant is calculated as the mean over all repetitions. Figure 7.11 shows the median and interquartile range of the CR over all participants per MCA setting for all consistently rating participants and for the two participant groups ($\mathbf{w_{y_s}}$ preferred $N = 7$, $\mathbf{w_{y_p}}$ pre-

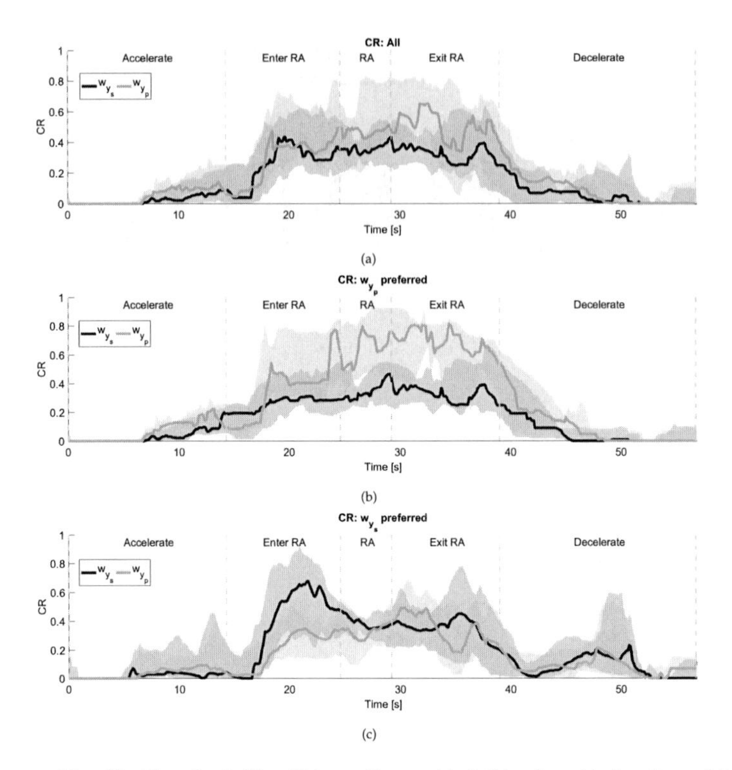

Figure 7.11: Mean CR with median (bold) and interquartile range (shaded) for all consistently rating participants (a) and those preferring $\mathbf{w}_{\mathbf{y}_\mathbf{s}}$ (b) and $\mathbf{w}_{\mathbf{y}_\mathbf{p}}$ (c), respectively.

ferred $N = 4$). The plots are divided in five sections, each showing the data belonging to a different section of the simulation, with RA referring to Roundabout.
The CR mainly increases during the roundabout section of the simulation for both MCA settings. The difference between MCA settings is relatively small. To better compare the two MCA settings with the CR, the rating of each participant is summarized with the mean CR over all time steps per MCA setting. Figure 7.12 shows the resulting boxplots containing the results for the eleven participants with a consistent CR per MCA setting.

No significant difference between the mean CR over time was found between conditions for all participants ($t(20) = 0.8735, p \geq 0.05$), nor for the participant group preferring $MCA\left(\mathbf{w}_{\mathbf{y}_\mathbf{s}}\right)$ ($t(12) = 1.5047, p \geq 0.05$) or $MCA\left(\mathbf{w}_{\mathbf{y}_\mathbf{p}}\right)$ ($t(6) = 0.5933, p \geq 0.05$).
To determine if specific parts of the simulation are rated differently between MCA setting, also the mean CR over all time steps for a specific section of the simulation is calculated per participant. Figure 7.13 shows the resulting boxplots per simulation section for all consistently rating participants, for those preferring $MCA\left(\mathbf{w}_{\mathbf{y}_\mathbf{s}}\right)$ and for those preferring $MCA\left(\mathbf{w}_{\mathbf{y}_\mathbf{p}}\right)$.

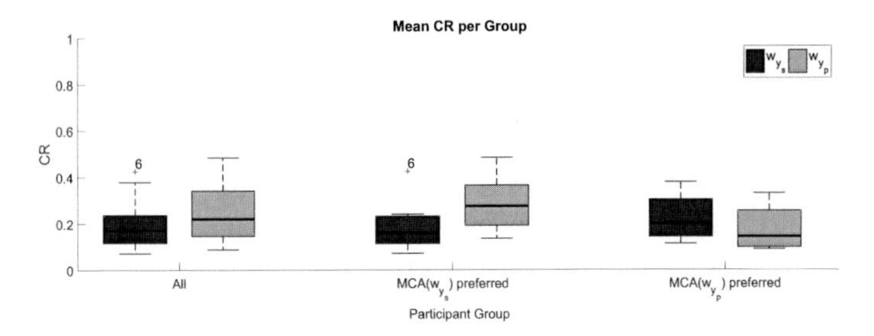

Figure 7.12: Boxplots of the mean CR over time and all three repetitions per MCA setting for participants with a consistent CR. The box shows the interquartile range and the median over all ratings.

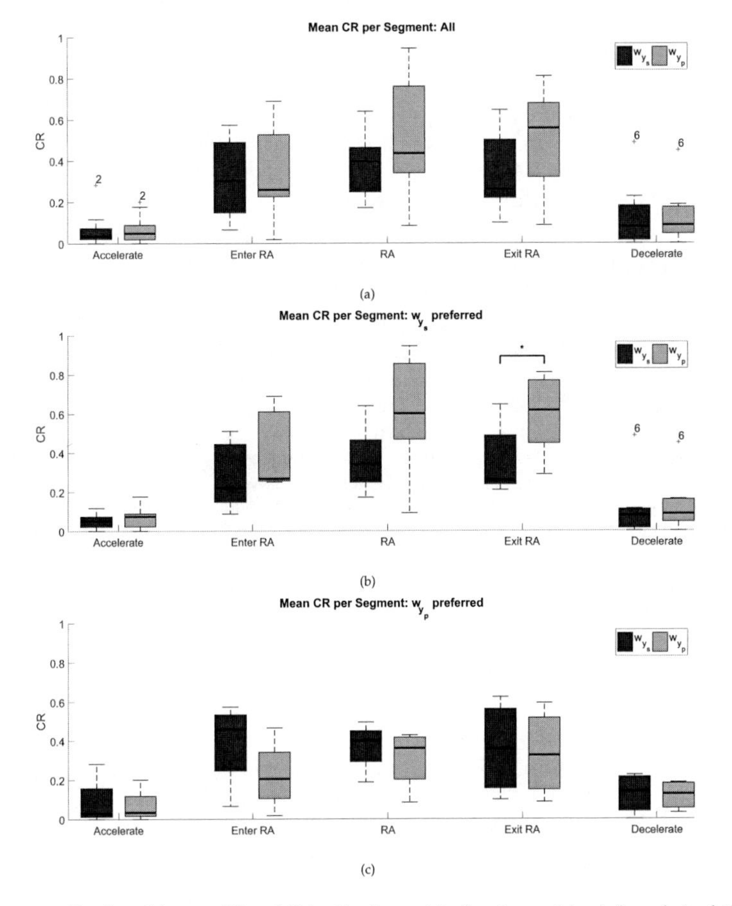

Figure 7.13: Boxplots of the mean CR per MCA setting for consistently rating participants for each simulation section. The boxes show the interquartile range and the median over all ratings.

No significant difference between the MCA settings was found for all participants combined (Accelerate: $z = 0.3951, p \geq 0.05$, Enter RA: $z = 0.3940, p \geq 0.05$, RA: $t(20) = 1.3891, p \geq 0.05$, Exit RA: $z = 1.3133, p \geq 0.05$, Decelerate: $z = 0.1315, p \geq 0.05$). For the participant group preferring $MCA\left(\mathbf{w}_{\mathbf{y}_\mathrm{s}}\right)$ the difference between MCA settings is more pronounced during and while exiting the roundabout, with only the latter showing a significant difference ($W = 36, p < 0.05$) between MCA settings (mean $MCA\left(\mathbf{w}_{\mathbf{y}_\mathrm{s}}\right) = 0.36$, mean $MCA\left(\mathbf{w}_{\mathbf{y}_\mathrm{p}}\right) = 0.60$). During all other sections no significant differences were found (Accelerate: $t(12) = 0.7012, p \geq 0.05$, Enter RA: $W = 40, p \geq 0.05$, RA: $t(12) = 1.9525, p \geq 0.05$, Decelerate: $W = 49.5, p \geq 0.05$). For the participant group preferring $MCA\left(\mathbf{w}_{\mathbf{y}_\mathrm{p}}\right)$ the difference seems instead more pronounced while entering the roundabout, but no significant differences were found. (Accelerate: $W = 17.5, p \geq 0.05$, Enter RA: $t(6) = 1.1445, p \geq 0.05$, RA: $t(6) = 0.6082, p \geq 0.05$, Exit RA: $t(6) = 0.1592, p \geq 0.05$, Decelerate: $t(6) = 0.1353, p \geq 0.05$).

7.5. DISCUSSION

In this chapter the effect of MCA optimization using a Motion Incongruence Rating (MIR) model based on previously collected rating data was investigated. An MPC-based MCA, using weights for linear acceleration and rotational velocity that were found using a baseline MIR model $MCA\left(\mathbf{w}_{\mathbf{y}_\mathrm{p}}\right)$, was compared to the same MCA using weights based on differences in variance between these motion channels for typical car manoeuvres, i.e., $MCA\left(\mathbf{w}_{\mathbf{y}_\mathrm{s}}\right)$. While the MIR predictions from dynamic MIR models from previous chapters indicated an improvement in MCA quality for $MCA\left(\mathbf{w}_{\mathbf{y}_\mathrm{p}}\right)$ compared to $MCA\left(\mathbf{w}_{\mathbf{y}_\mathrm{s}}\right)$, no significant differences between the conditions were found in the results of a performed human-in-the-loop validation experiment. A clear statement on the use of MIR models for MCA optimization can therefore not be made based on these results.

In the experiment, two different measurements of MCA quality were obtained, the OR and the CR, for each MCA setting. Before comparing these ratings between conditions, checks on the rating consistency over three repetitions were performed. The relatively large number of participants (2 for OR and 4 for CR) that did not provide consistent ratings for the three repetitions could be an indication that the difference between conditions was not large enough to be detected by the measurement methods used. Additionally, the large number of participants that provided inconsistent CR could be caused by the relatively short training time. The training included the same number of simulation repetitions as in previous motion cueing CR experiments [136, 137], but with two minutes per simulation part as compared to about ten minutes in previous experiments, the total training time was much shorter. To ensure rating consistency, it is therefore advised for future experiments to not only take into account the number of repetitions, but also consider the total training time.

The lack of significant differences between conditions was further analysed, through investigating the possible existence of different participant groups. This analysis showed that, while five participants preferred $MCA\left(\mathbf{w}_{\mathbf{y}_\mathrm{p}}\right)$, nine participants preferred $MCA\left(\mathbf{w}_{\mathbf{y}_\mathrm{s}}\right)$. Questionnaire results indicated that the main cause for a preference for $MCA\left(\mathbf{w}_{\mathbf{y}_\mathrm{s}}\right)$ were the strong motions present in the $MCA\left(\mathbf{w}_{\mathbf{y}_\mathrm{p}}\right)$ condition. When comparing the main vehicle motions (longitudinal and lateral acceleration) to the

simulator motions, however, both conditions show a reduction of the simulator motion as compared to the vehicle motion. The finding that most participants still judged the motion as "too strong" is in line with findings in previous research [180] and can possibly be caused by the relatively poor visual quality in the simulator [165]. The participants who, correctly, perceived the motions as too weak, might have been influenced by their prior knowledge of motion simulation.
The reason that the MIR models did not predict the preference for weaker motions is threefold. First, these models did not take into account the existence of cueing errors when one-to-one vehicle-simulator motion is obtained, i.e., all cueing errors are based on a difference between vehicle and simulator motions. For future research it could be useful to also investigate how to include the preference for below-unity gains between vehicle and simulator motions in the MIR models.
Secondly, the models were based on data from previous experiments, with other simulators that had a different visual quality from the simulator used in this study. A large FOV, for example, has been shown to improve speed estimation [95], while a small FOV can cause speed underestimation [94]. The small horizontal FOV (80°) of the Cyberpod used here, as compared to the larger much FOV of the CyberMotion (140°) and Daimler-Benz (360°) simulators used in the previous experiments, could have caused an underestimation of the vehicle speed, that was not, or to a lesser extent, present during the other experiments. This speed underestimation, in turn, could result in participants expecting *weaker* vehicle motions, corresponding to the incorrectly perceived vehicle speed.
Finally, the participants that partook in the experiments on which the MIR models were based differed considerably from the participants that took part in the experiment described here. For the current experiment, a participant database to which anyone could apply was used, while the previous experiments solely used MPI or Daimler employees as participants. The many differences (motion platform, visual system, participants, cueing errors, etc.) between the experiments used to develop the MIR models and the current experiment surely also influenced the accuracy of the prediction. For future research it is therefore advised to first investigate the use of MIR models for MCA optimization under more similar conditions, e.g., within the same experiment. From the results presented here, however, it can be concluded that the MIR models based on a limited dataset cannot directly be used for *any* MCA optimization.
Another challenge that was encountered, mainly when trying to use MIR models for the optimization of the MPC-based MCA for a more complex simulator than the CyberPod, is that during the optimization process, an infinite amount of different cueing errors are analysed using the MCA cost function. As MIR models can only predict cueing errors that are similar to those for which the MIR model was designed, not all predictions made by the MIR model during the optimization process are accurate. When a more limited simulator, such as a hexapod, is used, applying a MIR model-based cost function in the MCA optimization still gave reasonable results. However, when applying the same process to a simulator with *less* limitations, such as the CyberMotion Simulator at the MPI, very unrealistic motion profiles can result from such an optimization. This is mainly due to the many local minima that are possible with this type of simulator, in combination with the limited applicability of the MIR models. Using MIR models for the optimization of MCA for such simulators should therefore be done with extra care, and it is likely that additional constraints will need to be implemented to avoid local minimum solutions. Additionally, the MIR models should be based on a wider range of cueing errors when applying the process to these

type of simulators.
One of the key choices made in this chapter was to make the two conditions tested in this chapter comparable. For this purpose the output error weights $\mathbf{w}_{y_s}$ and $\mathbf{w}_{y_p}$ were scaled such that they had equal sums. It can be debated if this scaling procedure results in a fair comparison between conditions and whether other MPI MCA weight parameters influence the optimization in the two conditions similarly. For example, the absolute value of the lateral acceleration input error weight was fixed to one for both conditions, but its relative value compared to the corresponding output error weight, which for $\mathbf{w}_{y_p}$ was six times higher than for $\mathbf{w}_{y_p}$, differed per condition. Other scaling procedures could therefore be investigated in the future.
While the MIR models did not accurately predict the difference between the MCA settings for about two thirds of the participants, they did correctly predict that the main decrease in cueing quality would occur during the roundabout section of the simulation. The models also correctly predicted that hardly any decrease in cueing quality would be detected during the initial acceleration and final deceleration, even though the differences between vehicle and simulator motions during these sections were considerable. The fact that for one third of the participants the predicted MCA preference was indeed correct, makes it plausible that MIR models can aid in the tuning of cost function weights for optimization-based MCAs. However, further research, where, for example, the MIR model design and corresponding MCA optimization are performed within one experiment, needs to be conducted to investigate the usefulness and applicability of MIR models for MCA optimization.

7.6. Conclusion

In this chapter an example of how MIR models can be used for MCA optimization is explained and tested. The current results do not show a clear difference between the cueing quality of the MPI MCA optimized with the help of MIR models, or one heuristically-tuned based on the typical variance difference between linear acceleration and rotational velocity for typical car manoeuvres. Unexpectedly, more participants preferred the heuristic as compared to the optimization using MIR models. This preference seems to mainly be based on a preference for lower than unity gains between vehicle and simulator motions. The predicted preference for the optimization using MIR models did hold for about one third of the participants. Additionally, the low impact of cueing errors in longitudinal motions as compared to lateral motions was also predicted accurately.
The fact that for one third of the participants optimization using a MIR model was preferred indicates that, while not as broadly applicable as was done here, MIR models could be useful for MCA optimization. It is expected that MIR models developed using a much richer dataset will result in an MCA that is preferred by a wider range of participants. To fully understand the use of MIR models for MCA optimization, however, it is recommended to first further develop these models under conditions that are more similar to those which the resulting optimized MCA quality will be tested under, such as the same motion platform, visual system and participant group.

8

Conclusions and Recommendations

The research goal of this thesis was stated as follows:

To develop an MCA-independent off-line prediction method for time-varying perceived motion incongruence during vehicle motion simulation, to improve motion cueing quality

In this thesis, Perceived Motion Incongruence (PMI) refers to a feeling of decreased cueing quality due to the incongruence between visual and physical motion cues. It describes a novel approach to predict cueing quality and is divided in three parts. In Part I a continuous rating method to *measure* time-varying PMI was developed and experimentally validated. The data obtained with this measurement method were used to estimate and validate a PMI prediction *model* in Part II. In Part III a simple PMI prediction model was implemented in the cost function of an optimization-based MCA, to investigate whether this results in *optimized* PMI. The conclusions of each part are summarized in Sections 8.1 to 8.3, respectively. Section 8.4 presents the main recommendations for future research.

8.1. Measuring Perceived Motion Incongruence

The development of a time-varying PMI prediction model requires measurements of the PMI for both parameter estimation and validation of the model. As no such measurement method existed, in the first part of this thesis a subjective PMI measurement method was developed and tested. In **Chapter 2** this measurement method, based on continuous subjective ratings [107, 109], was presented and tested for reliability and validity in a human-in-the-loop experiment in the CyberMotion Simulator at the MPI for Biological Cybernetics, using simple motions and common (but exaggerated) cueing error types. The results show that participants can use the developed method to consistently rate the PMI due to common motion cueing errors (Cronbach's Alpha > 0.7). Furthermore, the relative magnitude of the time-averaged continuous rating for the different tested cueing error types was consistent with literature. For example, false cues were rated as much more detrimental than scaled cues [51], and roll rates were the main contributor to PMI for manoeuvres where this rate was above human perceptual threshold [68]. The continuous rating also correlated well (Pearson's $r > 0.8$) with the

more commonly used overall ratings, collected after each completed trial.
In **Chapter 3** the validated measurement method was used for the comparison of two MCAs during a realistic driving simulation scenario, in the Daimler Simulator. Unlike the experiment in **Chapter 2**, the vehicle motions were the result of a realistic drive along rural roads and through a city environment, while the simulator motions were obtained with two optimized MCAs. This experiment showed that participants could rate consistently under these more realistic conditions.
Subjective measures are, due to their high variance, often not preferred for engineering applications. From **Chapter 2** and **Chapter 3**, however, it can be concluded that highly relevant information can be obtained with the continuous subjective rating method presented in this thesis. Specifically, the time information obtained with this method allows for direct evaluation of the relative severity of different types of cueing errors, which is not possible with other evaluation methods. Additionally, unlike overall ratings, continuous ratings are not subjected to memory related biases such as the peak-end rule, which states that mainly the peak and the end experience, rather than the sum over all experiences, affect an overall rating [123, 181].
Other research where the continuous subjective rating method proposed in this thesis was used, confirms the effectiveness of this method for the evaluation of time-varying PMI during vehicle motion simulation [135, 182–184]. While the variance of the ratings between participants can be high, participants show to be capable of rating consistently and, moreover, the average rating over multiple participants compares well to the severity of different cueing errors as described in literature [51, 68]. These factors combined provide sufficiently good arguments to use the time-varying PMI measurement for modelling purposes.

8.2. Modelling Perceived Motion Incongruence

To model and predict the time-varying PMI, a model was developed using a data-driven approach. A key factor in PMI prediction is the relative weighing of different cueing errors. This includes weighting differences between similar cueing errors in different motion channels, but also cueing errors of different types, such as scaling, missing and false cues, within the same motion channel. To extract the different cueing error types from vehicle and simulator motion signals, a Cueing Error Detection Algorithm (CEDA) was developed in **Chapter 4**.
Unlike error detection as described in [139], where scale and shape errors are defined using the global scaling parameter of a washout filter, here the error detection is directly applied to the vehicle and simulator motion signals, making it independent of the applied cueing filter. This algorithm uses a wavelet-based semblance measure to distinguish between the relatively detrimental shape errors [41, 51, 144], i.e., missing and false cues, and the less detrimental scaling errors, and uses gain measures to further distinguish between missing and false cues.
The algorithm parameters were tuned using vehicle and simulator data from the experiment described in **Chapter 2** and the resulting algorithm was validated using data from a different simulator experiment, described in [135]. The algorithm correctly classified all cueing errors in scaling, missing and false cues and can thus facilitate relative cueing error type weighing for PMI modelling.
In **Chapter 5** a general form of a data-driven time-varying PMI prediction model was introduced. The proposed model consists of a non-linear part, that retrieves the different cueing errors from the measured vehicle and simulator motions, and a linear part,

that filters, weighs and combines these cueing errors into a single PMI output. The complexity of the non-linear part thus depends on the mathematical description of different cueing errors. A simple model would, for example, describe cueing errors as the absolute difference between simulator and vehicle motions for each motion channel. A more complex model could include the CEDA from **Chapter 4** to further differentiate between different cueing error types within each motion channel.

The PMI model was estimated and validated with the rating data from the measurement method described in **Chapter 2**. As this Motion Incongruence Rating (MIR) is an indirect measure of PMI, the proposed PMI model also includes a rating system that models the mapping of the PMI onto the MIR, in the linear-part of the model. The linear part of the model was chosen to have an ARX structure, which simplifies the parametrization of this part of the model.

Chapter 5 furthermore introduces a system identification process to identify and estimate the linear part of the model. This process is used to reduce model complexity by selecting only the subset of cueing errors that is truly needed for the PMI prediction of a given dataset. This input selection, for example, confirmed the finding in [124, 158] that linear jerk has a significant influence on the cueing quality. Additionally, the identification process is used to estimate the parameters/coefficients of the linear part of the model.

The identification process was used to design three PMI prediction models of different levels of complexity. All models were fitted to one part of the dataset from the experiment of **Chapter 2**, to analyse the explanatory power of these models, and applied to the remaining part of this dataset, to analyse the prediction power of these models. All models showed adequate explanatory (VAF values between 85% and 92.2%) and prediction power (VAF values between 68.7% and 85.9%), with both increasing with increasing model complexity. This shows that the proposed general PMI model and corresponding identification process can indeed be used to predict PMI ratings within one experiment.

In **Chapter 6** a process to estimate a Model Transfer Parameter (MTP) is described and validated. With the continuous rating method, the maximum rating within one experiment is given to the maximum PMI presented in that experiment. As this maximum PMI can differ between experiments, so do the corresponding PMI rating scales. With the MTP, rating data from one experiment can be rescaled and mapped onto the rating scale of a second experiment. This rescaling is, for example, needed when predicting PMI between different experiments or when combining rating data from different experiments into one dataset for model fitting. In [165] rating data from three different experiments are combined by matching the results of an identical experimental condition which was present in all three experiments. The unique combination of a time-varying quality measurement, a quality prediction model and the MTP scale linking method, renders replication of experimental conditions superfluous.

In **Chapter 6** the MTP between the datasets from the experiments described in **Chapter 2** and **Chapter 3** was estimated. First, the same three models of varying complexity as considered in **Chapter 5** were estimated using either dataset. Subsequently, the estimated MTP was used to analyse their prediction power when applied to the dataset from the other experiment. Additionally, the estimated MTP was used to combine the datasets from both experiments and the three models were again estimated using this combined dataset.

The results show that if a PMI prediction model is fitted to a sufficiently rich dataset, it can indeed be used to predict PMI from a different experiment (VAF values up to

77.1%). Results also show that especially models of low complexity benefit from fitting to the combined rather than the single datasets, in terms of explanatory power (with VAF improvements up to 6%). We can conclude from **Chapter 6** that the general PMI model and corresponding identification process proposed in this thesis, together with a sufficiently rich dataset, can also be used to *predict* general features of PMI ratings *between* experiments.
While many have focused on the design of physiological models for motion perception for decades [92, 93, 185–188], none of these models have managed to provide a single measure or prediction of cueing quality so far. The approach presented in this thesis shows that more research into data-driven approaches to predicting cueing quality is likely more efficient.

8.3. MINIMIZING PERCEIVED MOTION INCONGRUENCE

PMI prediction models open up many different opportunities for improving the perceived quality of cueing algorithms. Models can, for example, be used during off-line MCA parameter tuning. They can also be used to pinpoint troublesome parts of a simulation before an experiment is performed, such that these parts can be modified or removed. Analysing the correlation between cueing errors and the measured time-varying PMI, as is done during the development of PMI prediction models, can also aid in gaining a better understanding of which aspects of simulator-vehicle motion combinations are detrimental to motion cueing quality and possibly result in new insights into human motion perception.
Chapter 7 presents an initial attempt to utilize PMI prediction models for explicit MCA optimization. In this chapter a simple PMI prediction model, i.e., including only a weighted sum of absolute differences between vehicle and simulator motions in six motion channels, is implemented as part of the cost function of the optimization-based MCA developed in [42, 57]. The resulting PMI-based MCA is compared to the original MCA in a short human-in-the-loop experiment.
Results show that only a small group of participants, all with prior simulator experience, preferred the PMI-based MCA. The preference of the other, larger group seemed to mainly be based on a preference for "lower than unity gains" between vehicle and simulator motions. This preference has been found before in, for example, [189]. Overall, it was concluded in **Chapter 7** that for MCA optimization the model needs to be fitted to a much richer dataset, in terms of, among others, number and variety of participants, cueing errors and simulators.

8.4. RECOMMENDATIONS

This thesis describes a novel approach to improve perceived cueing quality of motion cueing algorithms. It provides a complete roadmap and describes how to *measure* and *model* PMI and how to apply such models to predict and with that *minimize* PMI in motion simulations. The results presented show the potential of this novel approach. For each step in the roadmap to improving cueing quality, however, several recommendations for future research can be made.
Regarding the *measurement* of PMI:

- The continuous rating method as presented in this thesis can only be used for passive driving simulations. It is recommended to investigate the necessary adjustments to this method such that it can also be used during active driving. Two main challenges for achieving this need to be addressed:

1. "Rating continuously while at the same time actively controlling a vehicle." It should be investigated if two such tasks can be done sufficiently well simultaneously or if an approach where participants first drive and afterwards rate their recorded drive should be adopted, such as proposed in [113] and [44].
2. "Obtaining sufficiently similar motion profiles for all participants such that ratings can still be averaged." Minimizing motion profile differences between participants could possibly be done with the help of haptic guiding forces to train participants to adopt predefined driving behaviour. Such training methods have previously been applied with regard to curve negotiation [190], eco-driving [191] and parking [192].

- While most participants in the experiments presented in this thesis managed to rate reliably, this was not the case for all tested participants. It is therefore recommended to investigate and improve the training part of the measurement method, such that optimal within-subject reliability is achieved. The dynamics of learning to provide reliable continuous ratings with respect to, for example, training duration, should be investigated.

Regarding the *modelling* of PMI:

- The thesis presents a general model for PMI prediction, that includes a non-linear part that allows for implementation of specialized cueing error calculation algorithms. It is recommended to both extend the Cueing Error Detection Algorithm (CEDA) presented in this thesis with more cueing error types, such as limiting and phase errors [97], as well as develop new algorithms to, for example, isolate the rotational angle, rate and acceleration errors above human perception thresholds. Additionally, known human absolute and discriminatory perception thresholds can be used to filter the translational motion channels.

- The linear part of the PMI prediction model is currently implemented as an ARX structure. The residuals of the predictions of several of the presented models showed autocorrelation, indicating that a different model structure might be more appropriate. Different model structures have been briefly tested, such as the ARMAX structure where a moving average is included, but these did not lead to any significant improvements in the model fit and prediction or reductions of the number of parameters. It is recommended to investigate the effects of different model structures on the accuracy of the models more thoroughly.

- This thesis showed that PMI models fitted to data from one experiment can be used to predict PMI in another experiment with different simulators and participants by simply using a Model Transfer Parameter (MTP) to map the two PMI rating scales onto one common scale. The between-experiment prediction of PMI was, however, less accurate than the within-experiment prediction. It is therefore recommended to investigate what the influence of differences between experiment set-ups are on the estimated PMI prediction models. For example, the models assume that vehicle motions can be derived from visuals and driving experience, but the visuals can differ greatly between simulators. It is recommended to investigate whether such differences between visuals can be captured in the non-linear part of the PMI prediction model via, for example, an algorithm that maps actual vehicle motions onto visually perceived vehicle motions. By capturing such known differences between simulators within the PMI prediction model, its between-experiment prediction power is expected to increase.

- The approach presented in this thesis shows that data-driven approaches to predicting cueing quality is likely to be more efficient than using physiological models for motion perception [92, 93, 185–188], as the latter do not provide one single measure or prediction of cueing quality. It is therefore recommended that research on MCA optimization focuses more on the development of these data-driven cueing quality models.
- Especially for automatic optimization of cueing quality, PMI prediction models need to be able to predict the PMI for all occurring cueing errors. Obtaining such models thus requires sufficiently rich datasets. It is therefore recommended to first investigate what a sufficiently rich dataset entails in this context, i.e., how to design input signals that are persistently exciting [193]. Secondly, a large database of diverse cueing error experiments, which can be combined using the process described in **Chapter 6**, should be established to estimate and validate such "universal" PMI models.

Regarding the *minimizing* of PMI:

- Before using the PMI prediction models for explicit MCA optimization, it is recommended to first use PMI for cueing quality research, such as off-line MCA tuning. Especially valuable will be these models' capacity for pinpointing the key troublesome parts of a simulation. Applying PMI prediction models in this way will also help to reveal the shortcoming of these models, which can in turn be used to fine-tune and improve them.
- **Chapter 7** describes how a simple PMI prediction model can be used as part of the cost function of an optimization-based MCA. **Chapter 5** and **Chapter 6** show, however, that the PMI prediction improves with increasing model complexity. It is therefore recommended to implement a more comprehensive model in future research and then determine whether this improves the cueing quality compared to a simple model.
- The results of **Chapter 7** only provide indications to why only a few participants preferred the PMI-based MCA. It is recommended to repeat this experiment under more controlled conditions, where the PMI prediction model is designed using data from a first experiment with the same simulator, participants and motions as a second experiment in which the PMI-based MCA is tested. Differences between these two experiments can then be introduced one by one to obtain a more thoroughly developed insight into the applicability of PMI prediction models for MCA optimization.

This thesis introduced a new way of looking at the long-standing problem of improving MCA quality. Hopefully these recommendations will provide a good starting point for other researchers to extend and improve the methods developed in this thesis.

Bibliography

[1] D. J. Allerton, *The impact of flight simulation in aerospace,* Aeronautical Journal **114**, 747 (2010).

[2] D. Stewart, *A Platform with Six Degrees of Freedom,* Institution of Mechanical Engineers **180**, 371 (1965).

[3] S. R. Dodd, J. Lancaster, A. Miranda, S. Grothe, B. DeMers, and B. Rogers, *Touch Screens on the Flight Deck,* Human Factors and Ergonomics Society Annual Meeting **58**, 6 (2014).

[4] O. Stroosma, H. J. Damveld, J. A. Mulder, R. Choe, E. Xargay, and N. Hovakimyan, *A Handling Qualities Assessment of a Business Jet Augmented with an L1 Adaptive Controller,* in *AIAA Guidance, Navigation, and Control Conference* (Portland, OR, USA, 2013) pp. 1–13.

[5] K. Bilimoria, E. R. Mueller, and C. R. Frost, *Handling Qualities Evaluation of Pilot Tools for Spacecraft Docking in Earth Orbit,* Journal of Spacecraft and Rockets **48**, 846 (2012).

[6] D. Tran and E. Hernandez, *Use of the Vertical Motion Simulator in Support of the American Airlines Flight 587 Accident Investigation,* in *AIAA Modeling and Simulation Technologies Conference* (Providence, RI, USA, 2012) pp. 1–7.

[7] I. G. Salisbury and D. J. Limebeer, *Optimal Motion Cueing for Race Cars,* IEEE Transactions on Control Systems Technology **24**, 200 (2016).

[8] I. G. Salisbury and D. J. Limebeer, *Motion cueing in high-performance vehicle simulators,* Vehicle System Dynamics **55**, 775 (2017).

[9] N. Reed, S. Cynk, and A. Parkes, *From Research to commercial fuel efficiency training for truck drivers using TruckSim,* in *Driver behavior and training,* edited by L. Dorn (Ashgate Publishing, Ltd, Farnham, United Kingdom, 2010) 4th ed., Chap. 20, pp. 257–268.

[10] M. J. Sullman, L. Dorn, and P. Niemi, *Eco-driving training of professional bus drivers - Does it work?* Transportation Research Part C: Emerging Technologies **58**, 749 (2015).

[11] H. E. Pettersson, J. Aurell, and S. Nordmark, *Truck driver behavior in critical situations and the impact of surprise,* in *Driving Simulation Conference* (Paris, France, 2006) pp. 285–294.

[12] A. Bolling, G. Sörensen, and J. Jansson, *Simulating the Effect of Low Lying Sun and Worn Windscreens in a Driving Simulator,* in *Driving Simulation Conference* (Paris, France, 2010) pp. 23–31.

[13] W. Brems, R. Uhlmann, A. Wagner, and J. Wiedemann, *Evaluation of Chassis Setups Using a Dynamic Driving Simulator,* in *Driving Simulation Conference* (Paris, France, 2016) pp. 85–92.

[14] G. Baumann, *Evaluation of steering feel and vehicle handling in the Stuttgart Driving Simulator (Bewertung von Lenkgefühl und Fahrverhalten im Stuttgarter Fahrsimulator),* in *International Munich Chassis Symposium* (Munich, Germany, 2014) pp. 201–215.

[15] A. Huesmann, D. Ehmanns, and D. Wisselmann, *Development of ADAS by Means of Driving Simulation,* in *Driving Simulation Conference* (Paris, France, 2006) pp. 131–141.

[16] M. A. Benloucif, C. Sentouh, J. Floris, P. Simon, S. Boverie, and J. Popieul, *Cooperation between the driver and an automated driving system taking into account the driver ' s state,* in *Driving Simulation Conference* (Paris, France, 2016) pp. 7–9.

[17] E. Zeeb, *Daimlers new full-scale, high-dynamic driving simulator - a technical overview,* in *Driving Simulation Conference* (Paris, France, 2010) pp. 157–165.

[18] M. Dagdelen, J.-c. Berlioux, F. Panerai, G. Reymond, and A. Kemeny, *Validation Process of the ULTIMATE High-Performance Driving Simulator,* in *Driving Simulator Conference* (Paris, France, 2006) pp. 37–47.

[19] T. Murano, T. Yonekawa, M. Aga, and S. Nagiri, *Development of High-Performance Driving Simulator,* SAE International Journal of Passenger Cars - Mechanical Systems **2**, 661 (2010).

[20] BMW Corporate Communications, *BMW Group builds new Driving Simulation Center in Munich,* (2018).

[21] B. L. Aponso, S. D. Beard, and J. A. Schroeder, *The NASA Ames Vertical Motion Simulator - A Facility Engineered for Realism,* in *Flight Simulation Conference* (London, UK, 2009) pp. 1–14.

[22] W. Bles, R. J. A. W. Hosman, and B. de Graaf, *Desdemona - Advanced disorientation trainer and (sustained-G) flight simulator,* in *AIAA Modeling and Simulation Technologies Conference* (Denver, CO, USA, 2000) pp. 1–4.

[23] F. M. Nieuwenhuizen and H. H. Bülthoff, *The MPI cybermotion simulator: A novel research platform to investigate human control behavior,* Journal of Computing Science and Engineering **7**, 122 (2013).

[24] P. Miermeister, M. Lächele, R. Boss, C. Masone, C. Schenk, J. Tesch, M. Kerger, H. Teufel, A. Pott, and H. H. Bülthoff, *The CableRobot simulator large scale motion platform based on Cable Robot technology,* in *IEEE Conference on Intelligent Robots and Systems*, Vol. 2016-Novem (Daejeon, Korea, 2016) pp. 3024–3029.

[25] B. Haycock, J. Campos, B. Keshavarz, G. Fernie, M. Potter, and S. K. Advani, *DriverLab : A Standards-Setting Simulator for Driving in Challenging Conditions,* in *Driving Simulation Conference* (Paris, France, 2016) pp. 229–231.

[26] P. Ray, *Quality flight simulation cueing - Why?* in *AIAA Flight Simulation Technologies Conference* (San Diego, CA, USA, 2013) pp. 138–147.

[27] D. J. Allerton, *Principles of Flight Simulation*, 1st ed. (John Wiley & Sons, Ltd, Chichester, UK, 2009) pp. 1–471.

[28] E. Mitsopoulos-Rubens, M. G. Lenné, and P. M. Salmon, *Effectiveness of simulator-based training for heavy vehicle operators : What do we know and what do we still need to know ?* in *Australasian Road Safety Research, Policing & Education Conference* (Brisbane, Australia, 2013) pp. 1–12.

[29] I. Siegler, G. Reymond, A. Kemeny, and A. Berthoz, *Sensorimotor integration in a driving simulator: contributions of motion cueing in elementary driving tasks*, in *Driving Simulation Conference* (Sophia Antipolis, France, 2001) pp. 1–12.

[30] J. Hogema, M. Wentink, and G. Bertollini, *Effects of Yaw Motion on Driving Behaviour , Comfort and Realism*, in *Driving Simulation Conference* (Paris, France, 2012) pp. 1–8.

[31] P. R. Lakerveld, H. J. Damveld, D. M. Pool, K. van der El, M. M. van Paassen, and M. Mulder, *The Effects of Yaw and Sway Motion Cues in Curve Driving Simulation*, IFAC-PapersOnLine **49**, 500 (2016).

[32] J. Bürki-Cohen and T. Go, *The Effect of Simulator Motion Cues on Initial Training of Airline Pilots*, in *AIAA Modeling and Simulation Technologies Conference* (Institute of Aeronautics and Astronautics, San Francisco, CA, USA, 2012) pp. 516–527.

[33] P. R. Grant, B. Yam, R. J. A. W. Hosman, and J. A. Schroeder, *Effect of Simulator Motion on Pilot Behavior and Perception*, Journal of Aircraft **43**, 1914 (2006).

[34] R. S. Kennedy, M. G. Lilienthal, K. S. Berbaum, D. R. Baltzley, and M. E. McCauley, *Simulator sickness in U.S. Navy flight simulators.* Aviation, space, and environmental medicine **60**, 10 (1989).

[35] W. H. Levison, R. E. Lancraft, and A. M. Junker, *Effects of Simulator Delays on Performance and Learning in a Roll-Axis Tracking Task*, in *Fifteenth Annual Conference on Manual Control* (Dayton, OH, USA, 1979) pp. 168–186.

[36] A. R. Valente Pais, M. Wentink, M. M. van Paassen, and M. Mulder, *Comparison of three motion cueing algorithms for curve driving in an urban environment*, Presence: Teleoperators and Virtual Environments **18**, 200 (2009).

[37] R. J. Telban and F. M. Cardullo, *Motion Cueing Algorithm Development: Human-Centered Linear and Nonlinear Approaches*, Tech. Rep. (National Aeronautics and Space Administration, Hampton, Virginia, USA, 2005).

[38] B. J. Correia Grácio, M. Wentink, J. E. Bos, M. van Paassen, and M. Mulder, *An Application of the Canal-Otolith Interaction Model for Tilt-Coordination During a Braking Maneuver*, in *AIAA Modeling and Simulation Technologies Conference* (Minneapolis, MN, USA, 2013) pp. 1–13.

[39] P. Pretto, J. Venrooij, A. Nesti, and H. H. Bülthoff, *Perception-Based Motion Cueing: A Cybernetics Approach to Motion Simulation*, in *Recent Progress in Brain and Cognitive Engineering*, Vol. 5, edited by S.-W. Lee, H. H. Bülthoff, and K.-R. Müller (Springer, Dordrecht, The Netherlands, 2015) Chap. 9, pp. 131–152.

[40] A. R. Naseri and P. R. Grant, *An Improved Adaptive Motion Drive Algorithm*, in *AIAA Modeling and Simulation Technologies Conference* (San Francisco, CA, USA, 2012) pp. 1–9.

[41] I. A. Qaisi and A. Traechtler, *Human in the loop: Optimal control of driving simulators and new motion quality criterion*, in *IEEE Conference on Systems, Man and Cybernetics* (IEEE, Seoul, Korea, 2012) pp. 2235–2240.

[42] M. Katliar, F. M. Drop, H. Teufel, M. Diehl, and H. H. Bülthoff, *Real-Time Nonlinear Model Predictive Control of a Motion Simulator Based on a 8-DOF Serial Robot*, in *European Control Conference* (Limassol, Cyprus, 2018) pp. 1529–1535.

[43] A. Beghi, M. Bruschetta, and F. Maran, *A real-time implementation of an MPC-based motion cueing strategy with time-varying prediction*, in *IEEE Conference on Systems, Man, and Cybernetics* (IEEE, Manchester, UK, 2013) pp. 4149–4154.

[44] F. M. Drop, M. Olivari, M. Katliar, and H. H. Bülthoff, *Model Predictive Motion Cueing: Online Prediction and Washout Tuning*, in *Driving Simulator Conference* (Antibes, France, 2018) pp. 71–78.

[45] P. R. Grant, Y. Papelis, C. Schwarz, and A. Clark, *Enhancements to the NADS motion drive algorithm for low-speed urban driving*, in *Driving Simulation Conference* (Paris, France, 2004) pp. 67–77.

[46] S. F. Ko and P. R. Grant, *Development and Testing of an Adaptive Motion Drive Algorithm for Upset Recovery Training*, in *AIAA Modeling and Simulation Technologies Conference* (Minneapolis, MN, USA, 2013) pp. 1–34.

[47] S. F. Schmidt and B. Conrad, *The Calculation of Motion Drive Signals for Piloted Flight Simulators*, Tech. Rep. CR-73375, Report No. 69-17, NABS-4869 (National Aeronautics and Space Administration, Washington, DC, USA, 1969).

[48] B. Conrad, S. F. Schmidt, and J. G. Douvillier, *Washout circuit design for multi-degrees-of-freedom moving base simulators.* in *AIAA Visual and Motion Simulation Conference* (Palo Alto, CA, USA, 1973).

[49] E. L. Groen and W. Bles, *How to use body tilt for the simulation of linear self motion.* Journal of vestibular research : equilibrium & orientation **14**, 375 (2004).

[50] J. A. Schroeder, *Helicopter Flight Simulation Motion Platform Requirements*, Tech. Rep. (National Aeronautics and Space Administration, Moffett Field, CA, USA, 1999).

[51] P. R. Grant and L. D. Reid, *Motion Washout Filter Tuning: Rules and Requirements*, Journal of Aircraft **34**, 145 (2008).

[52] M. Kurosaki, *Optimal Washout for Control of a Moving Base Simulator,* IFAC on A Link Between Science and Applications of Automatic Control **11**, 1311 (1978).

[53] R. Sivan, J. Ish-Shalom, and J. K. Huang, *An Optimal Control Approach to the Design of Moving Flight Simulators,* IEEE Transactions on Systems, Man and Cybernetics **12**, 818 (1982).

[54] R. V. Parrish, J. E. Dieudonne, R. L. Bowles, and D. J. Martin Jr., *Coordinated Adaptive Washout for Motion Simulators,* Journal of Aircraft **12**, 44 (1975).

[55] M. Dagdelen, G. Reymond, A. Kemeny, M. Bordier, and N. Maïzi, *Model-based predictive motion cueing strategy for vehicle driving simulators,* Control Engineering Practice **17**, 995 (2009).

[56] A. Beghi, M. Bruschetta, and F. Maran, *A real time implementation of MPC based Motion Cueing strategy for driving simulators,* in *IEEE Conference on Decision and Control* (Ieee, Maui, Hawaii, USA, 2012) pp. 6340–6345.

[57] M. Katliar, J. Fischer, G. Frison, M. Diehl, H. Teufel, and H. H. Bülthoff, *Nonlinear Model Predictive Control of a Cable-Robot-Based Motion Simulator,* IFAC-PapersOnLine **50**, 9833 (2017).

[58] L. D. Reid and M. A. Nahon, *Flight Simulation Motion-Base Drive Algorithms: Part 3 - Pilot Evaluations,* Tech. Rep. (Institute for Aerospace Studies, University of Toronto, Toronto, Canada, 1986).

[59] R. J. A. W. Hosman, H. C. van der Vaart, and G. A. J. van de Moesdijk, *Optimalization and Evaluation of Linear Motion Filters,* in *Conference on Manual Control* (Dayton, OH, USA, 1979) pp. 213–242.

[60] B. Gouverneur, J. A. Mulder, M. M. van Paassen, O. Stroosma, and E. Field, *Optimisation of the SIMONA Research Simulator's Motion Filter Settings for Handling Qualities Experiments,* in *AIAA Modeling and Simulation Technologies Conference* (Austin, Texas, USA, 2012) pp. 1–11.

[61] M. Bruenger-Koch, *Motion Parameter Tuning and Evaluation for the DLR Automotive Simulator,* in *Driving Simulation Conference* (Orlando, FL, USA, 2005) pp. 262–270.

[62] F. A. M. van der Steen, *Self-Motion Perception,* Phd thesis, Delft University of Technology (1998).

[63] A. R. Valente Pais, *Perception Coherence Zones in Vehicle Simulation,* Phd thesis, Delft University of Technology (2013).

[64] P. R. Grant and P. T. Lee, *Motion-Visual Phase-Error Detection in a Flight Simulator,* Journal of Aircraft **44**, 927 (2007).

[65] A. R. Valente Pais, M. M. van Paassen, M. Mulder, and M. Wentick, *Perception Coherence Zones in Flight Simulation,* in *AIAA Modeling and Simulation Technologies Conference* (Chicago, IL, USA, 2009) pp. 1–13.

[66] B. J. Correia Grácio, A. R. Valente Pais, M. M. van Paassen, M. Mulder, L. C. Kelly, and J. A. Houck, *Optimal and Coherence Zone Comparison Within and Between Flight Simulators,* Journal of Aircraft **50**, 493 (2013).

[67] K. N. De Winkel, M. Katliar, and H. H. Bülthoff, *Forced fusion in multisensory heading estimation,* PLoS ONE **10**, 1 (2015).

[68] A. Nesti, S. A. E. Nooij, M. Losert, H. H. Bülthoff, and P. Pretto, *Roll rate perceptual thresholds in active and passive curve driving simulation,* Simulation **92**, 417 (2016).

[69] C. M. Oman, *Sensory conflict in motion sickness: an Observer Theory approach,* in *Pictorial communication in real and virtual environments,* edited by S. Ellis (Taylor & Francis, London, UK, 1991) pp. 362–367.

[70] J. E. Bos, W. Bles, and E. L. Groen, *A theory on visually induced motion sickness,* Displays **29**, 47 (2008).

[71] K. N. De Winkel, M. Katliar, D. Diers, and H. H. Bülthoff, *Causal Inference in the Perception of Verticality,* Scientific Reports **8**, 1 (2018).

[72] C. R. Fetsch, A. H. Turner, G. C. DeAngelis, and D. E. Angelaki, *Dynamic Reweighting of Visual and Vestibular Cues during Self-Motion Perception,* Journal of Neuroscience **29**, 15601 (2009).

[73] D. M. Pool, *Objective Evaluation of Flight Simulator Motion Cueing Fidelity Through a Cybernetic Approach,* Phd thesis, Delft University of Technology (2012).

[74] L. Pariota, G. N. Bifulco, G. Markkula, and R. Romano, *Validation of driving behaviour as a step towards the investigation of Connected and Automated Vehicles by means of driving simulators,* in *Models and Technologies for Intelligent Transportation Systems* (Naples, Italy, 2017) pp. 274–279.

[75] M. Brünger-Koch, S. Briest, and M. Vollrath, *Do you feel the difference ? A motion assessment study,* in *Driving Simulation Conference* (Tsukuba, Japan, 2006) pp. 1–10.

[76] L. R. Young, *On Adaptive Manual Control,* Ergonomics **12**, 635 (1969).

[77] D. M. Pool, P. M. T. Zaal, M. M. Van Paassen, and M. Mulder, *Effects of Heave Washout Settings in Aircraft Pitch Disturbance Rejection,* Journal of Guidance, Control, and Dynamics **33**, 29 (2009).

[78] H. J. Damveld, M. Wentink, P. M. van Leeuwen, and R. Happee, *Effects of Motion Cueing on Curve Driving,* in *Driving Simulation Conference* (Paris, France, 2012) pp. 1–9.

[79] D. P. Jang, I. Y. Kim, S. W. Nam, B. K. Wiederhold, M. D. Wiederhold, and S. I. Kim, *Analysis of Physiological Response to Two Virtual Environments: Driving and Flying Simulation,* CyberPsychology & Behavior **5**, 11 (2002).

[80] S. E. Reardon, M. Field, and S. D. Beard, *Effects of motion filter parameters on simulation fidelity ratings simulations branch,* in *AHS 70th Annual Forum* (Montreal, Canada, 2014) pp. 1–12.

[81] P. R. Grant and B. Haycock, *The Effect of Jerk and Acceleration on the Perception of Motion Strength*, in *AIAA Modeling and Simulation Technologies Conference and Exhibit* (Keystone, CO, USA, 2006) pp. 1–11.

[82] P. R. Grant, M. Blommer, B. Artz, and J. Greenberg, *Analysing classes of motion drive algorithms based on paired comparison techniques*, Vehicle System Dynamics **47**, 1075 (2009).

[83] S. K. Advani and R. J. A. W. Hosman, *Towards Standardising High-Fidelity Cost-Effective Motion Cueing in Flight Simulation*, in *Royal Aeronautical Society Conference on: Cutting Costs in Flight Simulation. Balancing Quality and Capability* (London, UK, 2006).

[84] R. J. A. W. Hosman and S. K. Advani, *Design and evaluation of the objective motion cueing test and criterion*, Aeronautical Journal **120**, 873 (2016).

[85] N. J. I. Garrett and M. C. Best, *Driving simulator motion cueing algorithms–a survey of the state of the art*, in *Symposium on Advanced Vehicle Control* (Loughborough, UK, 2010) pp. 183–188.

[86] M. Bruschetta, F. Maran, and A. Beghi, *A nonlinear, MPC-based motion cueing algorithm for a high-performance, nine-DOF dynamic simulator platform*, IEEE Transactions on Control Systems Technology **25**, 686 (2017).

[87] M. Katliar, K. N. de Winkel, J. Venrooij, P. Pretto, and H. H. Bülthoff, *Impact of MPC Prediction Horizon on Motion Cueing Fidelity*, in *Driving Simulation Conference* (Tübingen, Germany, 2015) pp. 219–222.

[88] M. Bruschetta, F. Maran, C. Cenedese, A. Beghi, and D. Minen, *An MPC-Based Motion Cueing implementation with Time-Varying Prediction and drivers skills characterization*, in *Driving Simulation & Virtual Reality Conference* (Paris, France, 2016) pp. 7–9.

[89] Y. H. Chang, C. S. Liao, and W. H. Chieng, *Optimal motion cueing for 5-DOF motion simulations via a 3-DOF motion simulator*, Control Engineering Practice **17**, 170 (2009).

[90] L. R. Young, *Perception of the body in space: Mechanisms*, in *Handbook of Physiology - The nervous system III*, edited by I. Darian-Smith (American Physiological Society, 1984) Chap. 22, pp. 1023–1066.

[91] M. Baseggio, A. Beghi, M. Bruschetta, F. Maran, and D. Minen, *An MPC approach to the design of motion cueing algorithms for driving simulators*, in *IEEE Conference on Intelligent Transportation Systems, Proceedings* (Washington, DC, USA, 2011) pp. 692–697.

[92] J. E. Bos, W. Bles, and R. J. A. W. Hosman, *Modeling human spatial orientation and motion perception*, in *AIAA Modeling and Simulation Technologies Conference* (Montreal, Canada, 2013) pp. 1–11.

[93] J. Venrooij, P. Pretto, M. Katliar, S. A. E. Nooij, A. Nesti, M. Lächele, K. N. de Winkel, D. Cleij, and H. H. Bülthoff, *Perception-Based Motion Cueing : Validation in Driving Simulation*, in *Driving Simulation Conference* (Tübingen, Germany, 2015) pp. 153–161.

[94] P. Pretto, M. Ogier, H. H. Bülthoff, and J. P. Bresciani, *Influence of the size of the field of view on motion perception*, Computers and Graphics (Pergamon) **33**, 139 (2009).

[95] S. Durkee and N. Ward, *Effect of Driving Simulation Parameters Related to Ego-Motion on Speed Perception*, Driving Symposium on Human Factors in Driver Assessment, Training and Vehicle Design , 358 (2017).

[96] B. J. Correia Grácio, J. E. Bos, M. M. Van Paassen, and M. Mulder, *Perceptual scaling of visual and inertial cues: Effects of field of view, image size, depth cues, and degree of freedom*, Experimental Brain Research **232**, 637 (2014).

[97] T. D. van Leeuwen, *Simulator motion cueing error detection using a wavelet-based algorithm*, Master thesis, Delft University of Technology (2017).

[98] D. H. Weir, *Application of a driving simulator to the development of in-vehicle human-machine-interfaces*, IATSS Research **34**, 16 (2010).

[99] Advisory Group for Aerospace Research and Development, *Fidelity of Simulation for Pilot Training*, Tech. Rep. (North Atlantic Treaty Organization, Neuilly sur Seine, France, 1980).

[100] Y. J. Brown, F. M. Cardullo, and J. B. Sinacori, *Need-based evaluation of simulator force and motion cuing devices*, in *Flight Simulation Technologies Conference and Exhibit* (Boston, MA, USA, 2013) pp. 78–85.

[101] A. R. Valente Pais, M. M. Van Paassen, M. Mulder, and M. Wentick, *Perception Coherence Zones in Flight Simulation*, Journal of Aircraft **47**, 2039 (2010).

[102] P. Jonik, A. R. Valente Pais, M. M. van Paassen, and M. Mulder, *Phase Coherence Zones in Flight Simulation*, in *AIAA Modeling and Simulation Technologies Conference* (Portland, OR, USA, 2012) pp. 1–17.

[103] R. J. V. Bertin, C. Collet, S. Espié, and W. Graf, *Optokinetic or simulator sickness: objective measurement and the role of visual-vestibular conflict situations*, in *Driving Simulation Conference* (Orlando, FL, USA, 2005) pp. 280–293.

[104] D. M. Pool, P. Zaal, H. J. Damveld, M. M. van Paassen, and M. Mulder, *Evaluating Simulator Motion Fidelity using In-Flight and Simulator Measurements of Roll Tracking Behavior*, in *AIAA Modeling and Simulation Technologies Conference* (Minneapolis, MN, USA, 2013) pp. 1–25.

[105] P. Zaal and D. M. Pool, *Multimodal Pilot Behavior in Multi-Axis Tracking Tasks with Time-Varying Motion Cueing Gains*, in *AIAA Modeling and Simulation Technologies Conference* (National Harbor, MD, USA, 2014) pp. 174–190.

[106] D. Cleij, J. Venrooij, P. Pretto, D. M. Pool, M. Mulder, and H. H. Bülthoff, *Continuous Rating of Perceived Visual-Inertial Motion Incoherence During Driving Simulation,* in *Driving Simulator Conference,* edited by H. H. Bülthoff, A. Kemeny, and P. Pretto (Tübingen, Germany, 2015) pp. 191 – 198.

[107] ITU-R, *Methodology for the subjective assessment of the quality of television pictures,* Tech. Rep. (International Telecommunications Union, Geneva, Switzerland, 2002).

[108] S. S. Stevens, *The direct estimation of sensory magnitudes-loudness.* The American journal of psychology **69**, 1 (1956).

[109] M. Lambooij, W. A. Ijsselsteijn, and I. Heynderickx, *Visual discomfort of 3D TV: Assessment methods and modeling,* Displays **32**, 209 (2011).

[110] J. Freeman, S. E. Avons, D. E. Pearson, and W. A. Ijsselsteijn, *Effects of sensory information and prior experience on direct subjective ratings of presence,* Presence: Teleoperators and Virtual Environments **8**, 1 (1999).

[111] T. Eerola, P. Toiviainen, and C. Krumhansl, *Real-time prediction of melodies: Continuous predictability judgments and dynamic models,* in *Conference on Music Perception and Cognition,* Vol. 0 (Sydney, Australia, 2002) pp. 473–476.

[112] R. Cowie, A. Camurri, and D. Glowinski, *Results from the first series of experiments and first evaluation report,* Tech. Rep. (SIEMPRE Project Consortium, 2011).

[113] C. Schießl, *Subjective strain estimation depending on driving manoeuvres and traffic situation,* IET Intelligent Transport Systems **2**, 258 (2008).

[114] J. M. Girard, M. Wilczyk, Y. Barloy, P. Simon, and J. C. Popieul, *Towards an on-line assessment of subjective driver workload,* in *Driving Simulator Conference* (Orlando, FL, USA, 2005) pp. 382–391.

[115] T. Liu, G. Cash, N. Narvekar, and J. Bloom, *Continuous Mobile Video Subjective Quality Assessment Using Gaming Steering Wheel,* in *International Workshop on Video Processing and Quality Metrics for Consumer Electronics* (Scottsdale, AZ, 2012).

[116] E. G. Carmines and R. A. Zeller, *Reliability and Validity Assessment,* 9th ed. (Sage Publications, Inc., Beverly Hills, CA, USA, 1987).

[117] L. J. Cronbach, *Coefficient alpha and the internal structure of tests,* Psychometrika **16**, 297 (1951).

[118] J. F. jr. Hair, W. C. Black, B. J. Babin, and R. E. Anderson, *Multivariate Data Analysis,* 7th ed. (Pearson Prentice Hall, Upper Saddle River, NJ, USA, 2010).

[119] D. C. Howell, *Statistical Methods for Psychology,* 7th ed. (Wadsworth Cengage Learning, Belmont, CA, USA, 2010).

[120] D. S. Messinger, T. D. Cassel, S. I. Acosta, Z. Ambadar, and J. F. Cohn, *Infant Smiling Dynamics and Perceived Positive Emotion.* Journal of nonverbal behavior **32**, 133 (2008).

[121] L. D. Reid and M. A. Nahon, *Flight Simulation Motion-Base Drive Algorithms: Part 1 - Developing and Testing the Equations.*, Tech. Rep. 296 (University of Toronto, Institute for Aerospace Studies, Toronto, Canada, 1985).

[122] E. L. Groen, M. Wentink, A. R. Valente Pais, M. Mulder, and M. van Paassen, *Motion Perception Thresholds in Flight Simulation*, in *AIAA Modeling and Simulation Technologies Conference* (Keystone, CO, USA, 2012) pp. 1–11.

[123] D. S. Hands and S. E. Avons, *Recency and Duration Neglect in Subjective Assessment of Television Picture Quality*, Applied Cognitive Psychology **15**, 639 (2001).

[124] P. R. Grant and B. Haycock, *Effect of Jerk and Acceleration on the Perception of Motion Strength*, Journal of Aircraft **45**, 1190 (2008).

[125] S. Winkler and R. Campos, *Video quality evaluation for Internet streaming applications*, Human Vision and Electronic Imaging VIII **5007**, 104 (2003).

[126] S. Buchinger, W. Robitza, M. Nezveda, E. Hotop, P. Hummelbrunner, M. C. Sack, and H. Hlavacs, *Evaluating feedback devices for time-continuous mobile multimedia quality assessment*, Signal Processing: Image Communication **29**, 921 (2014).

[127] G. E. Cooper and R. P. Harper, *The use of pilot rating in the evaluation of aircraft handling qualities*, Tech. Rep. (National Aeronautics and Space Administration, Washington, DC, USA, 1969).

[128] L. L. Thurstone, *A law of comparative judgment*, Psychological Review **34**, 273 (1927).

[129] M. Mayrhofer, B. Langwallner, R. Schlüsselberger, W. Bles, and M. Wentink, *An Innovative Optimal Control Approach for the Next Generation Simulator Motion Platform DESDEMONA*, in *AIAA Modeling and Simulation Technologies Conference* (Hilton Head, SC, USA, 2012) pp. 1–14.

[130] N. J. Garrett and M. C. Best, *Model predictive driving simulator motion cueing algorithm with actuator-based constraints*, Vehicle System Dynamics **51**, 1151 (2013).

[131] J. B. Rawlings and D. Q. Mayne, *Advanced Model Predictive Control*, 5th ed. (Nob Hill Pub., 2012).

[132] F. Maran, M. Bruschetta, A. Beghi, and D. Minen, *Improvement of an MPC-based Motion Cueing Algorithm with Time-Varying Prediction and Driver Behaviour Estimation*, in *Driving Simulation Conference* (Tübingen, Germany, 2015) pp. 16–18.

[133] J. Andersson, *A General-Purpose Software Framework for Dynamic Optimization*, Phd thesis, KU Leuven (2013).

[134] A. Wächter and L. T. Biegler, *On the implementation of an interior-point filter line-search algorithm for large-scale nonlinear programming*, Mathematical Programming **106**, 25 (2006).

[135] T. D. van Leeuwen, D. Cleij, D. M. Pool, M. Mulder, and H. H. Bülthoff, *Time-varying perceived motion mismatch due to motion scaling in curve driving simulation,* Transportation Research Part F: Traffic Psychology and Behaviour **61**, 84 (2019).

[136] D. Cleij, J. Venrooij, P. Pretto, D. M. Pool, M. Mulder, and H. H. Bülthoff, *Continuous subjective rating of perceived motion incongruence during driving simulation,* IEEE Transactions on Human-Machine Systems **48**, 17 (2018).

[137] D. Cleij, J. Venrooij, P. Pretto, M. Katliar, H. H. Bülthoff, D. Steffen, F. W. Hoffmeyer, and H. P. Schöner, *Comparison between filter- and optimization-based motion cueing algorithms for driving simulation,* Transportation Research Part F: Traffic Psychology and Behaviour **61**, 53 (2019).

[138] M. Fischer, *Motion-Cueing-Algorithmen für eine realitätsnahe Bewegungssimulation,* Tech. Rep. (Deutsches Zentrum für Luft- und Raumfahrt e.V., Braunschweig, Germany, 2009).

[139] P. R. Grant, B. Artz, M. Blommer, L. Cathey, and J. Greenberg, *A Paired Comparison Study of Simulator Motion Drive Algorithms,* in *Driving Simulation Conference* (Paris, France, 2002) pp. 1–13.

[140] H. Asadi, S. Mohamed, K. Nelson, S. Nahavandi, D. R. Zadeh, and M. Oladazimi, *An Optimal Washout Filter Based on Genetic Algorithm Compensators for Improving Simulator Driver Perception,* in *Driving Simulation Conference* (Tübingen, Germany, 2015) pp. 173–182.

[141] S. Casas, I. Coma, J. V. Riera, and M. Fernández, *Motion-cuing algorithms: Characterization of users' perception,* Human Factors **57**, 144 (2015).

[142] M. Baarspul, *Flight simulation techniques with emphasis on the generation of high fidelity 6 DOF motion cues,* in *Congress of International Council of the Aeronautical Sciences* (London, UK, 1986).

[143] H. M. Heerspink, W. R. Berkouwer, O. Stroosma, M. M. van Paassen, M. Mulder, and J. A. Mulder, *Evaluation of Vestibular Thresholds for Motion Detection in the SIMONA Research Simulator,* in *AIAA Modeling and Simulation Technologies Conference* (San Diego, CA, USA, 2012) pp. 1–20.

[144] M. Fischer and J. Werneke, *The New Time-Variant Motion Cueing Algorithm For The DLR Dynamic Driving Simulator,* in *Driving Simulation Conference* (Monaco, 2008) pp. 57–67.

[145] C. Ramirez, R. Sanchez, V. Kreinovich, and M. Argaez, $\sqrt{x^2+\mu}$ *is the Most Computationally Efficient Smooth Approximation to* $|x|$*: a Proof,* Journal Of Uncertain Systems **8**, 205 (2014).

[146] G. R. J. Cooper and D. R. Cowan, *Comparing time series using wavelet-based semblance analysis,* Computers and Geosciences **34**, 95 (2008).

[147] J. M. Lilly and S. C. Olhede, *On the analytic wavelet transform,* IEEE Transactions on Information Theory **56**, 4135 (2010).

[148] A. R. Naseri and P. R. Grant, *Human discrimination of translational accelerations*, Experimental Brain Research **218**, 455 (2012).

[149] C. Torrence and G. P. Compo, *A Practical Guide to Wavelet Analysis*, Bulletin of the American Meteorological Society **79**, 61 (1998).

[150] G. P. Papaioannou, C. Dikaiakos, G. Evangelidis, P. G. Papaioannou, and D. S. Georgiadis, *Co-movement analysis of Italian and Greek electricity market wholesale prices by using a wavelet approach*, Energies **8**, 11770 (2015).

[151] S. K. Advani, R. J. A. W. Hosman, and M. Potter, *Objective Motion Fidelity Qualification in Flight Training Simulators*, in *AIAA Modeling and Simulation Technologies Conference* (Hilton Head, SC, USA, 2012) pp. 1–13.

[152] I. A. Qaisi and A. Treachtler, *Constrained Linear Quadratic Optimal Controller for Motion Control of ATMOS Driving Simulator*, in *Driving Simulation Conference* (Paris, France, 2012) pp. 1–8.

[153] J. Monen and E. Brenner, *Detecting changes in one's own velocity from the optic flow.* Perception **23**, 681 (1994).

[154] M. Lappe, F. Bremmer, and A. V. van den Berg, *Perception of self-motion from visual flow*, Trends in Cognitive Sciences **3**, 329 (1999).

[155] F. Festl, F. Recktenwald, C. Yuan, and H. A. Mallot, *Detection of linear ego-acceleration from optic flow*, Journal of Vision **12**, 10 (2012).

[156] I. P. Howard and L. Childerson, *The contribution of motion, the visual frame, and visual polarity to sensations of body tilt.* Perception **23**, 753 (1994).

[157] C. Fernández and J. M. Goldberg, *Physiology of peripheral neurons innervating otolith organs of the squirrel monkey. III. Response dynamics.* Journal of neurophysiology **39**, 996 (1976).

[158] F. Soyka, H. Teufel, K. Beykirch, P. Robuffo Giordano, J. Butler, F. M. Nieuwenhuizen, and H. H. Bülthoff, *Does Jerk Have to be Considered in Linear Motion Simulation?* in *AIAA Modeling and Simulation Technologies Conference* (Chicago, IL, USA, 2012) pp. 1381–1388.

[159] L. Ljung and K. Glover, *Frequency domain versus time domain methods in system identification*, Automatica **17**, 71 (1981).

[160] D. T. Westwick, E. A. Pohlmeyer, S. A. Solla, L. E. Miller, and E. J. Perreault, *Identification of multiple-input systems with highly coupled inputs: Application to EMG prediction from multiple intracortical electrodes*, Neural Computation **18**, 329 (2006).

[161] I. J. Myung, *Tutorial on maximum likelihood estimation*, Journal of Mathematical Psychology **47**, 90 (2003).

[162] F. E. Guedry, *Psychophysics of Vestibular Sensation*, in *Handbook of Sensory Physiology Volume VI/2*, edited by H. H. Kornhuber (Springer-Verlag, Berlin, 1974) 1st ed., Chap. I, pp. 3–154.

[163] S. A. Nooij, A. Nesti, H. H. Bülthoff, and P. Pretto, *Perception of rotation, path, and heading in circular trajectories,* Experimental Brain Research **234**, 2323 (2016).

[164] P. Perfect, E. Timson, M. D. White, G. D. Padfield, R. Erdos, and A. W. Gubbels, *A rating scale for the subjective assessment of simulation fidelity,* Aeronautical Journal **118**, 953 (2014).

[165] A. Berthoz, W. Bles, H. H. Bülthoff, B. J. Correia Grácio, P. Feenstra, N. Filliard, R. Hühne, A. Kemeny, M. Mayrhofer, M. Mulder, H. G. Nusseck, P. Pretto, G. Reymond, R. Schlüsselberger, J. Schwandtner, H. Teufel, B. Vailleau, M. M. Van Paassen, M. Vidal, and M. Wentink, *Motion scaling for high-performance driving simulators,* IEEE Transactions on Human-Machine Systems **43**, 265 (2013).

[166] F. Savona, A. M. Stratulat, E. Diaz, V. Honnet, G. Houze, P. Vars, S. Masfrand, V. Roussarie, and C. Bourdin, *The Influence of Lateral , Roll and Yaw Motion Gains on Driving Performance on an Advanced Dynamic Simulator,* in *International Conference on Advances in System Simulation* (Nice, France, 2014) pp. 113–119.

[167] N. J. Dorans, M. Pommerich, and P. W. Holland, *Statistics for Social and Behavioral Sciences,* 1st ed., Statistics for Social and Behavioral Sciences (Springer New York, New York, NY, USA, 2007).

[168] M. H. Pinson and S. Wolf, *An objective method for combining multiple subjective data sets,* in *SPIE 5150, Visual Communications and Image Processing* (Lugano, Switzerland, 2003) pp. 583–592.

[169] W. H. Chen, D. A. Revicki, J. S. Lai, K. F. Cook, and D. Amtmann, *Linking Pain Items from Two Studies Onto a Common Scale Using Item Response Theory,* Journal of Pain and Symptom Management **38**, 615 (2009).

[170] D. Thissen, J. W. Varni, B. D. Stucky, Y. Liu, D. E. Irwin, and D. A. DeWalt, *Using the PedsQLTM 3.0 asthma module to obtain scores comparable with those of the PROMIS pediatric asthma impact scale (PAIS),* Quality of Life Research **20**, 1497 (2011).

[171] F. Ott, L. Pohl, M. Halfmann, G. Hardiess, and H. A. Mallot, *The perception of ego-motion change in environments with varying depth: Interaction of stereo and optic flow,* Journal of Vision **16**, 4 (2016).

[172] L. Ljung, *The Matlab user's guide,* 7th ed., Vol. 7 (The MathWorks, Inc, Natick, M, 2007) p. 237.

[173] Z. Fang and A. Kemeny, *Explicit MPC motion cueing algorithm for real-time driving simulator,* in *Power Electronics and Motion Control Conference* (Harbin, China, 2012) pp. 874–878.

[174] P. R. Grant, *The Development of a Tuning Paradigm for Flight Simulator Motion Drive Algorithms,* Phd thesis, University of Toronto (1996).

[175] D. Mayne, *Control of Constrained Dynamic Systems,* European Journal of Control **7**, 87 (2007).

[176] J. E. Bos, S. N. MacKinnon, and A. Patterson, *Motion sickness symptoms in a ship motion simulator: effects of inside, outside, and no view.* Aviation, space, and environmental medicine **76**, 1111 (2005).

[177] J. Ruscio, *A Probability-Based Measure of Effect Size: Robustness to Base Rates and Other Factors,* Psychological Methods **13**, 19 (2008).

[178] J. C. H. Li, *Effect size measures in a two-independent-samples case with nonnormal and nonhomogeneous data,* Behavior Research Methods **48**, 1560 (2016).

[179] K. O. McGraw and S. P. Wong, *A Common Language Effect Size Statistic,* Psychological Bulletin **111**, 361 (1992).

[180] E. L. Groen, M. S. V. Valenti Clari, and R. J. A. W. Hosman, *Evaluation of Perceived Motion During a Simulated Takeoff Run,* Journal of Aircraft **38**, 600 (2008).

[181] B. L. Fredrickson and D. Kahneman, *Duration neglect in retrospective evaluations of affective episodes.* Journal of personality and social psychology **65**, 45 (1993).

[182] J. R. van der Ploeg, *Sensitivity Analysis of a Model Predictive Control-based Motion Cueing Algorithm,* Master thesis, Delft University of Technology (2018).

[183] F. Ellensohn, M. Breyer, M. Schwienbacher, J. Venrooij, and D. Rixen, *A Filter-Based Motion Cueing Algorithm for a Redundant Driving Simulator,* Pamm **18**, e201800445 (2018).

[184] F. Ellensohn, J. Venrooij, M. Schwienbacher, and D. Rixen, *Experimental evaluation of an optimization-based motion cueing algorithm,* Transportation Research Part F: Traffic Psychology and Behaviour **62**, 115 (2019).

[185] G. L. Zacharias and L. R. Young, *Influence of combined visual and vestibular cues on human perception and control of horizontal rotation,* Experimental Brain Research **41**, 159 (1981).

[186] R. J. Telban and F. M. Cardullo, *An integrated model of human motion perception with visual-vestibular interaction,* in *AIAA Modeling and Simulation Technologies Conference* (Montreal, Canada, 2013).

[187] R. J. A. W. Hosman, F. M. Cardullo, and J. E. Bos, *Visual-Vestibular interaction in motion perception,* in *AIAA Modeling and Simulation Technologies Conference* (Portland, OR, USA, 2012) pp. 1–13.

[188] M. Newman, B. Lawson, A. Rupert, and B. McGrath, *The Role of Perceptual Modeling in the Understanding of Spatial Disorientation During Flight and Ground-based Simulator Training,* in *AIAA Modeling and Simulation Technologies Conference* (Minneapolis, MN, USA, 2013) pp. 1–14.

[189] B. J. Correia Gracio, *The Effects of Specific Force on Self-Motion Perception in a Simulation Environment,* Phd thesis, Delft University of Technology (2013).

[190] M. Mulder, D. A. Abbink, and E. R. Boer, *The effect of haptic guidance on curve negotiation behavior of young, experienced drivers*, in *IEEE Conference on Systems, Man and Cybernetics* (Singapore, Singapore, 2008) pp. 804–809.

[191] H. Jamson, D. L. Hibberd, and N. Merat, *The Design of Haptic Gas Pedal Feedback to Support Eco-Driving*, in *International Driving Symposium on Human Factors in Driver Assessment, Training, and Vehicle Design* (Bolton Landing, NY, USA, 2017) pp. 264–270.

[192] M. Hirokawa, N. Uesugi, S. Furugori, T. Kitagawa, and K. Suzuki, *Effect of haptic assistance on learning vehicle reverse parking skills*, IEEE Transactions on Haptics **7**, 334 (2014).

[193] L. Ljung, *System Identification: Theory for the User* (Prentice Hall PTR, Upper Saddle River, NJ, USA, 1999).

[194] G. E. P. Box and D. R. Cox, *An Analysis of Transformations*, Journal of the Royal Statistical Society: Series B (Methodological) **26**, 211 (2018).

[195] M. Rosenblatt, *Remarks on Some Nonparametric Estimates of a Density Function*, The Annals of Mathematical Statistics **27**, 832 (1956).

[196] R. M. Mallery, O. U. Olomu, R. M. Uchanski, V. A. Militchin, and T. E. Hullar, *Human discrimination of rotational velocities*, Experimental Brain Research **204**, 11 (2010).

[197] H. Akaike, *Information Theory and an Extension of the Maximum Likelihood Principle*, in *International Symposium on Information Theory* (Budapest, Hungary, 1973) pp. 267–281.

[198] F. M. Drop, D. M. Pool, M. M. Van Paassen, M. Mulder, and H. H. Bülthoff, *Objective model selection for identifying the human feedforward response in manual control*, IEEE Transactions on Cybernetics **48**, 2 (2018).

[199] S. M. Lynch, *Introduction to Applied Bayesian Statistics and Estimation for Social Scientists*, 1st ed. (Springer Science+Business Media, LLC, New York, NY, USA, 2007).

[200] P. J. Brockwell and R. A. Davis, *Introduction to Time Series and Forecasting.*, 2nd ed. (Springer-Verlag, New York, NY, USA, 2002).

[201] M. Jansen, *Lecture notes in statistics*, 1st ed., Vol. 161 (Springer-Verlag, New York, NY, USA, 2001) p. 191.

The Model: MIR Averaging

During a continuous rating experiment, multiple rating signals are obtained from the different participants and trials. In this study the main goal is to predict the *average* Motion Incongruence Rating (MIR) from such an experiment. In this appendix several ways of combining the participant ratings into one average MIR are discussed.
Standard methods for averaging data include taking the mean, mode or median of a set of data points. The mean is an often used method, but only really describes the average for normally distributed data and is less suitable to describe the average for non-normally distributed datasets [119]. The mode does not seem appropriate for continuous data, where values are hardly ever repeated exactly. To obtain a value for the mode of the dataset, the data first needs to be discretized, such that the mode can be calculated over specific bins with equal range. In this form, the mode, just as the median, can be used to describe the average MIR at one time step.
A downside of using either of these methods is that, over time, they result in a non-smooth signal as can be seen from Figure A.1 where these methods were applied to the continuous ratings obtained in the experiment described in Chapter 2. Because at some time steps several modes are found, at those time steps the mean over these modes is taken instead.
Two other averaging techniques involve analyzing the actual distribution of the data points at each time step. The first technique involved calculating Box-Cox power transformations [194] with the goal of obtaining normal distributions for the transformed dataset. The Box-Cox power transformations were calculated for several lambda's at each time step, as shown in the top plot of Figure A.1. The lambda value of the transformation resulting in the most time steps with normally distributed data was used to transform the ratings at all time steps. The mean of these transformed ratings was taken and transformed back to the rating unit using the same transformation. This results in a signal equally smooth to directly taking the mean over all trials and participants.
The last technique involved the Kernel Density Estimation (KDE) [195], which was used to obtain a most likely distribution of the dataset at each time step. The peak of this distribution (mode) was then taken as the average of the ratings at this time step. Since this method involves analyzing the distributions of the data points at each time step separately, it is assumed that with this technique the most accurate average of the participant ratings can be obtained.

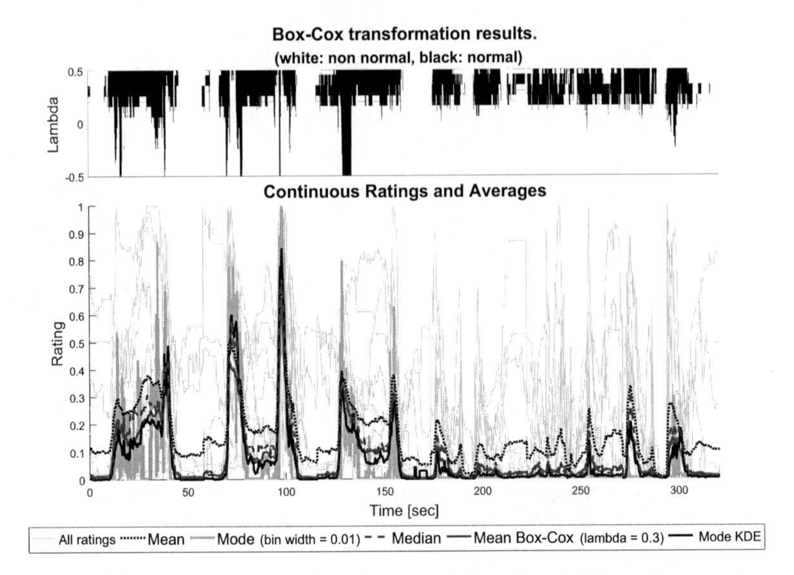

Figure A.1: Results for different averaging techniques for continuous rating data.

In Figure A.1 the effects of the different averaging techniques is shown. It shows that using the mean and mode can result in very different ratings, while the median, Box-Cox and KDE transformations give similar results. Using the mean leads to much higher average ratings, due to outliers in the data. The mode, on the other hand, results in a very non-smooth signal because the distributions are somewhat flat at some moments in time. From this analysis we conclude that it is best to use the median to describe the average MIR over time.

B

The Model: Optional Non-linear Subsystems

The model described in Section 5.2 of this thesis consists of a non-linear part and a linear part, where the former can be seen as a container for several non-linear subsystems that extract different cueing errors from the simulator and vehicle motions. In this appendix two of such possible subsystems, related to perceived rotations, are described. It should be noted that these subsystems serve as examples, and were not validated using experiments.

MCA tuning experts often take into account that rotations with a rate below the perceptual thresholds are not perceived by the participants. Hence, rotational rates resulting from tilt-coordination do not have to be minimized further than these thresholds. To account for this in the model, the rotational rates in both the vehicle and simulator motions can also be transformed into perceived rates and used to calculate additional cueing errors with the subscript ε_{perc}. Additionally, an attempt is made to calculate the perceived angle based on these thresholded rotational rates, which are then used to obtain additional cueing errors also indicated with the subscript ε_{perc}. These transformations of rotational rate and angle are further explained here.

B.1. Rotational Rate

To determine whether the rotational rate is perceived, a distinction is made between detection thresholds, below which a motion is not perceived, and discrimination thresholds, below which the *difference* between two motions is not perceived. To calculate which part of the rotational motion is above *both* thresholds, multiple threshold functions based on Equation 4.1 in Chapter 4 are used. The simulator ($F_{det_{sim}}$) and the vehicle ($F_{det_{veh}}$) motion threshold functions are used to determine if the motion mismatch is caused by perceivable simulator or vehicle motions, by comparing to motion to the detection threshold th_{det} for a given motion channel. A third threshold function (F_{det}) is then used to determine whether the sum of the first two threshold functions is above 0.5, i.e., whether either the simulator or vehicle motion is above the detection threshold. Next, a fourth threshold function (F_{discr}) is used to determine if the motion mismatch is above the discrimination threshold th_{discr}. The discrimination threshold is calculated with equation B.1, which is based on Steven's Power law for rotational motions and the parameters for rotational velocity found by [196]:

$$th_{discr} = 0.88 \cdot I^{0.37}, \tag{B.1}$$

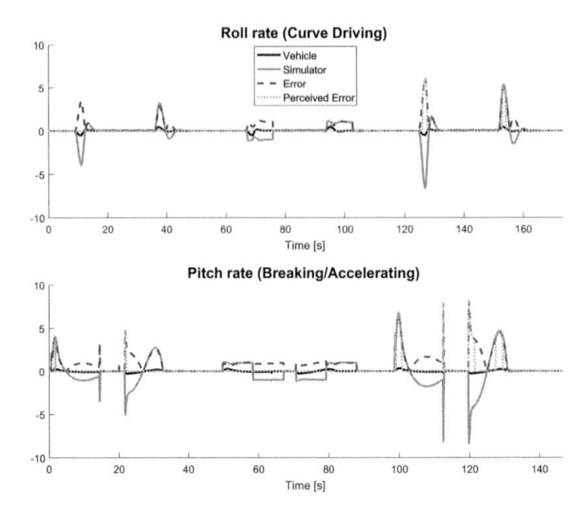

Figure B.1: Actual and perceived roll (top) and pitch (bottom) rate errors for curve driving and breaking/accelerating manoeuvres, respectively, using a perceptual threshold of 4 deg/s.

where I, the stimulus intensity, is equal to the simulator rotational rate in pitch, roll or yaw. Finally, a fifth threshold function (F_{perc}) is used to determine whether the motion mismatch could be perceived based on both the detection and discrimination thresholds.

The final threshold function is used to set the rotational rate errors that are assumed to not be perceived to zero, while all other errors keep their original values. Applying this transformation to roll and pitch rate signals from the experiment described in Chapter 2, using a discrimination threshold of three and four degrees per second for roll and pitch rate respectively, results in perceived rotational rate errors for the curve driving and breaking/accelerating manoeuvres, respectively, as shown in Figure B.1. The simulator motion shown in Figure B.1 was obtained using three different MCAs, resulting in different simulator motions for the same manoeuvres. As car driving results in only very limited rotational motion, the actual cueing errors are very similar to the absolute value of the simulator motion signals. Using the transformation explained before, the actual cueing error is transformed in the perceived cueing error, which only occurs during the last curve driving manoeuvre and the first and last breaking/accelerating manoeuvres.

B.2. Rotational Angle

The human vestibular and somatosensory systems cannot distinguish between changes in specific force due to rotation or due to linear acceleration. In motion simulation this tilt/translation ambiguity is used to simulate linear acceleration via rotation. By combining this rotation with visual cues of linear acceleration, the central nervous system (incorrectly) resolves the ambiguity and interprets the perceived force as linear acceleration. However, when the rotational rate is above the perceptual threshold, this

additional information could be used by the CNS to (correctly) interpret the perceived specific force as the result of rotation. The effect this has on motion cueing quality was shown in [92], where, for longitudinal acceleration maneuvers, the MCA quality strongly depended on mismatches in rotational angle. This makes including the perceived rotational angle as a cueing error in the model desirable. How exactly different sensory signals, such as skin pressure, specific force and rotational rate, are combined into a perceived rotation angle over time, however, is still unknown. In this paragraph an initial attempt is made to calculate the time-varying perceived rotational angle from rotational rate signals with a simple algorithm.
There are two apparent processes that can be considered: 1) leaky integration of the perceived rotational rate and 2) a switching process based on perceived rotational rate for attributing the perceived specific force to either rotation or to linear acceleration. Integration of the perceived rotational rate results in perceived rotation angles when both the skin pressure and specific force signals indicate no rotation is present. Instead the focus is therefore put on basing the algorithm on the switching process.
The algorithm is based on three assumptions:

1. Attributing a proportion of the specific force to rotation when the rotational rate is above threshold always occurs without error, e.g., rotations above threshold result in a correctly perceived rotation angle.

2. Decreasing the rotational angle, and with that the specific force, will decrease the perceived rotational angle, e.g., either the actual rotation angle or the perceived rotational angle at discrete time step $t-1$ to be perceived, depending on which is smaller.

3. Other rotational rates do not influence the perceived rotational angle, e.g., the perceived angle at t is equal to that at $t-1$

Currently, these three assumptions are implemented in Matlab using a for loop and if/else statements in combination with threshold functions for detecting above threshold rotational rates and the direction of changes in rotational rate. The perceived rotational angle error is then calculated as the difference between the perceived angle in the vehicle and the simulator. For a brief initial analysis of the algorithm, its outcome is compared with the rating results presented first in Chapter 2. For this comparison the rating during the braking/accelerating manoeuvres is used, as the longitudinal acceleration errors were small and the rated perceived motion incongruence is assumed to instead be caused by the rotational errors.
In Figure B.2 the actual and perceived pitch angle and rate during the braking/accelerating maneuvers using three different MCAs is compared to the motion incongruence rating at this time. To better understand the rating, other relevant model inputs are also shown in the bottom plot in this figure. In all plots the motion signals are multiplied with an estimated gain for better comparison with the rating signal. Figure B.2 shows that the algorithm predicts that during the second manoeuvre, where the pitch rate is below threshold, no rotational angle is perceived. During this manoeuvre peaks in the ratings nevertheless can be observed, but they do not seem to fit either the perceived or the actual rating angles. Instead, these peaks could be the result of, for example, errors in the longitudinal or pitch acceleration.
In the other two manoeuvres the algorithm predicts that the initial rotational angles during deceleration (around $t=10$ and $t=105$) are perceived as being larger than the second set of rotational angles during breaking (around $t=30$ and $t=125$). This

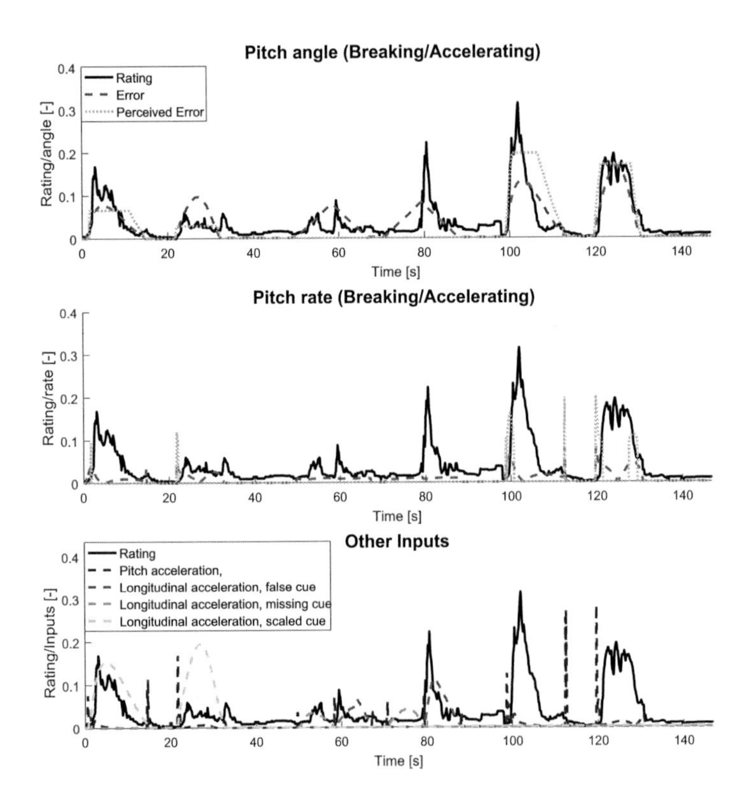

Figure B.2: MIR and actual and perceived pitch angle (top) rate (middle) unitless signals during the braking accelerating manoeuvre using three different MCAs. MIR and additional unitless model inputs are shown in the bottom plot.

perceived angle is the exact opposite of the actual rotational angles. Comparing both angles to the rating suggests that the perceived angle fits the observed rating better in this respect.

It would be very interesting to further investigate the validity of this algorithm, as its outcome can be very helpful in analyzing the *cause* for perceived motion incongruence. With showing this preliminary result, the author hopes to convince and encourage researchers with a stronger background in human perception to investigate such an algorithm further.

SI Process: Estimating Criteria Thresholds

In Chapter 5 the thresholds th_{uic} and th_{uoc} are introduced. These thresholds are used in the system identification process for input selection and order reduction, respectively, and should be set depending on the goal of the model. In this study, the goal is to develop a simple model that can be used to optimize MCAs. The model should thus capture the main features of the MIR, and not be too complex. To determine the appropriate value of the thresholds th_{uic} and th_{uoc}, Steps 1-3 of the SI process were applied to synthetic data generated with known artificial MIR models.

C.1. Synthetic Datasets

In Figure C.1 a short overview is given on how the synthetic data and models are generated. To ensure that the particular characteristics of the model inputs are represented in the synthetic data, the actual measured inputs are used in this process. The model output, instead, is synthetic and not based on the measured ratings.
For a pre-specified number of inputs I and maximum polynomial order n_{max}, which are selected from the sets I_{all} and $n_{max_{all}}$ respectively, a random ARX model is generated to represent the linear part of the MIR model. For the generation of this model the orders of polynomials A and B_i are set at a random integer in the range $[1 - n_{max}]$. The ARX parameters are then set at random values, while the gain for a certain input does not exceed a gain of 2 at any frequency and the gain at zero frequency is higher than 0.5. These boundaries are set to ensure reasonable filters for each input with each selected input having a detectable influence on the synthetic output.

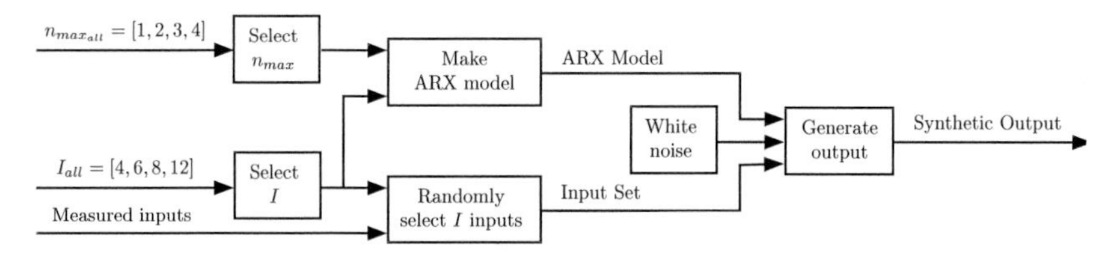

Figure C.1: Generation of synthetic data and model for analysis of threshold value influence on model identification.

The I inputs to this system are randomly selected from the total input set as described in Section 5.2.1. These inputs, together with a white noise signal, are passed through the ARX model to generate the synthetic output. One synthetic dataset thus consists of the I selected inputs and the corresponding synthetic output and is generated by an ARX model with known parameters and a maximum polynomial order of n_{max}.
In total $4 \cdot 4 \cdot 7 \cdot (6+1) \cdot 5 = 3{,}920$ synthetic datasets were generated for four different number of inputs (I_{all} = [4,6,8,12]), four different maximum polynomial orders ($n_{max_{all}}$ = [1,2,3,4]), seven different input selection thresholds ($th_{uic_{all}}$ = [0.01,0.05,0.1, 0.2,0.3,0.6,1]) and six different order reduction thresholds ($th_{uoc_{all}}$ = [0.01,0.05,0.1,0.2, 0.3,0.6]) plus the use of the Akaike Information Criterion (AIC) [197] instead of SC_{ord_p}, that each were repeated five times. For each dataset the inputs, polynomial orders and ARX parameters were randomly chosen and randomly generated Gaussian white noise was used.

C.2. RESULTS

Using the first three steps of the SI process, the inputs and model orders were estimated and compared to the original values. Figure C.2 shows the input selection (a) and order reduction (b) results for all synthetic datasets. Each colored line in Figure C.2 shows the results for a specific number of inputs and maximum model order in the synthetic dataset. Each data point on a line shows the average result over five datasets, using the corresponding threshold th_{uic} or th_{uoc} on the horizontal-axis. The black lines indicate the average over all results at each threshold value. The dotted vertical line indicates the chosen thresholds, which are discussed in the next paragraph.
Figure C.2(a) shows, from left to right, as a percentage of the actual number of inputs in the synthetic data: the inputs that were correctly selected, the inputs that were missing and the inputs that were incorrectly selected. The figure shows that increasing the value of th_{uic} results in an increasing number of inputs that is considered irrelevant after the input selection process, even though they were actually used to generate the synthetic output. On the other hand, very low values of th_{uic} result in considering certain inputs relevant, even though they were not used to generate the synthetic output. In this case, the ARX model probably used these inputs to fit the white noise. Increasing the number of inputs to generate the synthetic data also increases the chance on missing inputs after the input selection process and so does increasing the maximum model order.
Figure C.2(b) shows, from left to right, as a percentage of one (polynomial A) plus the number of correctly selected inputs (polynomials B_i): the polynomials orders that were correctly estimated, the polynomials orders that were estimated too low and the polynomials orders that were estimated too high. In these figures, the results when using the AIC instead of th_{uoc} to reduce the polynomial order are also shown. The model orders are generally estimated too low for n_{max} higher than one. A likely explanation for this is that these model orders influence frequency ranges that are not represented in the input signals and thus do not influence the model fit. While increasing th_{uoc} results in even more polynomial orders to be estimated too low, for low values of this threshold the polynomial orders are estimated too high. Interestingly, using the AIC instead of th_{uoc} to reduce the polynomial order results in an even higher percentage of polynomial orders being estimated too high. A similar finding was observed by Drop [198].

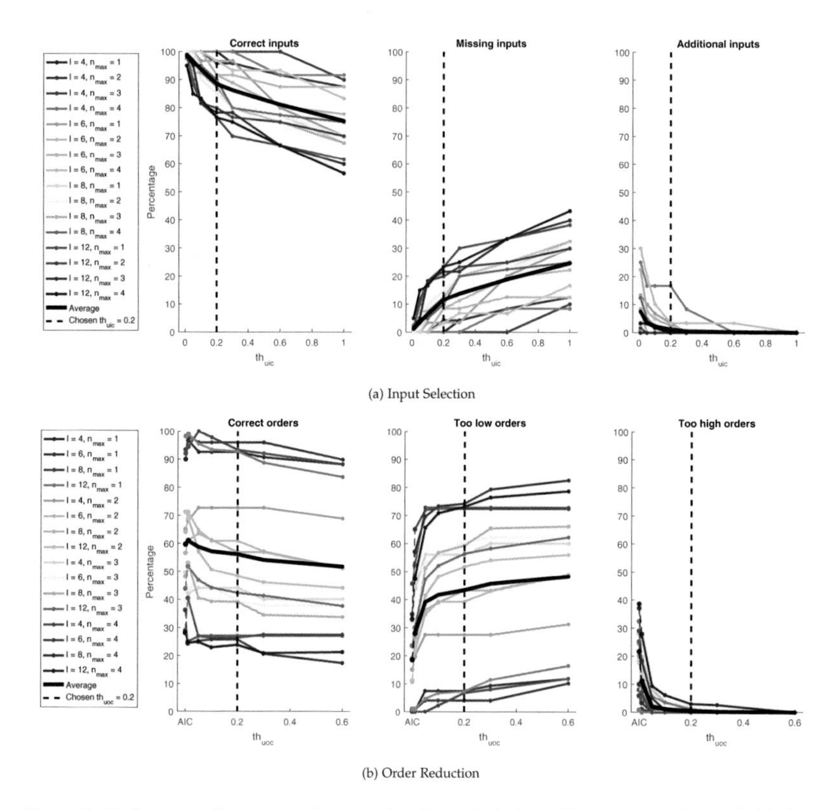

Figure C.2: Performance of input and order estimations for synthetic data with varying criterion thresholds th_{uic} and th_{uoc}, respectively.

C.3. THRESHOLD CHOICE

Setting the value for th_{uic} too low reduces the model fit because relevant inputs might not be selected. On the other hand, setting this value too high can result in an unnecessarily complex model and with that a reduction of its predictive power. A value of $th_{uic} = 0.2$ is chosen such that in Figure C.2(a) almost 90% of the inputs is selected correctly, while less than hardly any ($< 1\%$) additional inputs are selected.
Even though using the AIC is a widely accepted method for model selection, the synthetic data results show that this criterion is not strict enough for the type of data used here, resulting in unnecessarily high polynomial orders in more than 20% of the cases. Using the threshold method with $th_{uoc} = 0.2$ model orders are almost never ($< 0.5\%$) overestimated while the same percentage of correct polynomial orders (59% vs. 56% correct orders) only decreases marginally. Instead therefore the threshold method with $th_{uoc} = 0.2$ is used in this study. The model orders that are underestimated can be corrected manually during the model explanatory analysis that is performed in Step 5 of the SI process.

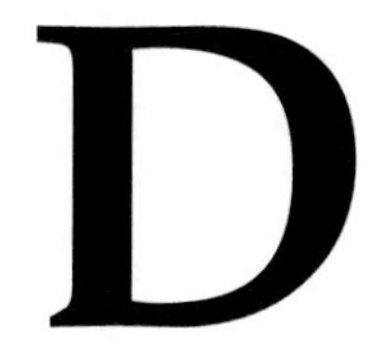

SI Process: Selection Criteria Parameters

In Section 5.3.2 of Chapter 5 the criterion for input selection and order reduction is discussed and the parameters W (time step weight vector) and P_{ANIC} (Avoid Negative Input Contributions parameter) are introduced. To show the effects of these parameters on the models resulting from the SI process, the initial SI parameters of the CEDA model are used to show the influence of W and P_{ANIC} on the goodness of fit of the final model.

D.1. Influence of W

In Equation 5.1 the selection criterion used for input selection and order reduction is described. The weight vector W is typically set to one for each time step ($W = W_1$), indicating that the error between modelled and measured MIR is equally important at all time steps. This weight vector, however, can be used to improve the predictive power of the model, by increasing the weight at certain important time steps.
One way of selecting important time steps is by increasing the weight at time steps that describe important features of the output signal. As was discussed in [136], the off-line overall rating of a maneuver was best described by the maximum continuous rating during this maneuver. This implies that minimizing peaks in the continuous rating is the most important, when trying to improve the overall MCA quality. To more accurately model these peaks in particular, a peak detection algorithm (such as Matlab function 'findpeaks') can be used to detect the peaks and valleys in the measured MIR. The corresponding time steps, and the surrounding ones, can then be given and increased weight during the criterion calculation by setting $W = W_{peaks}$. For the dataset used in this study, however, adjusting W to W_{peaks} did not give a significantly different result.
A more commonly used criterion for selecting important time steps is by varying the weight according to the level of agreement between participants ($W = W_{std}$). When the spread between participants is large, the weights for these time steps should be small, while the weights should increase with a decreasing spread. In Figure D.1 the weights W_{std}, which are inversely proportional to the standard deviation of the rating at each time step, for the evaluation dataset z_e are shown. Figure D.1 shows that the highest agreement between participants is found in the middle of maneuver BA and MCA_{NL} where the median rating is zero. In general, the agreement is highest at times when the

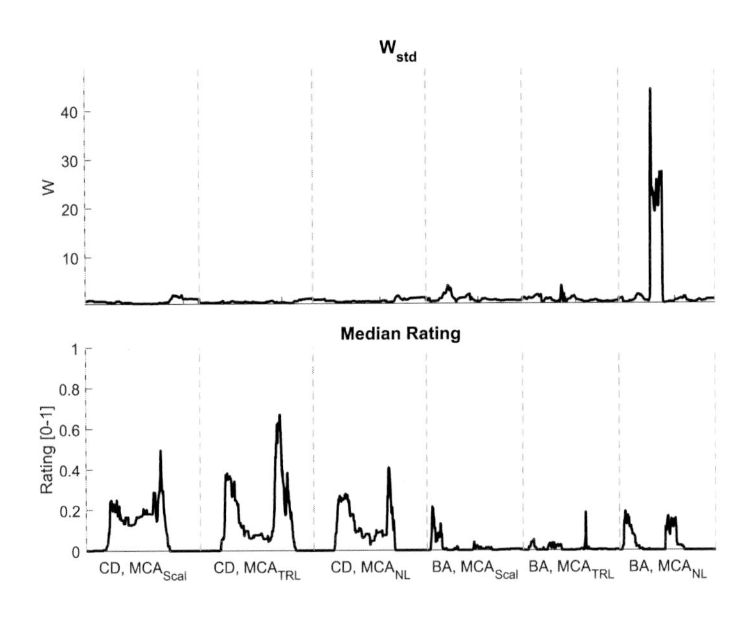

Figure D.1: Weights W_{std} for dataset z_e.

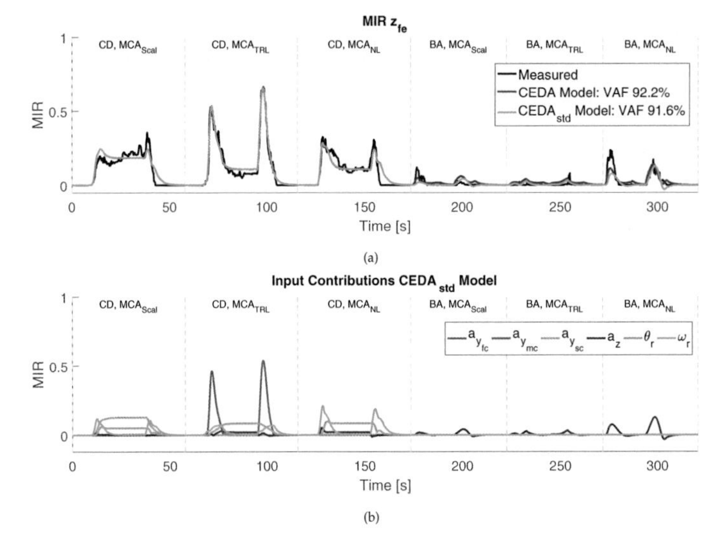

Figure D.2: Model $CEDA_{std}$: Goodness of fit and input contributions for fit to dataset z_{fe}.

median rating is zero.

A new model, $CEDA_{std}$, is obtained using the same initial parameters and choices as the $CEDA$ model described in Paragraph 5.4.1, but using $W = W_{std}$ instead of the original $CEDA$ model setting $W = W_1$. In Figure D.2 the effects of setting $W = W_{std}$, compared to $W = W_1$ for the CEDA model are shown for the parameter estimation dataset z_{fe}. The main difference between models $CEDA$ and $CEDA_{std}$ is found during the BA maneuvers. The input selection process resulted in leaving out ω_p when the weight parameter was changed from W_1 to W_{std}. After the order reduction step, the $CEDA_{std}$ models therefore contain one parameter less than the $CEDA$ model. Adjusting W_1 to W_{std} decreases the total fit with a decrease in VAF of 0.6%.

In Figure D.3 the prediction for dataset z_p is shown for both the $CEDA$ and $CEDA_{std}$ model, as well as the input contributions of the $CEDA_{std}$ model. Figure D.3(a) shows that the prediction power is increased minimally with 0.2%. Taking into account the agreement between participants is generally assumed to improve the prediction power of the model, as it reduces the chance of fitting the model to noise in the dataset. The prediction results shown in Figure D.3 imply that this only has a very marginal effect on the prediction power. Additionally, the spread between participants caused such disproportionately large weights during maneuver BA with MCA_{NL}, that it makes the model less suitable for prediction of different maneuvers.

Because no clear indication was found that variable weights significantly improve the model, for simplicity the weights in this study are all set to W_1. However, when more data become available, it is advised to investigate if weights based on participant agreement can improve the prediction power. The relation between weights and agreement can be adjusted and possibly the removal of negative input contributions can be improved.

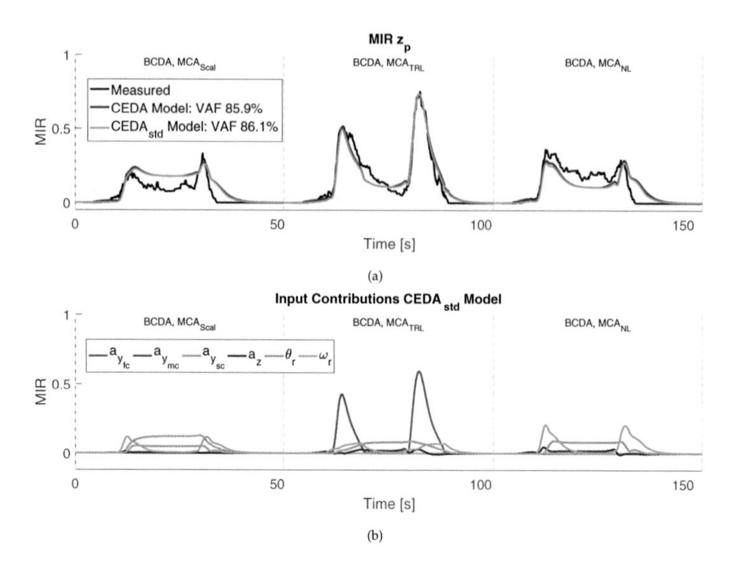

Figure D.3: Model $CEDA_{std}$: Goodness of fit and input contributions for prediction of dataset z_p.

D.2. Influence of ANIC

The CEDA model was obtained by setting the parameter P_{ANIC} to one, such that the removal of inputs causing higher negative input contributions are mostly avoided. To show the influence of this parameter a model was obtained by setting P_{ANIC} to zero. In Figure D.4(a) the goodness of fit to dataset z_{fe} and in Figure D.4(b) the corresponding input contributions are shown. The prediction of this model for dataset z_p is shown in Figure D.5(a) and the corresponding input contributions are shown in Figure D.5(b).
The resulting model has fourteen parameters, obtained by adding one extra input, compared to thirteen parameters in the CEDA model. Figures D.4(b) and D.5(b) show that the contribution of input α_y is fully negative, e.g., 180 degrees phase shift. While this additional input improves the model by increasing both the fit and the prediction with an absolute increase in VAF of 0.5% and 1% respectively, the negative input contribution would suggest that cueing errors are *improving* the cueing quality. To make the SI process more robust against such illogical choices, the ANIC parameter should thus be set to one. However, if this is not a requirement for the use of the model, setting P_{ANIC} to zero likely results in a higher goodness of fit, because the input selection is less constrained.

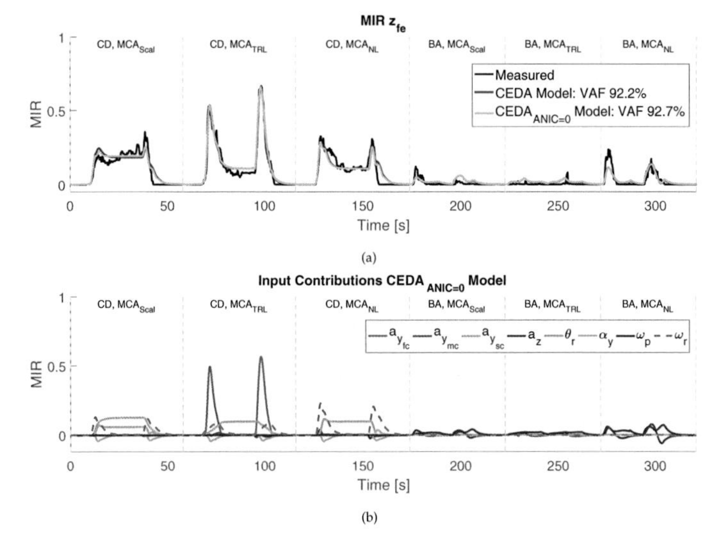

Figure D.4: $CEDA_{ANIC=0}$ model fits to dataset z_{fe}.

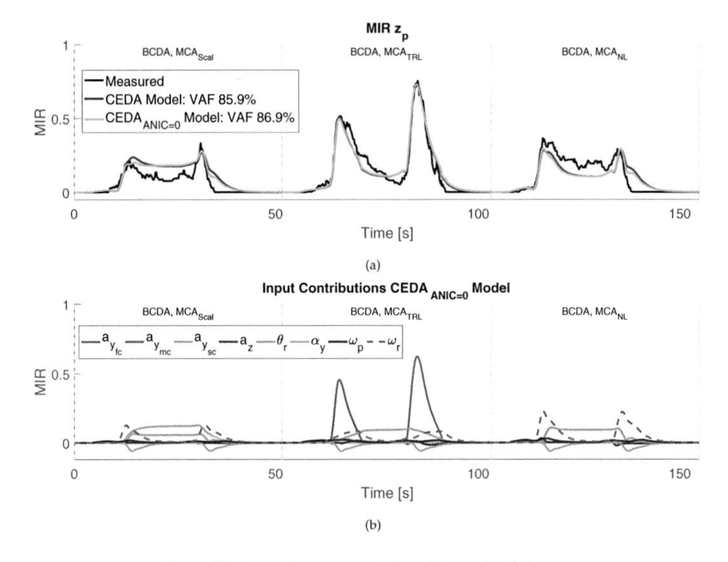

Figure D.5: $CEDA_{ANIC=0}$ model predictions for dataset z_p.

SI Process: Parameter Estimation

Parameter estimation is a technique that is used to estimate the most likely parameters of system. In this appendix first a general overview is given on the background and different implementations of maximum likelihood estimation. We then discuss which the relation between this estimation technique, often used in the human sciences, with the prediction error method for parameter estimation, often used in engineering. After this introduction to the subject of parameter estimation, the influence of the noise characteristics observed in this study on the prediction error parameter estimation is analysed.

E.1. Maximum Likelihood Estimation

To calculate the parameters of a model, the system response to a known set of inputs can be compared to the measured system output. The parameters can then be calculated by equating the difference between measured and modelled system outputs to zero and solving for the parameters. However, in the real world, measurements of system outputs are a combination of the system output plus noise. This relation for linear systems can be described by:

$$y_{meas} = y + \eta = X \cdot B + \eta, \tag{E.1}$$

where y_{meas} is the measured system output, y the true system output, X are the system inputs, B is the polynomial with the true system parameters and η is the noise. Without exact knowledge of the noise, the true system parameters can thus no longer be directly calculated. Instead, parameters can be estimated by maximizing the probability of them being the true parameters of the system, given the measurements. Such a rule describing how to obtain the most probable model parameters is called an estimator.
The many different estimators described in literature can generally be derived from the Bayes' estimator. This estimator is based on Bayes' Rule for continuous random variables [199] which states that the probability density of the parameters given the measurements, e.g., the posterior probability density, is defined with:

$$f\left(\beta|y_{meas}\right) = \frac{f\left(y_{meas}|\beta\right) f\left(\beta\right)}{f\left(y_{meas}\right)} \tag{E.2}$$

The probability density of certain parameters being the true system parameters given the measurements thus depends on three probability densities:

1. Probability density of the measurements occurring given the parameters: $f(y_{meas}|\beta)$
   - where $f(y_{meas}|\beta)$ is called the likelihood function and depends on the noise distributions.
2. Prior distributions of the parameters: $f(\beta)$
   - this contains information on the parameters before any evidence, e.g., measurements, are taken into account
3. Marginal distribution of the measurements: $f(y_{meas})$
   - the mean probability of the measurements over all values of β

The Bayes' estimator tries finding parameter estimates that maximize the posterior probability. A point estimator based on this rule is the maximum a posteriori (MAP) estimator, which assumes that the most likely value for β can be found at the mode (e.g., the maximum) of the posterior probability distribution. For the relatively simple case where $f(\beta|y_{meas})$ is a unimodal distribution, this is a reasonable assumption.
The MAP estimator is described with:

$$\hat{\beta}_{MAP} = \underset{\beta}{\arg\max}\, f(\beta|y_{meas}) = \underset{\beta}{\arg\max}\, f(y_{meas}|\beta)\, f(\beta) \tag{E.3}$$

However, in many applications the prior distributions $f(\beta)$ are unknown and instead are assumed to be uniform. As the marginal distributions are independent of the parameters, maximizing the posterior probability then becomes proportional to maximizing the likelihood function $f(y_{meas}|\beta)$. For linear systems as described in equation E.1, this likelihood function represents the distribution of the noise η. E.g., given certain parameters $\hat{B}$ and the corresponding modelled output y_{mod}, the likelihood of certain measurements y_{meas} to occur depends on the noise distribution.
For many applications it is easier to use the negative of the log likelihood, rather than the likelihood, such that minimization and summation instead of maximization and products can be used. The maximum likelihood estimator is then described by the following rule:

$$\hat{\beta}_{MLE} = \underset{\beta}{\arg\min} \quad -\ln \quad f(y_{meas}|\beta) \tag{E.4}$$

The maximum likelihood estimator can be further simplified if more is known about the noise distributions described by $f(y_{meas}|\beta)$. In many cases these simplified estimators use the following definition of the measurement error ε:

$$\varepsilon = y_{meas} - y_{mod} = y_{meas} - X \cdot B \tag{E.5}$$

If the modelled output y_{mod} represents the true system output y, then the measurement error ε will equal the system noise η. In Table E.1 an overview is given for some of the most common distribution assumptions and the corresponding simplified MLEs.
In this table, IID stands for Independent Identically Distributed. This means that the

noise distribution locations (i.e., mean) and shape (i.e., variance) are identical for all noise measures and should not depend on any other system variable [200]. Identically distributed variables are thus also homoscedastic variables, e.g., all noise distributions have the same variance [201].
The generalized least squares method allows for a certain degree of correlation between the noise and system variables described by the matrix V, e.g., IID noise is not required. For independent, but heteroscedastic noise the weighted least squares, with weights w_i corresponding to the noise variances for each measurement, can be used. When the IID noise assumption applies, the ML estimator simplifies to the ordinary least squares estimator, which solely depends on the measurement error. Due to its simplicity and the available analytical solution, this estimator is used in many applications, including modelling of dynamical systems.
When, instead of the normal distribution, the IID noise follows a Laplace, or double exponential, distribution, the Least Absolute Deviation estimator can be used. This distribution type has an excess kurtosis of three, meaning that the distribution peak is much sharper than that of a normal distribution.

Table E.1: Maximum likelihood estimation methods

Estimation Method	Estimator	Analytical Solution	Assumptions	Features
Generalized Least Squares (GLS)	$\hat{\beta}_{GLS} = \arg\min_{\beta} \varepsilon^T V \varepsilon$	Yes	Normal Distributions	Sensitive to outliers
Weighted Least Squares (WLS)	$\hat{\beta}_{WLS} = \arg\min_{\beta} \sum_{i=1}^{n} w_i \varepsilon_i^2, \, with \, w_i = \frac{1}{\sigma_i^2}$	Yes	Independent Normal Distributions	Sensitive to outliers
Ordinary Least Squares (OLS)	$\hat{\beta}_{OLS} = \arg\min_{\beta} \sum_{i=1}^{n} \varepsilon_i^2$	Yes	IID Normal Distributions	Sensitive to outliers. Implemented in Matlab ARX method
Least Absolute Deviation (LAD)	$\hat{\beta}_{LAD} = \arg\min_{\beta} \sum_{i=1}^{n} \lvert\varepsilon_i\rvert$	No	IID Laplace Distributions	Resistant to outliers and robust to departures from the normality assumption

E.2. Prediction Error Method for ARX model structures

For dynamical systems with an ARX model structure, the relation between system output y_{meas}, inputs X and noise η is described by:

$$Ay_{meas} = X \cdot B + \eta, \tag{E.6}$$

where A and B represent the output and input polynomials, respectively. The simple OLS estimator described in the previous section, depends on the equality of system noise η and measurement error ε for the correct model. With this model structure, however, the measurement error ε also includes system noise from previous time steps and is thus never equal to η. Instead, the prediction error is introduced. This prediction error is calculated with the measured and predicted, rather than modelled, outputs for the model, y_{meas} and y_{pred}, respectively. The measured output for an ARX model structure is:

$$y_{meas} = X \cdot B_{real} + \eta - a_{1_{real}} \cdot y_{meas} \cdot z^{-1} - a_{2_{real}} \cdot y_{meas} \cdot z^{-2} \ldots - a_{n_{real}} \cdot y_{meas} \cdot z^{-n} \tag{E.7}$$

Here, X are the system inputs, B is the input polynomial, η indicates the system noise and $a_{i_{real}}$ are the parameters of the output polynomial A. The predicted output is then calculated with:

$$y_{pred} = X \cdot B - a_1 \cdot y_{meas} \cdot z^{-1} - a_2 \cdot y_{meas} \cdot z^{-2} \ldots - a_{n_a} \cdot y_{meas} \cdot z^{-n_a}, \tag{E.8}$$

where a_i are the parameters the modelled output polynomial A. The prediction error is then calculated with:

$$\varepsilon_{pred} = y_{meas} - y_{pred} \tag{E.9}$$

If the system noise is IID Gaussian noise, the ML estimator reduces to the OLS estimator:

$$\hat{\beta}_{OLS} = \arg\min_{\beta} \sum_{i=1}^{n} \varepsilon^2_{pred_i} \tag{E.10}$$

E.3. Influence of non-IDD noise

In this study the initial guess of the model structure of the linear part of the perceived motion incongruence model resembles the ARX model structure. For this structure, the parameter estimation is based on the prediction error, rather than the measurement error, because the measurement error depends on passed instances of the modelled output. If the system noise is assumed to be IID Gaussian noise, an analytical solution for the model parameters can be found, which significantly decreases the SI process calculation time. As this process is often repeated many times for one model, any time improvement can increase the number of different non-linear choices and SI process parameter combinations that can be tried in a certain time span, eventually leading to more optimal models.

In this study, however, the system noise is not IID Gaussian noise. The analytical solution obtained from the OLS estimation is therefore not necessarily the maximum likelihood estimate. In this appendix an analysis of the effects of this departure from IID Gaussian noise on the parameter estimation is performed. First, the system noise distribution is modelled in Section E.3.1, after which this noise model is used in Section E.3.2 for back estimation of ARX model parameters. The latter leads to the conclusion that the observed noise distribution does not significantly influence the parameter estimation, and the OLS estimator can still be considered as a maximum likelihood estimator.

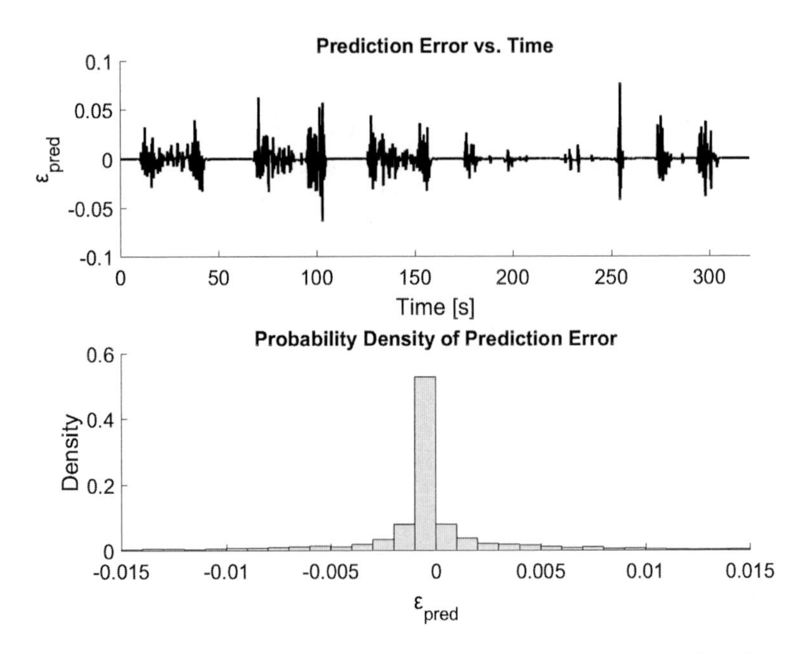

Figure E.1: Prediction error ε_{pred} versus time (top) and its probability density function (bottom).

E.3.1. Noise Modelling

The system noise in ARX model structures is described by the prediction error. If this system noise is IID Gaussian noise the model parameters can be estimated with the OLS estimator. In Figure E.1 the resulting prediction error for the data used in this study is shown. Figure E.1 shows the prediction error over time in the top plot and the total prediction error probability density distribution in the bottom plot. Clearly, the variance of the system noise is not constant over time, e.g., heteroscedasticity occurs. The bottom plot additionally shows that the total system noise is not normally distributed. As the prediction error only provides one sample from the noise distribution at each time step, it is not possible to model the heteroscedasticity accurately from the prediction error. Instead, here we try to estimate the system noise from the measurements.

The measurements in this study included 48 time signals (16 participants each rating 3 trials), leading to 48 measurement points at each time step. These measurements show the distribution of the noise around a true underlying average rating that we wish to model. The relation between the measured and true system output y, e.g., the average rating, is shown in:

$$y_{meas} = y + \eta_{meas}(y), \tag{E.11}$$

where η_{meas} differs from the system noise η in equation E.7 as it includes system noise from previous time steps as well. The true system output that we aim to model is described with:

$$y = BX - a_1 y \quad z^{-1} - a_2 y \quad z^{-2} - \ldots - a_n y \quad z^{-n} \tag{E.12}$$

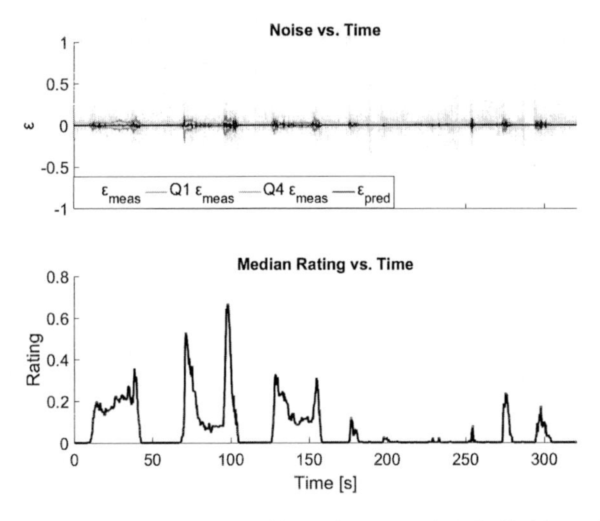

Figure E.2: Top: system noise estimates ε_{pred} (red line) and ε_{meas} (gray dots). The black lines indicate the inter quartile range of the noise estimate ε_{meas} at each time. Bottom: median over all rating measurements.

Combining equations E.7, E.11 and E.12, the measurement noise η_{meas} can thus be described as the difference between the measured and the system output:

$$y_{meas} - y = \eta_{meas}(y) = \eta - a_1(y_{meas} - y)z^{-1} - a_2(y_{meas} - y)z^{-2} - \dots - a_n(y_{meas} - y)z^{-n} \tag{E.13}$$

This results in the relation between system noise and measurement noise:

$$\eta = \eta_{meas} + a_1(\eta_{meas})z^{-1} + a_2(\eta_{meas})z^{-2} + \dots + a_n(\eta_{meas})z^{-n} \tag{E.14}$$

The system noise distribution can thus be derived from the measurement noise distribution and the polynomial A parameters. As the polynomial A parameters and the true system output are not known, to determine the system noise distribution we need estimates of these values.
Here we start by estimating the parameters of polynomial A. For this estimation we assume that the measured system output is the median rating at each time step and that the system noise is IID Gaussian noise. Using the prediction error method with OLS estimation, the system polynomials A and B are estimated.
To estimate system noise, the true system output y is assumed to be the median of the measured outputs. The latter is hereby defined as the measured system noise estimate ε_{meas}:

$$\varepsilon_{meas} = \eta_{meas} + \hat{a}_1(\eta_{meas})z^{-1} + \hat{a}_2(\eta_{meas})z^{-2} + \dots + \hat{a}_n(\eta_{meas})z^{-n} \tag{E.15}$$

In Figure E.2 the system noise estimates using equation E.9 for ε_{pred} and equation E.15 for ε_{meas} are shown versus time. Figure E.2 shows that the prediction error is for the most part within the interquartile range of the measured system errors. Indicating that

they are both representing the similar underlying system noise. It is also clear that both estimates show heteroscedasticity over time, e.g., the variance is not constant, and seems to correlate with the median of the measurements, i.e., the lower the median value, the smaller the variance.
To model this heteroscedasticity related to the median rating, distributions were fitted to the noise estimates ε_{meas} corresponding to a specific value of the median rating. Normal, logistic and the Student's t location-scale distribution types, each type having increasing kurtosis, were used. In Figure E.3 the fits of the different distributions to the histograms of the measured noise estimates ε_{meas} is shown. The average negative log likelihood over all fitted Student's t location-scale distributions, shown in the legends, is lower than for the normal and logistic distributions. In Figure E.4 the median and inter quartile range (IQR) of the measured noise estimates ε_{meas} and the fitted distributions are shown. As expected from the average negative log likelihoods for the Student's t location-scale distribution type, the median and IQR for this distribution type also fit best to the measured noise estimates ε_{meas} distributions, and is therefore used to describe the distribution of ε_{meas}.
To describe the heteroscedasticity of ε_{meas} over system output y, the parameters of the Student's t location-scale distributions, location μ, scale σ and shape parameter ν, are modelled using a seventh order polynomial. In Figure E.5 these parameters vs. y and the seventh order polynomial fits are shown. The dotted line indicates an extrapolation of these parameters for values of y up to one, which, for simplicity, is assumed to be the median of the parameters from $y = 0.5$ and higher. This extrapolation is needed to assure reasonable parameters for the complete system output, i.e., rating, range $[0 - 1]$. The polynomial fit takes into account that for some values of y many more measurement samples are used than for others, by using the number of samples as weights in the optimization cost function. The final noise model as a function of the true system output is now described as:

$$f_{\varepsilon_{mod}}(y) = f_{tLocationScale}(poly7_{\mu}(y), poly7_{\sigma}(y), poly7_{\nu}(y)) \tag{E.16}$$

To validate this noise model $f_{\varepsilon_{mod}}(y)$, random samples are drawn from it with y as the median rating and compared to ε_{meas}. The top plot in Figure E.6 shows the resulting measured and modelled noise (ε_{meas} and ε_{mod} respectively) over time, as well as their median and first and third quartiles. The resulting overall distributions of ε_{meas} and ε_{mod} over all time steps is shown in the bottom plot. Figure E.6 shows that the proposed noise model can describe the measured noise relatively well and can be used to determine the influence of the heteroscedasticity and non-normality of the noise on the arx parameter estimation. It should be noted, however, that even though the interquartile ranges are very similar, the measured noise does contain more outliers than the noise sampled from the noise model. For future work it is therefore recommended to improve this noise model by, for example, varying distribution type as well as distribution parameters per median rating, by modelling additional dependencies of the distributions aside from the dependency on median rating, or by replacing the polynomial functions with more functions that fit better to the observed distribution parameters.

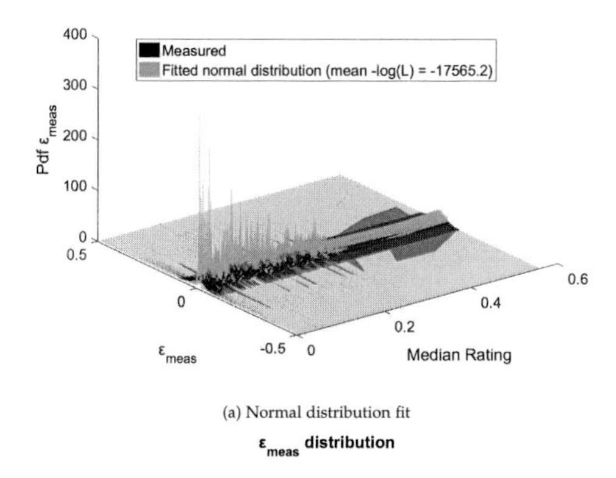

(a) Normal distribution fit

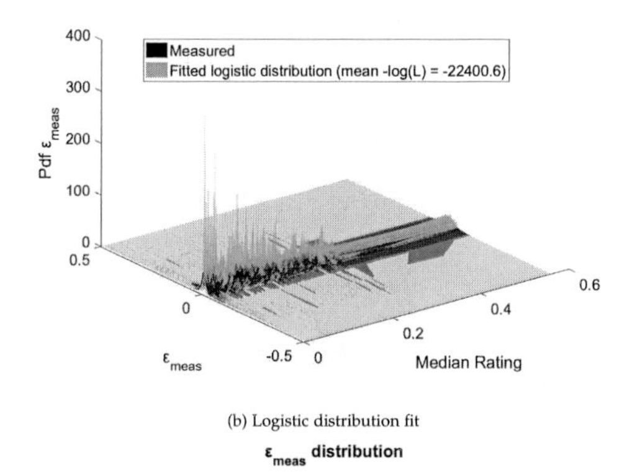

(b) Logistic distribution fit

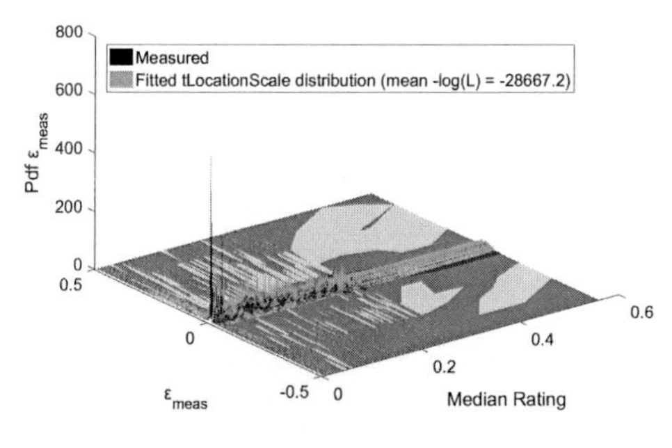

(c) Student's t location-scale distribution fit

Figure E.3: Standard distribution fits to measured noise estimates ε_{meas}.

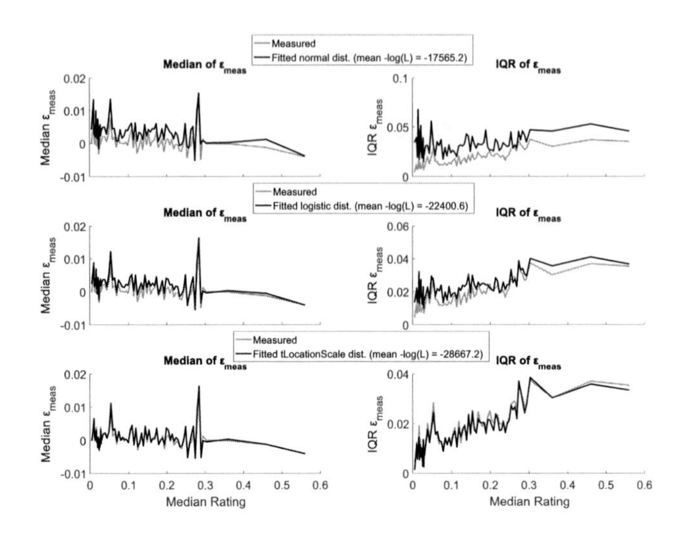

Figure E.4: Median (left column) and inter quartile range (IQR) (right column) of the measured noise estimates ε_{meas} (green) and the fitted distributions (red).

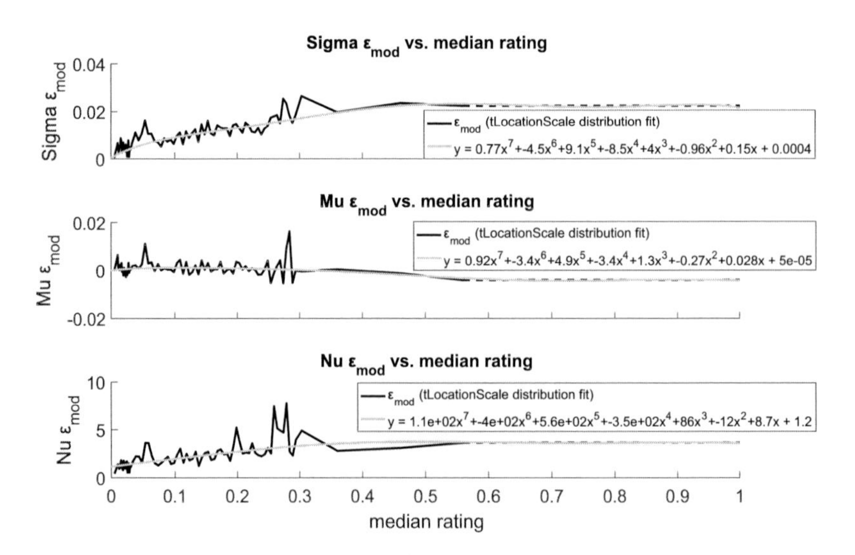

Figure E.5: From top to bottom: the Student's t location-scale distributions, location μ, scale σ and shape parameter ν and their fit.

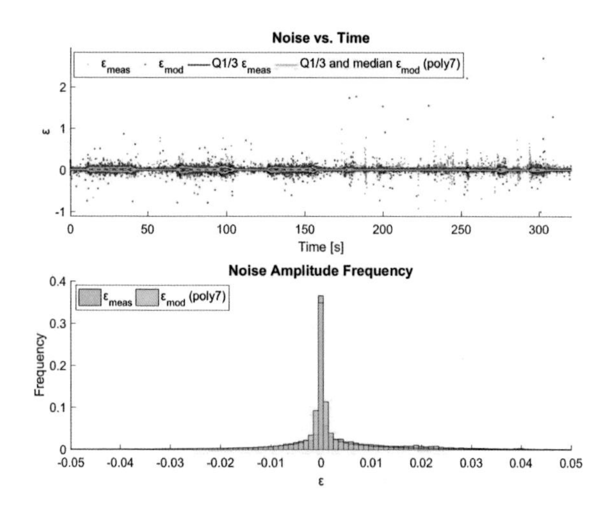

Figure E.6: Measured and modelled system noise estimates over time (top) and their probability density distributions (bottom).

E.3.2. Back Estimation of ARX parameters

Using the noise model from the previous section, synthetic outputs are generated using random ARX model polynomials and the inputs in a similar way as was described in Section C.1.

Synthetic datasets were made using both distributions of white Gaussian noise and the noise model described in the previous section. Furthermore, the number of inputs, between three and nine, and the maximum polynomial order, between one and four, were varied and each combination was repeated twenty times with randomly chosen parameters. In total this led to $2x6x4x20 = 960$ different synthetic models that were back-estimated using the ARX function of Matlab.

In Figure E.7 the errors between the actual and the estimated system parameters are shown for all conditions (top row), per number of inputs (middle row) and per maximum polynomial order (bottom row), for system errors from both white Gaussian noise and noise model distributions. The right hand plots show the boxplots without outliers, for clarity.

To determine significant differences in parameter estimation errors between the two noise distributions, t-tests were performed for each pair shown in Figure E.7. No significant differences between distributions were found.

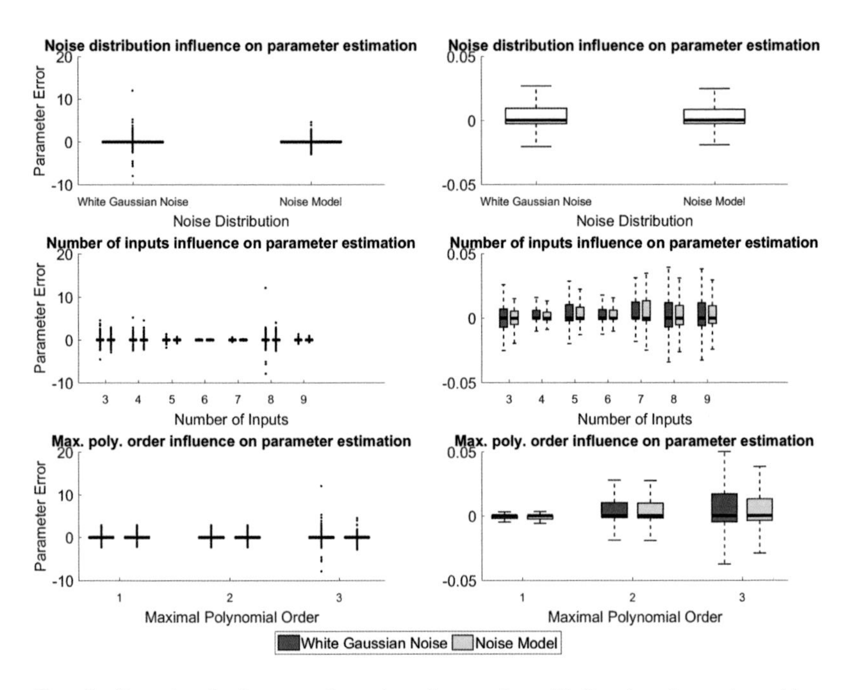

Figure E.7: Parameter estimation errors when system noise comes from white Gaussian noise or noise model distributions.

E.3.3. CONCLUSION

Since no significant differences between system noise from white Gaussian and actual fitted noise model distributions were found, the effects of non IID Gaussian noise as observed from the rating measurements is reasonably assumed to not influence the least squares parameter estimation for ARX models significantly. The parameters found with this estimation method can therefore be regarded as to be close to the maximum likelihood estimations of the system parameters.
However, small differences between the noise model and the measured noise might were observed. For future work it is therefore recommended to improve the noise model and confirm that the noise characteristics do not significantly influence the parameter estimation.

MTP Estimation

When using the continuous rating method to rate Perceived Motion Incongruence (PMI), a rating of this incongruence, the MIR, is obtained. This MIR is rated on a scale that is anchored to the minimum and maximum PMI perceived during the experiment. Comparing the rating obtained from two different experiments thus requires one to transform both MIRs to have the same scale. To do this, the model transfer parameter (MTP) is introduced in Chapter 6. This constant maps the rating scale of one of the experiments onto the rating scale of the other experiment.

To calculate the MTP, two methods have been tested and compared: a serial and parallel modelling-scaling method. In this appendix both methods are explained and analysed. To validate the methods and determine whether they can successfully estimate the MTP, data from a single experiment are split in two sub-datasets and the MTP between them is calculated. As both subsets originate from the same experiment, they were rated on the same rating scale and the estimated MTP should thus, by definition, be unity. The method validation thus amounts to comparing the resulting MTP estimates to the true MTP of one.

The unity MTP estimation is done twice for each estimation method: once using the CMS and once using the Daimler experiment dataset. The subsets that are derived from each experiment are used to fit and apply a PMI model to. For a robust estimation, the model fit to both subsets should be reasonably good. To accomplish this, the two subsets should include segments, i.e., combinations of manoeuvre and MCA, that have similar input power. To this end, the overall datasets are split using Table F.1, where the first subset contains the Basis group and at least one of the segments from each of the Choices groups. The second subset contains at least one segment from each of the Choices groups.

For each estimation method, the MTP is estimated for different combinations of segments in each subset and for three different models, each time using a randomized initial MTP value between 0.1 and 2, as explained in Section 6.3.3. For the CMS experiment in total 8 different subset pairs are possible, while for the Daimler experiment, which has many more segments, a total of 48 different subset pairs is used. Each method thus performed 24 and 144 MTP estimations for the CMS and Daimler experiment, respectively.

Table F.1: Grouping of experiment sections with similar input power.

Exp.	Basis	Choices			
1	BA_{Scal}, BA_{TRL}, BA_{NL}	CD_{Scal}, $BCDA_{Scal}$	CD_{TRL}, $BCDA_{TRL}$	CD_{NL}, $BCDA_{NL}$	
2	$TLW_{M/D}$, $TLA_{M/D}$	$RuC_{M/D}$, $OvT_{M/D}$	$Rou_{M/D}$, $TuL_{M/D}$	$SID_{M/D}$, $TLD_{M/D}$	$Ci1_{M/D}$, $Ci2_{M/D}$, $Ci3_{M/D}$

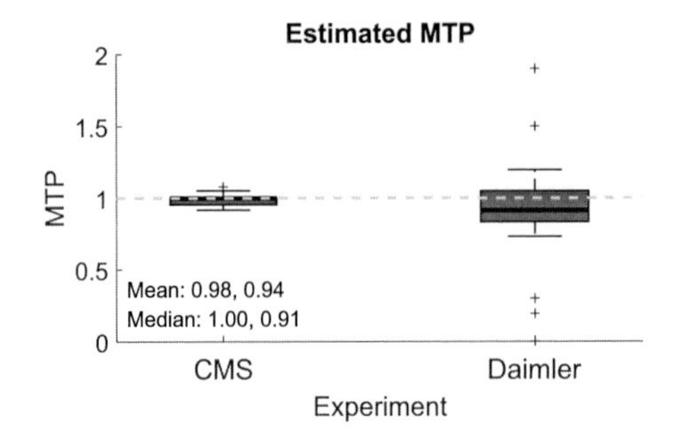

Figure F.1: Boxplots showing median and interquartile range for serial unit MTP estimation for CMS (left) and Daimler (right) experiments.

F.1. SERIAL

For the serial MTP estimation method, first a model is fitted to the first subset, after which the MTP is estimated by applying this model, multiplied with the MTP, to the second subset. In this case, no data from the second subset are thus used to fit the model. The optimal MTP is found by minimizing the one-step-ahead prediction error for the second subset.

In Figure F.1 the results of the serial MTP estimations for the different subset pairs of both experiments is shown in boxplots, where the CMS boxplots describes 24 MTP estimates and the Daimler boxplot describes 144 MTP estimates. For both experiments, the unity MTP falls within the interquartile ranges of both boxplots. The final MTP estimate should be calculated as the average over all these estimates.

In Figure F.2 the histograms and normality plots of the estimates for both experiments are shown. This figure shows that the estimates of both experiments do not exactly follow a normal distribution. The Lilliefors test for normality confirms this (CMS: $D = 0.26674$, $p < 0.001$, Daimler: $D = 0.14017$, $p < 0.001$) and the mean is therefore not a valid descriptor of the average MTP estimate. Instead the median should be used. With the serial MTP estimation method, the unity MTP is thus estimated as 1.00 and 0.91 for the CMS and the Daimler experiments, respectively. This estimation method can thus be perfect, in case of the CMS experiment, but has an error from unity of about 9%, in the case of the Daimler experiment.

F.2. PARALLEL

For the parallel MTP estimation method, the MTP is estimated using an iterative process of alternately fitting a model to both subsets and estimating the MTP. Before combining both subsets, the inputs of the second subset are multiplied with the MTP, while

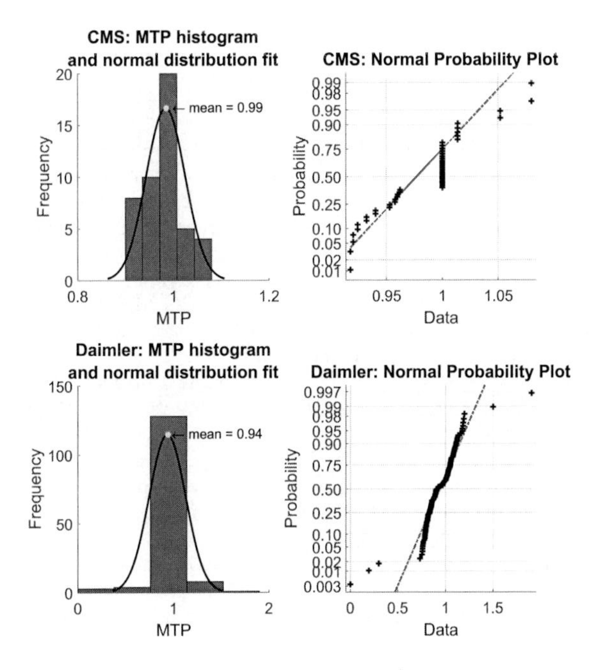

Figure F.2: Histograms and normal distribution fits (left) and normality plots (right) for serial unit MTP estimations for CMS (top) and Daimler (bottom) experiments.

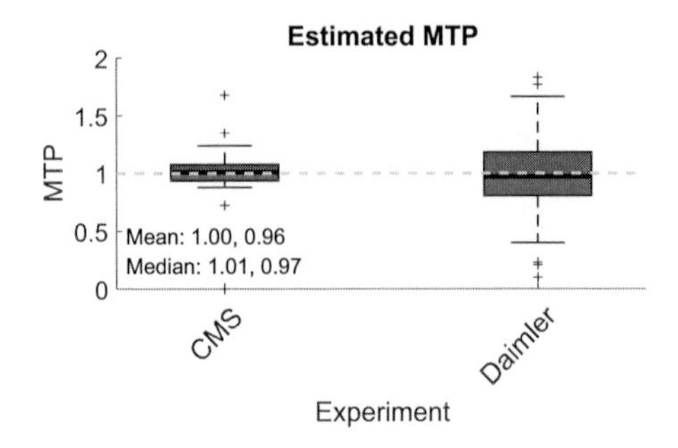

Figure F.3: Boxplots showing median and interquartile range for parallel unit MTP estimation for CMS (left) and Daimler (right) experiments.

the inputs of the first subset remain unchanged. The optimal MTP is found using a gradient-descent method by minimizing the one-step-ahead prediction error for the complete dataset. In this case the model is thus fitted to data from both subsets.

In Figure F.3 the results of the parallel MTP estimation for the different subset pairs of both experiments is shown in boxplots, where the CMS boxplot describes 24 MTP estimates and the Daimler boxplot describes 144 MTP estimates. For both experiments, unity MTP is a likely candidate for the MTP as it falls well inside the interquartile ranges of both boxplots. The final MTP estimation contains the additional step of averaging over these different MTP estimation results.

Figure F.4 shows that, the results do not exactly follow a normal distribution. This is mainly caused by local minima being found when the initial MTP estimate given to the optimization is close to zero. The Lilliefors test for normality confirmed that the MTP estimations are not normally distributed (CMS: $D = 0.2510$, $p < 0.001$, Daimler: $D = 0.085938$, $p < 0.05$), indicating that the mean value over all MTP estimations is not a good descriptor of the average MTP estimate. Instead, therefore, the median over all MTP estimates is used as the overall MTP estimate.

Using the median to describe the final MTP estimate for the CMS and Daimler experiments, MTP estimates of 1.01 and 1.03 are found, respectively. The error from unity using the parallel MTP estimation estimate is thus 1% and 3%, for the CMS and Daimler experiments, respectively.

F.3. Discussion and Conclusion

The serial MTP estimation process has the advantage of a low computation time compared to the parallel estimation process. However, the results of the serial process are much less accurate (9% error) than those of the parallel process (3% error). Also this estimation process seems less reliable as there is a large difference in error between the two experiments.

The relatively low accuracy and reliability can be explained by the use of only one subset to fit a model. With small subsets, there are many possible models that can describe the subset equally well. However, not all of these models would show the same prediction performance. For example, the roll angle and the yaw rate cueing errors have a similar shape in the CMS experiment, and either of them might provide a good fit. However, in the Daimler experiment, these two signals are not similar and the predic-

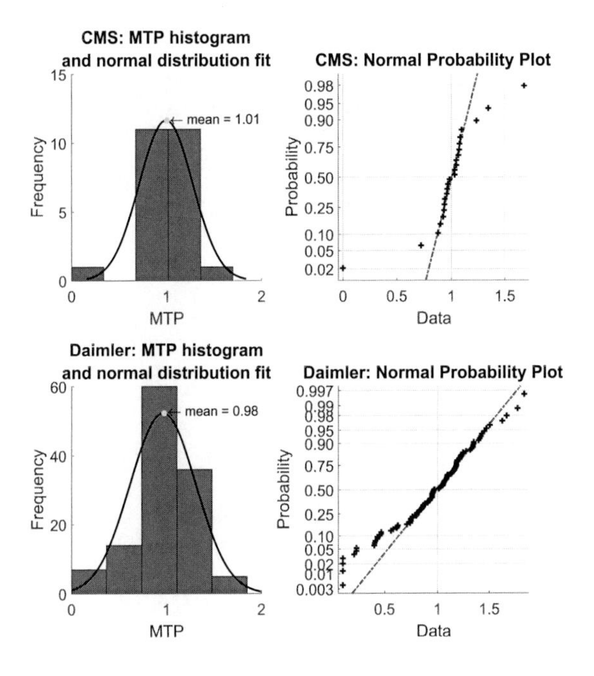

Figure F.4: Histograms and normal distribution fits (left) and normality plots (right) for parallel unit MTP estimations for CMS (top) and Daimler (bottom) experiments.

tion power of a model using the roll angle instead of the yaw rate would thus differ. If the chosen model does not fit the second subset well enough, the MTP estimate could adversely be used to partially correct for the bad predictive power of the model. Using both subsets for model fitting, makes sure that the model is most optimal for both subsets. The resulting MTP estimate is therefore less likely to account for poor descriptive power of the model.

The results found in this study are supported by [169], where a similar comparison between a serial and parallel scale transformation estimation was made. In their study also the parallel estimation process proved to provide more accurate results.

When more data are available for model fitting, and a more accurate model can thus be obtained, the difference between results of the serial and parallel MTP estimation processes are expected to reduce. For this study, however, the parallel MTP estimation process is used, as it results in more accurate MTP estimates.

While the final MTP estimation is quite accurate, the spread in the results of the parallel MTP estimation method is quite large. This spread is most likely caused by the simplicity of the search algorithm that often finds local minima when the initial MTP is far from the actual MTP. This spread can be reduced by improving the search algorithm, but a more elaborate search algorithm would also likely result in a higher computation time. As currently the MTP calculation for the Daimler experiment already took several days, and resulted in a sufficiently accurate MTP estimate, further improvement of the search algorithm currently seems unpractical.

Model Structure, Residual and Uncertainty Analysis

In this appendix the model structure, residual and uncertainty analysis for the models fitted to datasets from the CMS and the Daimler experiment, as well as the combined dataset, are shown. Additionally, the residual analysis for the prediction of the MIR in the other dataset is presented.

G.1. CMS Data Results

In this section the results of the models, fitted to the dataset of the CMS experiment, and applied to the dataset of the Daimler experiment are shown.

G.1.1. Model Structure

Using the SI process described in Chapter 5 the inputs, polynomial orders for output and input polynomials A and B, input delays, DC gains and output contribution percentage (OCP) shown in Table G.1 for each model were selected. Table G.1 shows that the models have a similar number of parameters, with the Basic model having two parameters less than the AI and CEDA models. For the AI model, the cueing error in yaw rate, as selected for the Basic model, was replaced by a combination of lateral jerk and roll angle. For the CEDA model, the cueing error in yaw rate was replaced by the error in roll angle and the error in vertical acceleration was replaced by the errors in pitch and roll angle.

In the CEDA model, also the cueing error in lateral acceleration was split up in three different types of cueing errors, that each were weighted differently: the false and missing cues were given a more than three times higher DC gain than the scaled cue. The order for polynomial A was high for all three models, while the B polynomial for rotational rate is solely a gain for all three models. Delays are above one second only for the AI model for cueing errors in lateral acceleration and roll angle.

For the Basic model, most of the rating can be explained with the lateral acceleration. However, the gain of the vertical acceleration is higher than that of the lateral acceleration for this model. For the AI and CEDA models most of the rating can be explained with the rotational rate, followed by the lateral acceleration. For the CEDA model, most of the rating can be explained with the combined cueing errors in lateral acceleration,

Table G.1: Estimated Basic, AI and CEDA model parameters when fitting to the CMS experiment dataset.

(a) Polynomial orders A (MIR) and B (cueing errors). N indicates the number of free parameters.

	Signal											
Model	MIR	a_y	$a_{y_{sc}}$	$a_{y_{mc}}$	$a_{y_{fc}}$	a_z	j_y	θ_r	θ_p	ω_r	ω_y	**N**
Basic	4	2				2				1	2	11
AI	4	2				2	2	2		1		13
CEDA	4		2	1	2			1	2	1		13

(b) Input delays per cueing error

	Signal									
Model	a_y	$a_{y_{sc}}$	$a_{y_{mc}}$	$a_{y_{fc}}$	a_z	j_y	θ_r	θ_p	ω_r	ω_y
Basic	0.6				0.0				0.1	0.0
AI	1.2				0.0	0.0	1.7		0.4	
CEDA		0.0	0.3	0.7			0.0	0.0	0.2	

(c) DC Gain per cueing error

	Signal									
Model	a_y	$a_{y_{sc}}$	$a_{y_{mc}}$	$a_{y_{fc}}$	a_z	j_y	θ_r	θ_p	ω_r	ω_y
Basic	0.33				0.51				0.08	0.01
AI	0.20				0.24	0.22	0.01		0.07	
CEDA		0.12	0.44	0.36			0.01	0.003	0.06	

(d) OCP per cueing error

	Signal									
Model	a_y	$a_{y_{sc}}$	$a_{y_{mc}}$	$a_{y_{fc}}$	a_z	j_y	θ_r	θ_p	ω_r	ω_y
Basic	41				19				24	16
AI	28				11	8	31		21	
CEDA		10	13	12			37	8	20	

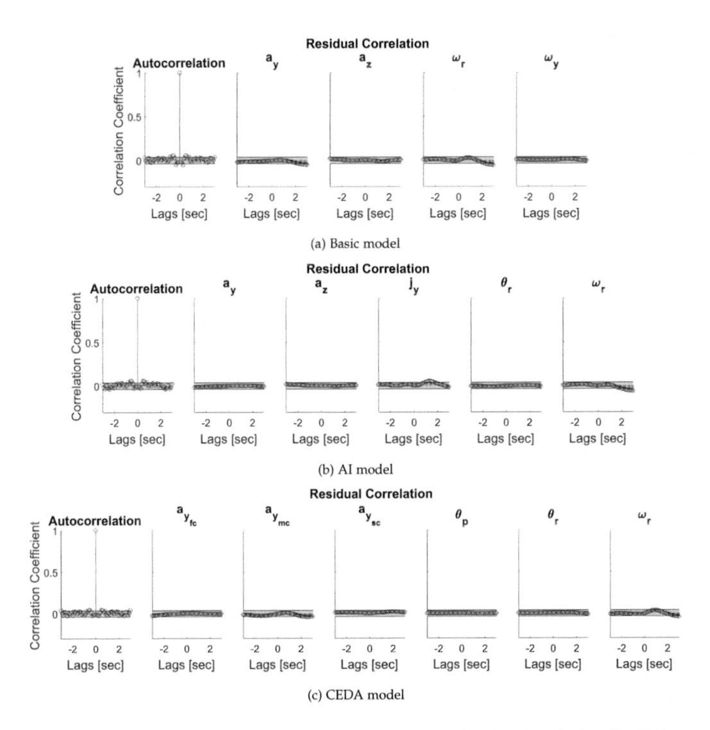

Figure G.1: Residual analysis of the Basic, AI and CEDA models when fitted and applied to the CMS experiment dataset.

followed by cueing errors in roll angle.

G.1.2. MODEL FIT

The residual and uncertainty analyses results of the models fitted to the dataset from the CMS experiment are discussed. The residual analyses for all models is shown in Figure G.1. The auto-correlations of the residuals, left subplot, and cross-correlations of the residuals with the inputs, all other subplots, for each model are shown in red, and the 95% confidence interval (CI) is indicated with the gray area.

Despite all models having a high A polynomial order, there is still some residual auto-correlation visible for the Basic and AI models. Increasing the order even more reduces this residual correlation, but hardly improves the model fit or prediction. To avoid overfitting, the A polynomial order is therefore not further increased. While the cross-correlation between residuals and most cueing errors do not significantly exceed the 95% CI, the correlation with the cueing error in ω_y does exceed the 95% CI, especially for the Basic and AI models. In future research it could be investigated whether this correlation is reduced when removing the rotation errors below human perception threshold from the cueing error.

The uncertainty analysis is illustrated in Figure G.2, showing the parameter uncertainty and covariance matrix, and Figure G.3, showing bode plots of the input/output polynomials and their uncertainty over frequency.

The correlation coefficients displayed in Figure G.2 show that no clear correlation exists between parameters related to different cueing errors or the MIR for any of the models. The parameter 99% CI in Figure G.2 show that the largest uncertainty is present in the parameters related to the vertical acceleration cueing error in models Basic and AI. Replacing this cueing error with the roll angle cueing error for the CEDA model, seems to reduce the parameter uncertainty.
Figure G.3 shows that for low frequencies, the parameters related to the yaw rate cueing error for the Basic model have a large uncertainty, while the parameters related to the vertical acceleration are somewhat uncertain over the whole frequency range. For the AI model, the parameters related to the lateral jerk cueing error are uncertain at high frequencies. For the CEDA model the parameters related to the scaled cues in lateral acceleration and the pitch angle are most uncertain across the entire frequency range displayed here.

G.1.3. Model Prediction

The residual analysis for the models fitted to the dataset from the CMS experiment and applied to the data from the Daimler experiment is shown. The residual analyses for all models is shown in Figure G.4.
Most residual auto and cross-correlation do not exceed the 95% confidence interval, however, some incorrect input delays might have caused the constant offset of the cross-correlations from zero for many of the cueing errors. Additionally, the B polynomials related to the lateral acceleration and jerk seem to not capture all the cueing error information in the Daimler dataset.

G.2. Daimler Data Results

Results of the models fitted to the dataset of the Daimler experiment and applied to the dataset of the CMS experiment are discussed here.

G.2.1. Model Structure

Using the SI process described in Chapter 5 the inputs, polynomial orders, input delays, DC gains and output contribution percentage (OCP) shown in Table G.2 for each model, fitted to the Daimler experiment dataset, were selected.

Table G.2 shows that fitting the models to the Daimler data results in the selection of less cueing error types than when fitting them to the CMS data. The number of model parameters does not vary much between the different models, with the CEDA model having two parameters more than the AI and one parameter more than the Basic model. The input delays are zero for all cueing errors except the false cue in lateral acceleration, where the delay is 0.1 seconds. The Basic model includes only cueing errors in linear acceleration, of which the vertical acceleration has the highest and the longitudinal acceleration the lowest DC gain. The CEDA model also shows higher DC gains for lateral than for longitudinal accelerations, where the false cues have a higher DC gain than the scaled and missing cues. All models mainly use the cueing errors in lateral acceleration to model the rating. The other cueing errors contribute to the modelled rating about equally.

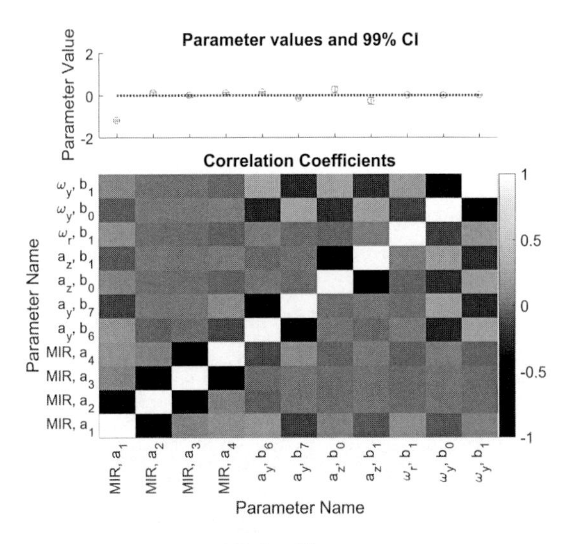

(a) Basic model.

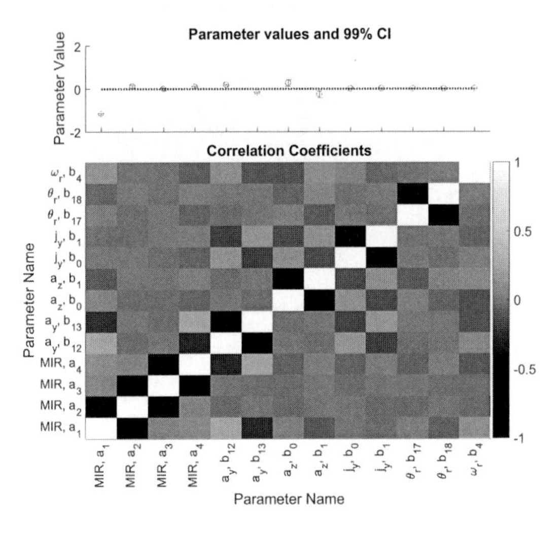

(b) AI model.

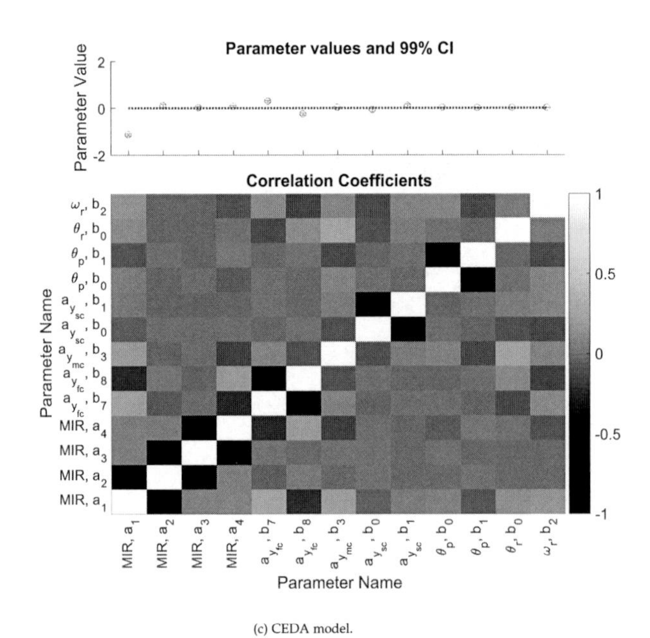

(c) CEDA model.

Figure G.2: Uncertainty analysis of the Basic, AI and CEDA models when fitted and applied to the CMS experiment dataset.

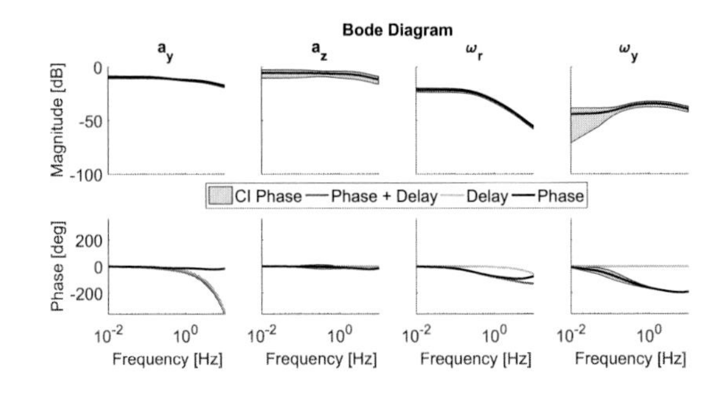

(a) Basic model.

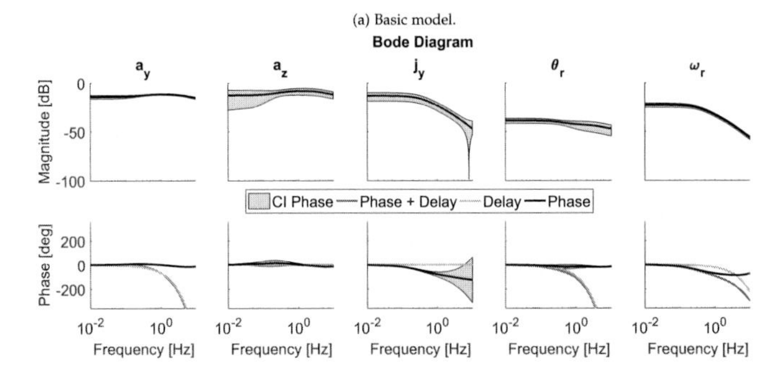

(b) AI model.

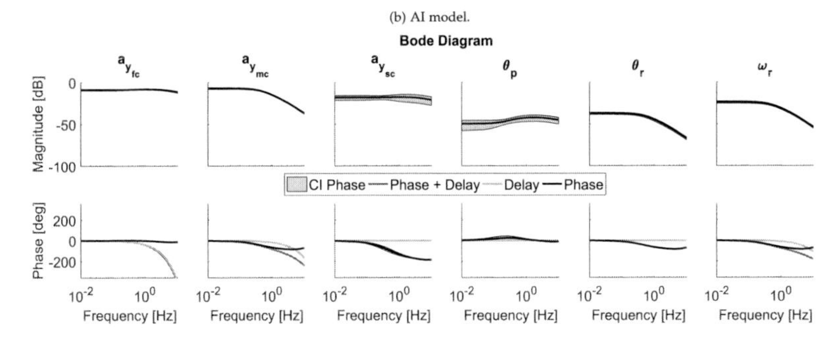

(c) CEDA model.

Figure G.3: Bode plots for each $\frac{B_i}{A}$ polynomial pair of the Basic, AI and CEDA models when fitted and applied to the CMS experiment dataset.

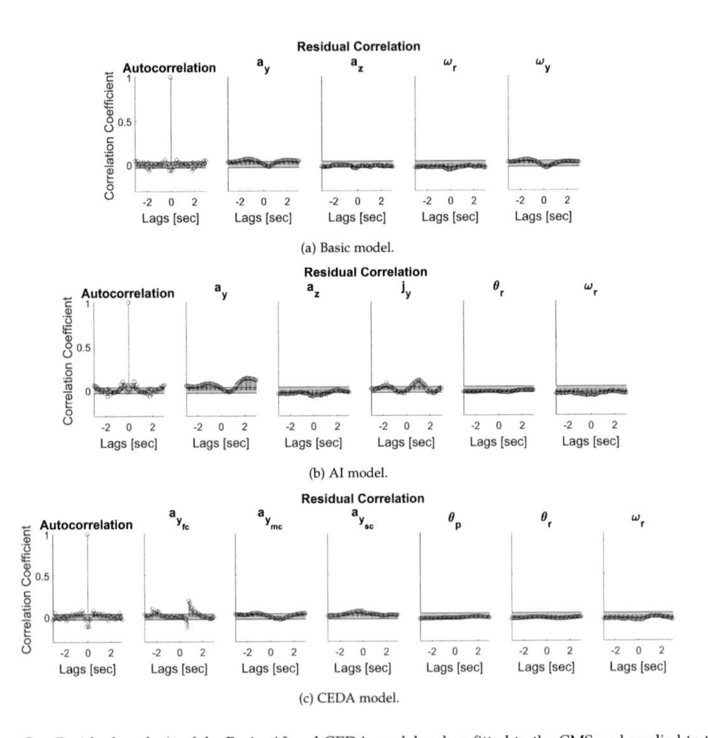

(a) Basic model.

(b) AI model.

(c) CEDA model.

Figure G.4: Residual analysis of the Basic, AI and CEDA models when fitted to the CMS and applied to the Daimler experiment dataset.

Table G.2: Estimated Basic, AI and CEDA model parameters when fitting to the Daimler experiment dataset.

(a) Polynomial orders A (MIR) and B (cueing errors). N indicates number of free parameters.

	Signal									
Model/Signal	MIR	a_x	$a_{x_{mc}}$	a_y	$a_{y_{sc}}$	$a_{y_{mc}}$	$a_{y_{fc}}$	a_z	θ_r	**N**
Basic	3	1		2				1		7
AI	3	1		1					1	6
CEDA	1		2		1	2	1		1	8

(b) Input delays per cueing error

	Signal							
Model/Signal	a_x	$a_{x_{mc}}$	a_y	$a_{y_{sc}}$	$a_{y_{mc}}$	$a_{y_{fc}}$	a_z	θ_r
Basic	0.0		0.0				0.0	
AI	0.0		0.0					0.0
CEDA		0.0		0.0	0.0	0.1		0.0

(c) DC Gain per cueing error

	Signal							
Model/Signal	a_x	$a_{x_{mc}}$	a_y	$a_{y_{sc}}$	$a_{y_{mc}}$	$a_{y_{fc}}$	a_z	θ_r
Basic	0.12		0.38				0.65	
AI	0.13		0.37					0.01
CEDA		0.14		0.32	0.38	0.57		0.01

(d) OCP per cueing error

	Signal							
Model/Signal	a_x	$a_{x_{mc}}$	a_y	$a_{y_{sc}}$	$a_{y_{mc}}$	$a_{y_{fc}}$	a_z	θ_r
Basic	13		76				11	
AI	13		72					15
CEDA		13		11	58	2		16

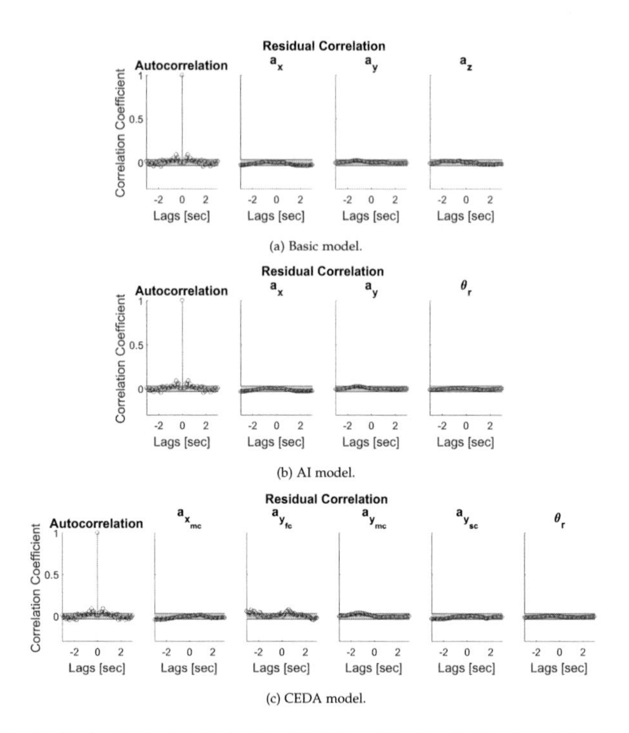

(a) Basic model.

(b) AI model.

(c) CEDA model.

Figure G.5: Residual analysis of the Basic, AI and CEDA models when fitted and applied to the Daimler experiment dataset.

G.2.2. Model Fit

The residual and uncertainty analyses results of the models fitted to the dataset from the Daimler experiment are shown. The residual analyses for all models are shown in Figure G.5.

All models show some residual autocorrelation, indicating that not all information in the output signal is captured correctly. Increasing the order of polynomial A to high values of around eight does decrease this autocorrelation, but does not improve the model fit and increases the uncertainty of the parameters. For the Basic and AI models, no considerable cross-correlation between residuals and cueing errors is found. For the CEDA model, the false cue in lateral acceleration does show a cross-correlation with the residuals, indicating that information in this cueing error is not fully captured by this model.

The uncertainty analysis is shown in Figure G.6, showing the parameter uncertainty and covariance matrix, and Figure G.7, showing bode plots of the input/output polynomials and their uncertainty over frequency.

The correlation coefficients in Figure G.6 show no considerable correlation between parameters related to different cueing for any of the models, indicating that the input selection process removed sufficient inputs. Likely due to the few number of parame-

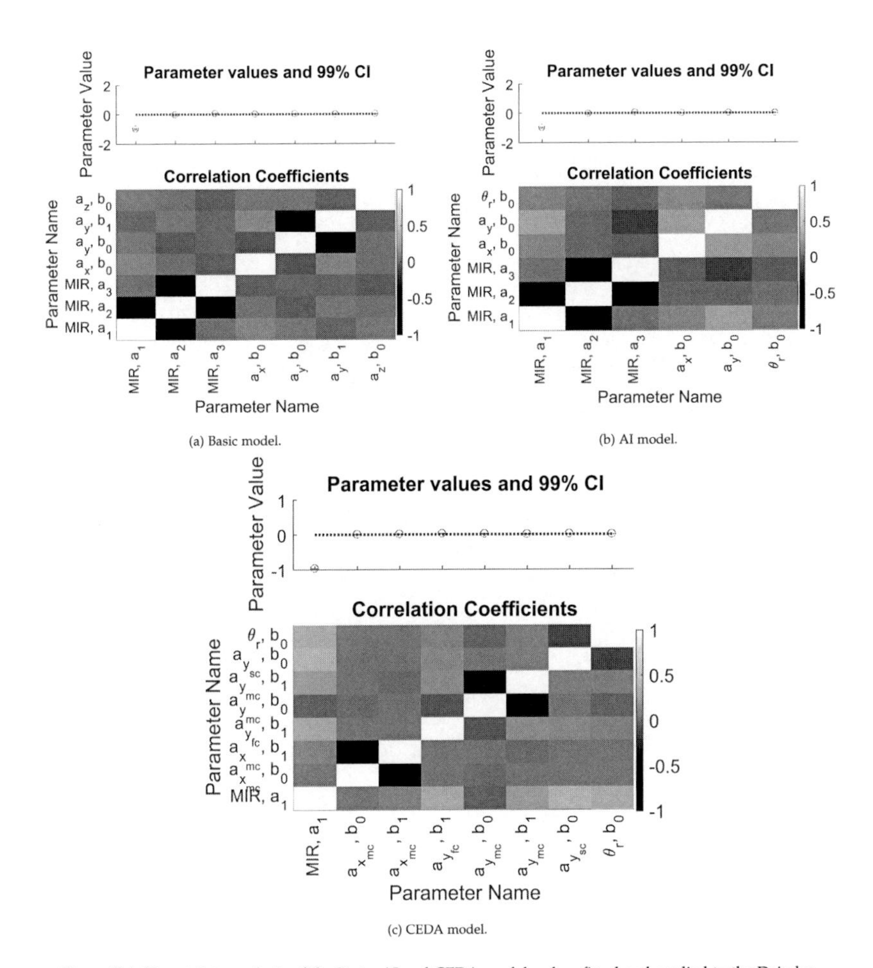

(a) Basic model.

(b) AI model.

(c) CEDA model.

Figure G.6: Uncertainty analysis of the Basic, AI and CEDA models when fitted and applied to the Daimler experiment dataset.

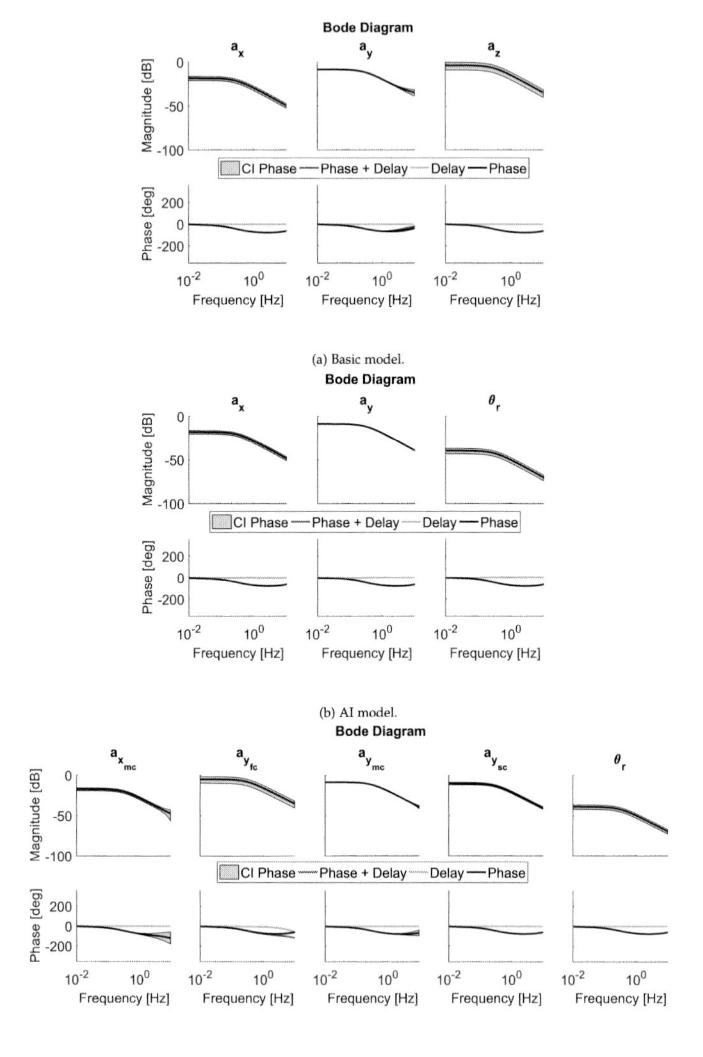

(a) Basic model.

(b) AI model.

(c) CEDA model.

Figure G.7: Bode plots for each $\frac{B_i}{A}$ polynomial pair of the Basic, AI and CEDA models when fitted and applied to the Daimler experiment dataset.

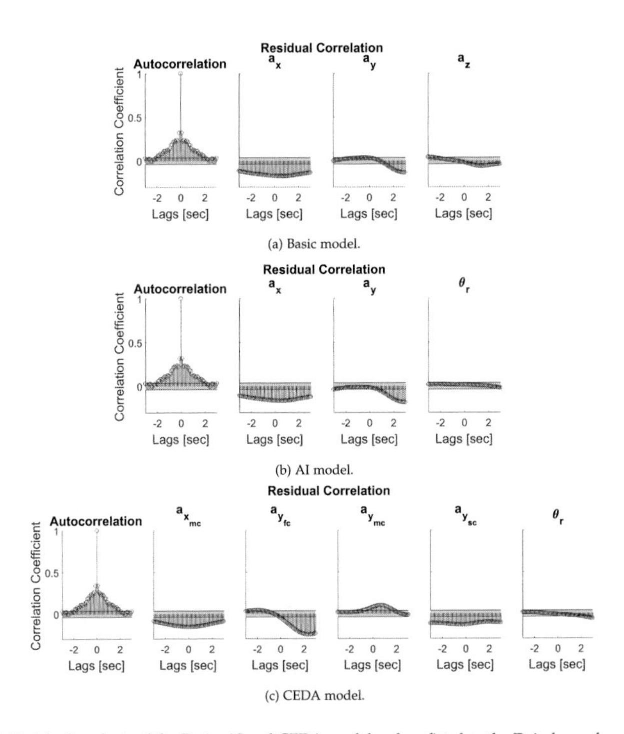

Figure G.8: Residual analysis of the Basic, AI and CEDA models when fitted to the Daimler and applied to the CMS experiment dataset.

ters in these models, the uncertainty for each parameter is relatively small.
Figure G.6 shows that the parameter estimates related to the lateral acceleration cueing errors are more uncertain higher frequencies. The parameter estimates related to the vertical acceleration in the Basic model, and the roll angle in the AI and CEDA model show a relatively constant uncertainty for the whole frequency range displayed here.

G.2.3. Model Prediction

The residual analysis for the models fitted to the dataset from the Daimler experiment and applied to the data from the CMS experiment is discussed. The residual analyses for all models are shown in Figure G.8.

All models show a large residual autocorrelation, indicating that information in the rating during the CMS experiment was not captured correctly by the model. Incorrect delays for the longitudinal acceleration cueing error could be responsible for the large cross-correlations between this error and the residuals. Overall the CEDA model, having the most inputs, shows the most cross-correlation between residuals and cueing errors, indicating that information in these cueing errors is not captured well by the model.

G.3. Combined Experiment Fitting

The results of the models fitted to a combination of the datasets from the CMS and the Daimler experiments are discussed.

G.3.1. Model Structure

Using the SI process described in Chapter 5 the inputs, polynomial orders, input delays, DC gains and output contribution percentage (OCP) shown in Table G.3 for each model were selected.
For all models the cueing errors in lateral and vertical acceleration, as well as roll rate, were selected. The AI and CEDA models also use cueing errors in roll angle. The AI model also uses the yaw acceleration, while the CEDA model instead includes the missing cue in longitudinal acceleration.
The polynomial A order is relatively high for all models, while most B polynomial orders are low. The Basic model has a much lower number of parameters than the CEDA model, with a difference of six parameters. This difference is mainly caused by the difference in number of inputs (three versus seven inputs). Most of the cueing errors have an estimated delay of zero, but for the AI model the roll rate is estimated to have a delay of 1.1 seconds, and for the CEDA model the missing cue in longitudinal acceleration has an estimated delay of 1 second and the roll rate of 0.8 seconds.
For the Basic model, the vertical acceleration has a higher DC gain than the lateral acceleration, while for the AI and CEDA models the opposite is true. These models instead have additional power in the roll angle. The CEDA model DC gain for the scaled cues in lateral acceleration are about half of the DC gains of the missing and false cues. The largest contribution to the modeled rating comes from the lateral acceleration for all models. For the AI and CEDA models, also the roll angle provides a relatively large contribution.

G.3.2. Model Fit

The residual analyses for all models is shown in Figure G.9. All models show some residual autocorrelations, indicating that not all information in the ratings is captured by the model. Cross-correlations are found between the residuals and cueing errors in lateral acceleration and roll rate for all models. In the CEDA model this mainly concerns the false cues in lateral acceleration.
The uncertainty analysis is shown in Figure G.10, showing the parameter uncertainty and covariance matrix, and Figure G.11, showing bode plots of the input/output polynomials and their uncertainty over frequency.
The correlation coefficients in Figure G.10 show that no considerable correlation is found between parameters related to different cueing errors, indicating that the input selection removed sufficient inputs. The estimated parameters shown in this figure have relatively small confidence intervals, apart from the parameters related to the cueing errors in vertical acceleration for the AI model. Figure G.11 shows that the uncertainty of the parameters related to vertical acceleration are relatively large and constant over the frequency range shown here for all models.

Table G.3: CMS-Daimler: Estimated Model Parameters

(a) Polynomial orders *A* (MIR) and *B* (cueing errors). N indicates number of free parameters.

	Signal										
Model/Signal	*MIR*	$a_{x_{mc}}$	a_y	$a_{y_{sc}}$	$a_{y_{mc}}$	$a_{y_{fc}}$	a_z	θ_r	ω_r	α_y	**N**
Basic	4		2				1		1		8
AI	4		1				4	1	1	1	12
CEDA	4	1		1	1	4	1	1	1		14

(b) Input delays per cueing error

	Signal								
Model/Signal	$a_{x_{mc}}$	a_y	$a_{y_{sc}}$	$a_{y_{mc}}$	$a_{y_{fc}}$	a_z	θ_r	ω_r	α_y
Basic		0.0				0.1		0.3	
AI		0.0				0.0	0.0	1.1	0.0
CEDA	0.4		0.5	0.0	0.6	0.0	0.0	0.0	

(c) DC Gain per cueing error

	Signal								
Model/Signal	$a_{x_{mc}}$	a_y	$a_{y_{sc}}$	$a_{y_{mc}}$	$a_{y_{fc}}$	a_z	θ_r	ω_r	α_y
Basic		0.45				0.56		0.06	
AI		0.41				0.30	0.01	0.03	0.01
CEDA	0.09		0.22	0.44	0.39	0.18	0.01	0.05	

(d) OCP per cueing error

	Signal								
Model/Signal	$a_{x_{mc}}$	a_y	$a_{y_{sc}}$	$a_{y_{mc}}$	$a_{y_{fc}}$	a_z	θ_r	ω_r	α_y
Basic		72				14		14	
AI		64				8	15	7	5
CEDA	6		11	40	5	4	22	12	

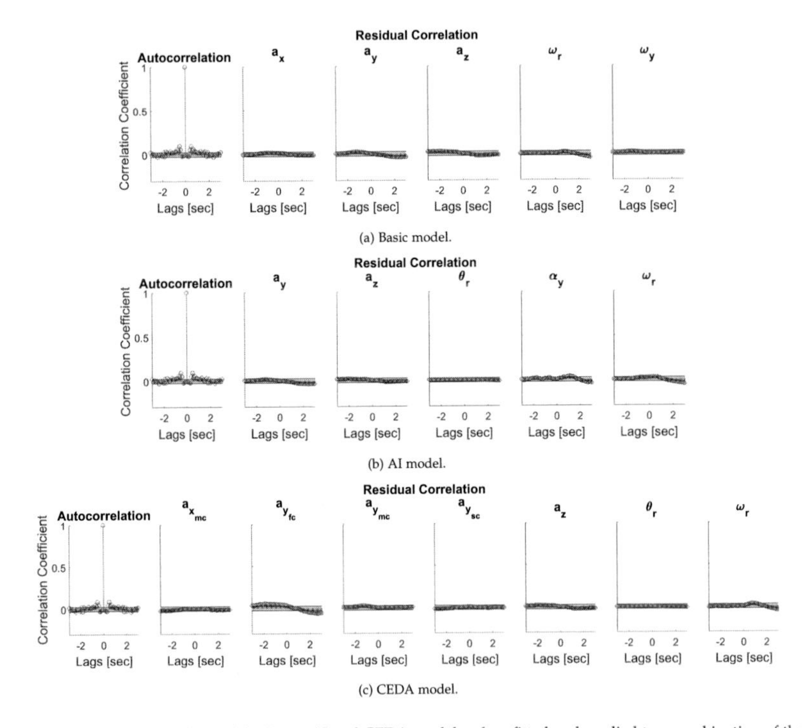

(a) Basic model.

(b) AI model.

(c) CEDA model.

Figure G.9: Residual analysis of the Basic, AI and CEDA models when fitted and applied to a combination of the datasets from the CMS and the Daimler experiments

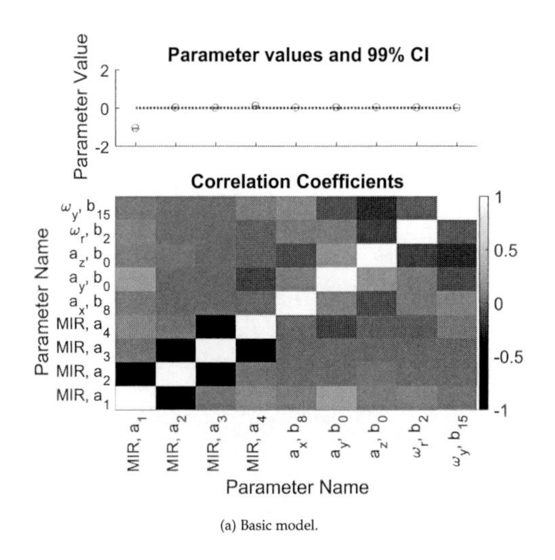

(a) Basic model.

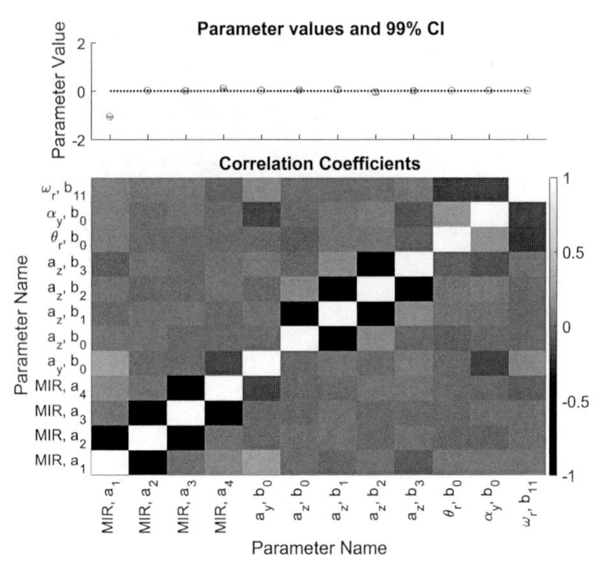

(b) AI model.

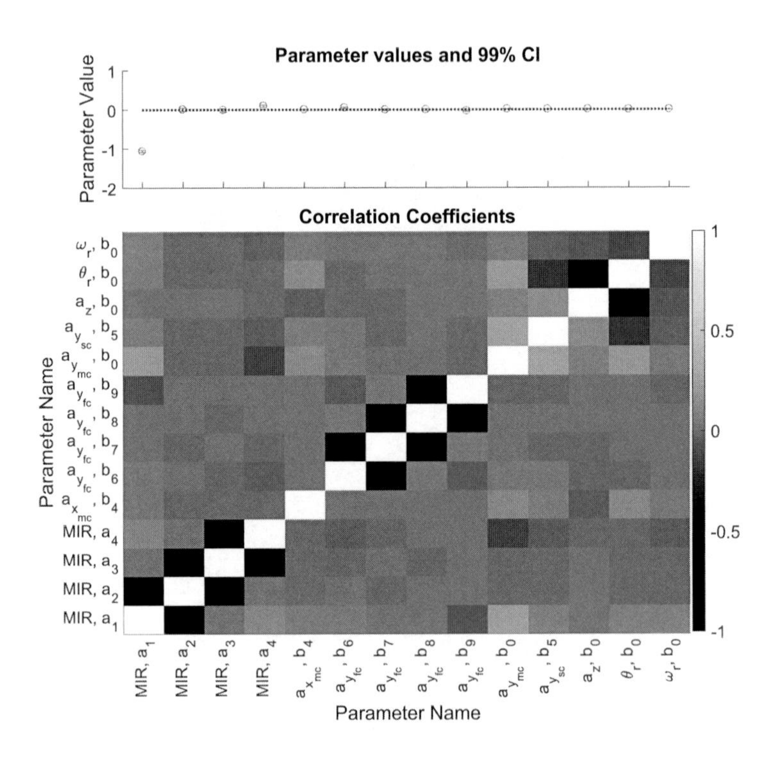

(c) CEDA model.

Figure G.10: Uncertainty analysis of the Basic, AI and CEDA models when fitted and applied to a combination of the datasets from the CMS and the Daimler experiments

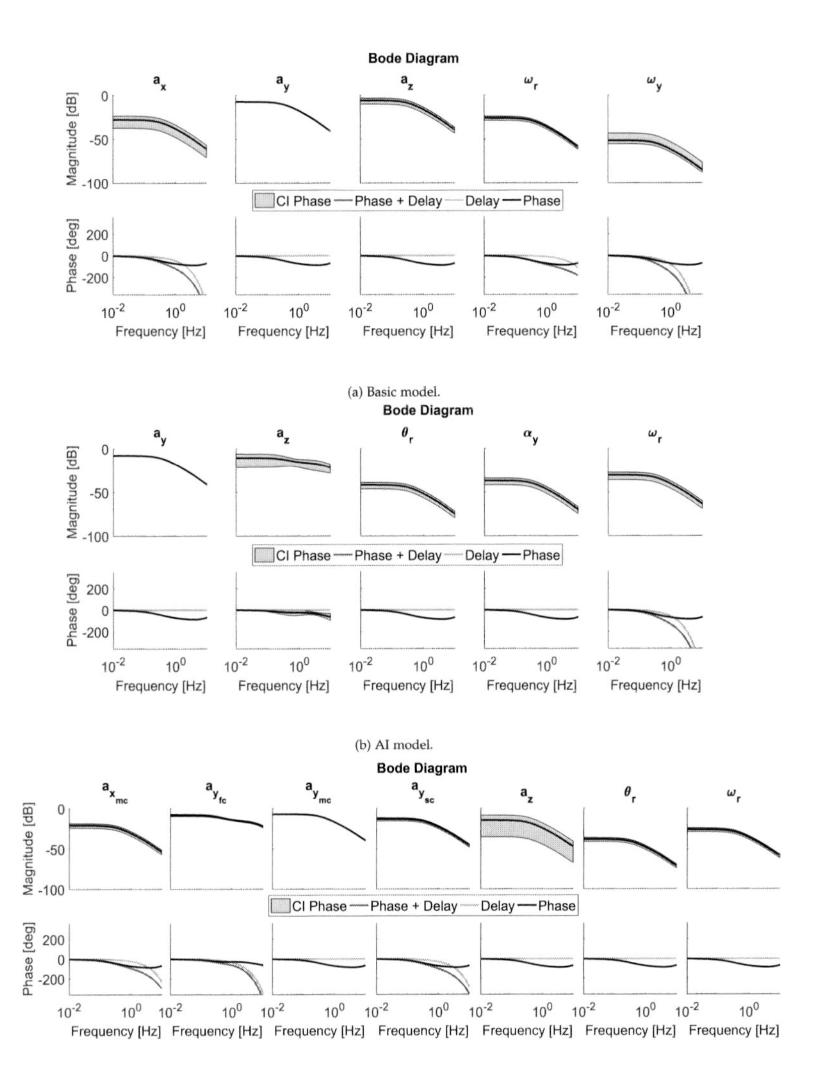

(a) Basic model.

(b) AI model.

(c) CEDA model.

Figure G.11: Bode plots for each $\frac{B_i}{A}$ polynomial pair of the Basic, AI and CEDA models when fitted and applied to a combination of the datasets from the CMS and the Daimler experiments

H Questionnaire

Figure H.1 shows the questionnaire that was given at the end of the experiment described in Chapter 7.
The responses for Question 2 are reported with levels 1 till 7, from low to high influence, and the different factors are summarized as follows:

- **stronger**: The motion I experienced was stronger than I expected
- **weaker**: The motion I experienced was weaker than I expected
- **direction**: The motion I experienced was in a wrong direction
- **delay**: The motion I experienced came later than I expected
- **advance**: The motion I experienced came earlier than I expected
- **icyRoad**: I felt like I was sliding on an icy road
- **falseCue**: I experienced motion when I did not expect any motion
- **missingCue**: I did not experience motion when I expected motion
- **jerky**: I experienced unrealistic jerky motions
- **rotationsCurves**: I experienced rotations during the curves
- **rotationsAccDec**: I experienced rotations during accelerating/decelerating
- **roadRumble**: I experienced unrealistic road rumble
- *Not used in analysis*: Other

Subject ID					Date					
MISC scores:										
Overall Rating										

Continuous rating of motion quality in the Hexapod Simulator - Questionnaire

1. How would you rate your current state?

	1	2	3	4	5	6	7	8	9	
Tired, Demotivated, Distracted, Weak, Ill					→					*Energetic, Motivated, Concentrated, Fit, Healthy*

2. Please give an indication on how much the following factors influenced your mismatch rating (cross one of the boxes):

	Low		Medium			High	
The motion I experienced was stronger than I expected							
The motion I experienced was weaker than I expected							
The motion I experienced was in a wrong direction							
The motion I experienced came later than I expected							
The motion I experienced came earlier than I expected							
I felt like I was sliding on an icy road							
I experienced motion when I did not expect any motion							
I did not experience motion when I expected motion							
I experienced unrealistic jerky motions							
I experienced rotations during the curves							
I experienced rotations during accelerating/decelerating							
I experiences unrealistic road rumble							
Other:							

3. Which (type of) car did you imagine being seated in during the experiment?

....

4. How would you rate your own driving style as compared to the driving style in the experiment?

-4	-3	-2	-1	0	1	2	3	4

Less aggressive → *More aggressive*

5. How often you do travel by car (either as passenger or driver)?

Never	*Yearly*	*Monthly*	*Weekly*	*Few days a week*	*Daily*

General comments:

....

....

Thank you for participating!

Figure H.1: Questionnaire

Acknowledgements

Reading a biography of Norbert Wiener, "inventer" of the field of Cybernetics, I knew that this was the field I wanted to do a PhD in. But where would I find a place where I can do that? Google: search cybernetics, phd, research. Output: Tuebingen, Germany. So maybe I should start there, thanks Google! Of course that was only the start, and there are many many more people that I owe a big thank you to for helping me through this PhD journey.
First of all I would like to sincerely thank Heinrich Buelthoff, for creating a place like the MPI for Biological Cybernetics. I have learned so much from the great international group of people from all different research fields. I would also like to thank both him and Max Mulder for giving me the opportunity and support to pursue my PhD. I am grateful to Heinrich for giving me lots of freedom in not only defining my PhD topic, but also in my research that followed.
I would like to thank Max for all his extremely useful insights in my work, helping me focus when I started drifting away from my main goals again, being so approachable and always showing a sincere interest in my work. You made sure to never be a cause for any delay in my research, by reviewing my lengthy drafts within days and replying to emails often instantly. You left me many a time impressed and somewhat guilty for not working equally hard as you.
Both Heinrich and Max I would also like to thank very much for their complete understanding of any delays caused by having my first child towards the end of my PhD work, and Max for even pointing out the relative unimportance of quickly finishing my PhD work in this case. As I am writing this, I again realize how much this meant to me.

Next to my two promotors, I also am very grateful to my team of daily supervisors. While taking direction from in total 5 different people could have caused problems, the advice of each one was incredibly valuable and complementary rather than contradictory, making it a joy to work with all of them.
I would like to thank Daan for his support and advice all throughout my research. I am sorry that you had to read through all those pages of initially often quite incoherent draft versions of my work. But I am extremely grateful that you did, and gave me your useful insights and suggestions in your always calm and friendly manner.
I would like to thank Joost and Paolo for giving me the opportunity to work in their research group. Being physically far from the TU Delft, Joost was my on-site engineering help desk and a great discussion partner, always managing to describe my messy thoughts and ideas in a more organized and logical manner. I also really appreciated your sense of appropriateness and not being afraid to speak up when you thought a line was crossed.
To Paolo I would like to show my gratitude for giving me a very valuable different point of view (yes, engineers need statistics too ;)) and for our long very helpful discussions structuring my work into a clear story.

I would also like to thank all my other colleagues in the MPSim group: Ksander,

Johannes, Daniel, Maria, Misha, Suzanne and Alessandro, thank you for the great work environment. I would like to thank my office mate Ksander in particular for helping me understand statistical models, for the great discussions and for lightening up the mood in our office together with Johannes. Thanks to you two I now always lock my screen ;).
I also owe a big thanks to Maria for her help with my experiments, I dont know how driving experiments can be ran on the CMS without you ;). At the end of my time at the MPI I also really appreciated our talks as two pregnant women and new mommy's. Of course I also owe a big thanks to Misha for helping me out so many times with all kinds of modelling/coding/maths challenges I encountered throughout my PhD work. But I also want to thank him for being his weird, attentive, curious, sweet self, cheering me up so many times.
While working in another building physically separated our group from the rest of the department, I am very happy to have gotten to know the other groups during our common lunches, outings and evenings in town. Thank you all for creating such a fun environment to be in, not only during work, but also afterwards. And of course also thanks for all your interesting Labmeeting talks which taught me a lot!

I also want to thank Senya, for being my "first" friend in Tuebingen and letting me tag along on his fun activities. And a big thanks to Frank for, among others, helping me get started on my modelling endeavours and generally being such a nice friend.
To Rabi and Stephan and their family I owe a big thanks for all the fun outside of work and reminding me once again how much fun having a family is. Rabi became a very close friend who, next to being a great support throughout my pregnancy knowing exactly what I was going through ;) and showing me the ins and outs of motherhood, is also just a great person to be around.
To my friends from "back home", Kirsten, Christina and Jon, thank you for your friendship and being oke with just a mere couch to crash on after travelling so far to come visit us in Tuebingen. :)

And then, to my family. First a special thanks to my mom, for taking care of Lukas every Thursday morning so I could finalize my thesis work during my mommy days. And a big thank you to my big brother for utilizing his exceptional graphics skills to design a thesis cover that reflects not only my work but also me. I could now try summing up all the other things I owe gratitude for to my family, but that would easily double the number of pages of even this thesis. Therefore just, thank you for everything. And finally, to my dear husband Sylvain and sweet son Lukas. To Lukas, thank you for not giving me a hard time in my belly, so I could still run my last experiment and for generally making my life so much better. And lastly, Sylvain, thank you for putting up with all my plans, for supporting me through my education, career and life, no matter the city or the country. If it wasn't for your ability and willingness to tackle new challenges head on, moving abroad, learning a new language twice, finding a new job etc., this thesis would not have happened. Thank you sweetheart, Ik houd van jou, je t'aime, I love you!

Curriculum Vitæ

Diane Cleij was born on May 27, 1986 in Rotterdam, The Netherlands. From 1998 to 2004 she attended the Libanon Lyceum in Rotterdam, where she obtained her VWO diploma.
In 2004 she enrolled at the Faculty of Aerospace Engineering at Delft University of Technology in Delft, the Netherlands. As part of her M.Sc. studies she interned at Entropy Control Inc. in La Jolla, CA, USA, where she analysed driving behaviour differences between elderly and young drivers. In 2011 she obtained her M.Sc. degree in Aerospace Engineering at the Control and Simulation Department at the Delft University of Technology in Delft, The Netherlands. For her final thesis work she developed a haptic shared steering wheel controller based on human neuromuscular models and analysed the corresponding driver-controller interaction.
In February 2012 she started as a Consultant Mechatronics at Alten Mechatronics in Eindhoven, The Netherlands, where she worked on several robotics related in-house projects. In October 2012 she was assigned to work at ASML as a Designer Mechatronics at the Stage Position Measurement Department in Eindhoven, The Netherlands. She was working in the interface between the Stage Position Measurement software and hardware, where she was, among other, designing and analysing several specification tests.
In February 2014 she moved to Germany to start her PhD at the Motion Perception and Simulation Department of the Max Planck Institute for Biological Cybernetics in Tübingen, Germany, in cooperation with the Control and Simulation Department of the Faculty of Aerospace Engineering at Delft University of Technology in Delft, The Netherlands. Her research, as described in this thesis, investigated the measuring, modelling and minimizing of perceived motion incongruence in vehicle motion simulation. After her first son was born, she moved back to The Netherlands and started working at the Dutch Institute for Road Safety Research (SWOV) in The Hague, The Netherlands, while finishing her PhD thesis. At SWOV she is working on several road safety research projects related to connected and automated vehicles.

List of Publications

D. Cleij, D. M. Pool, M. Mulder and H. H. Bülthoff, *Optimizing an Optimization-Based MCA usingPerceived Motion Incongruence Models* , *Driving Simulator Conference*, (Antibes, France, 2020), Abstract submitted.

J. R. van der Ploeg, **D. Cleij**, D. M. Pool, M. Mulder and H. H. Bülthoff, *Sensitivity Analysis of an MPC-based Motion Cueing Algorithm for a Curve Driving Scenario*, (Antibes, France, 2020), Abstract submitted.

D. Cleij, J. Venrooij, P. Pretto, M. Katliar, H. H. Bülthoff, D. Steffen, F. W. Hoffmeyer and H-P Schöner, *Comparison between filter- and optimization-based motion cueing algorithms for driving simulation*, `https://doi.org/10.1016/j.trf.2017.04.005`, in *Transportation Research Part F: Traffic Psychology and Behaviour*, **61**, (2019).

T. D. van Leeuwen, **D. Cleij**, D. M. Pool, M. Mulder and H. H. Bülthoff, *Time-varying perceived motion mismatch due to motion scaling in curve driving simulation*, `https://doi.org/10.1016/j.trf.2018.05.022`, in *Transportation Research Part F: Traffic Psychology and Behaviour*, **61**, (2019).

D. Cleij, J. Venrooij, P. Pretto, D. M. Pool, M. Mulder and H. H. Bülthoff, *Continuous Subjective Rating of Perceived Motion Incongruence During Driving Simulation*, `https://doi.org/10.1109/THMS.2017.2717884`, in *IEEE Transactions on Human-Machine Systems*, **48**, 1 (2018).

M. Grottoli, **D. Cleij**, P. Pretto, Y. Lemmens, R. Happee and H. H. Bülthoff, *Objective evaluation of prediction strategies for optimization-based motion cueing*, `https://doi.org/10.1177/0037549718815972`, in *SIMULATION*, **95**, 8, (2019).

T. D. van Leeuwen, **D. Cleij**, D. M. Pool, M. Mulder and H. H. Bülthoff, *Time-varying perceived visual-motion mismatch due to lateral specific force scaling during passive curve driving simulation*, in *Driving Simulator Conference*, (Stuttgart, Germany, 2017), pp. 121-122.

J. Venrooij, **D. Cleij**, M. Katliar, P. Pretto, H. H. Bülthoff, D. Steffen, F. W. Hoffmeyer and H-P Schöner, *Comparison between filter- and optimization-based motion cueing in the Daimler Driving Simulator*, in *Driving Simulator Conference*, (Paris, France, 2016), pp. 31-38.

D. Cleij, J. Venrooij, P. Pretto, D. M. Pool, M. Mulder and H. H. Bülthoff, *Continuous rating of perceived visual-inertial motion incoherence during driving simulation*, in *Driving Simulator Conference*, (Tübingen, Germany, 2015), pp. 191-198.

J. Venrooij, P. Pretto, M. Katliar, S. A. E. Nooij, A. Nesti, M. Lächele, K. N. de Winkel, **D. Cleij** and H. H. Bülthoff, *Perception-based motion cueing: validation in driving simulation*, in *Driving Simulator Conference*, (Tübingen, Germany, 2015), pp. 153-161

D. A. Abbink, **D. Cleij**, M. Mulder and M. M. Paassen, *The Importance of Including Knowledge of Neuromuscular Behaviour in Haptic Shared Control*, https://doi.org/10.1109/ICSMC.2012.6378309, in *IEEE International Conference on Systems, Man, and Cybernetics (SMC)*, (Seoul, South Korea, 2012), pp. 3350-3355.

E. Boer, **D. Cleij**, J. D. Dawson and M. Rizzo, *Serialization of vehicle control at intersections in older drivers*, https://www.ncbi.nlm.nih.gov/pubmed/24273755, in *Proceedings of the 6th International Driving Symposium on Human Factors in Driving Assessment, Training and Vehicle Design*, (Olympic Valley - Lake Tahoe, CA, USA, 2011), pp. 17-23.

MPI Series in Biological Cybernetics

ISSN 1618-3037

1	Volker H. Franz	The relationship between visually guided motor behavior and visual perception ISBN 978-3-89722-722-4 40.50 €
2	Markus von der Heyde	A Distributed Virtual Reality System for Spatial Updating. Concepts, Implementation, and Experiments ISBN 978-3-89722-781-1 40.50 €
3	Anastasia Sigala	Visual object categorization and representation in primates: Psychophysics and Physiology ISBN 978-3-89722-961-7 40.50 €
4	Isabell Berndt	Event-related Lateralizations of EEG activity during Pointing ISBN 978-3-8325-0180-8 40.50 €
5	Rainer Rosenzweig	Experimentelle Bestimmung der Verrechnungszeiten beim Stereosehen anhand der verzögert wahrgenommenen Tiefenumkehr von bewegten, teilweise verdeckten Objekten ISBN 978-3-8325-0236-2 40.50 €
6	Titus R. Neumann	Biomimetic Spherical Vision. Biomimetische Algorithmen zum sphärischen Sehen ISBN 978-3-8325-0410-6 40.50 €
8	Bernhard Riecke	How far can we get with just visual information? Path integration and spatial updating studies in Virtual Reality ISBN 978-3-8325-0440-3 40.50 €
9	Barbara Knappmeyer	Faces in Motion ISBN 978-3-8325-0637-7 40.50 €
10	Ulrike von Luxburg	Statistical Learning with Similarity and Dissimilarity Functions ISBN 978-3-8325-0767-1 40.50 €
11	Heiko Hoffmann	Unsupervised Learning of Visuomotor Associations ISBN 978-3-8325-0858-6 40.50 €
12	Thomas Navin Lal	Machine Learning Methods for Brain-Computer Interfaces ISBN 978-3-8325-1048-0 40.50 €
14	Gerald Franz	An empirical approach to the experience of architectural space ISBN 978-3-8325-1137-1 40.50 €

15	Elisabeth Huberle	Temporal and spatial properties of shape processing in the human visual cortex: Combined fMRI and MEG adaptation ISBN 978-3-8325-1453-2 40.50 €
16	Tobias Pfingsten	Machine Learning for Mass Production and Industrial Engineering ISBN 978-3-8325-1505-8 40.50 €
17	Daniel Berger	Sensor Fusion in the Perception of Self-Motion ISBN 978-3-8325-1555-3 40.50 €
18	Karin Pilz	The role of facial and body motion for the recognition of identity ISBN 978-3-8325-1579-9 40.50 €
19	Christian Wallraven	A computational recognition system grounded in perceptual research ISBN 978-3-8325-1662-8 40.50 €
20	Wolfram Schenck	Adaptive Internal Models for Motor Control and Visual Prediction ISBN 978-3-8325-1899-8 40.50 €
21	Xiaohai Sun	Causal Inference from Statistical Data ISBN 978-3-8325-1916-2 40.50 €
22	Tobias Meilinger	Strategies of Orientation in Environmental Spaces ISBN 978-3-8325-1997-1 38.50 €
23	Wolf Kienzle	Learning an Interest Operator from Human Eye Movements ISBN 978-3-8325-2021-2 34.00 €
24	Martin Schwaiger	Konstruktion und Entwicklung omnidirektionaler Laufplattformen ISBN 978-3-8325-2097-7 37.00 €
25	Marco Vitello	Perception of moving tactile stimuli ISBN 978-3-8325-2647-4 35.00 €
26	Nina Gaißert	Perceiving Complex Objects. A Comparison of the Visual and the Haptic Modalities ISBN 978-3-8325-2794-5 47.50 €
27	David Engel	Shape-Centered Representations: From Features to Applications ISBN 978-3-8325-2820-1 54.00 €
28	Lewis L. Chuang	Recognizing Objects from Dynamic Visual Experiences ISBN 978-3-8325-2842-3 39.00 €

29	Regine Armann	Faces in the Brain - a Behavioral, Eye-tracking and High-level Adaptation Approach to Human Face Perception ISBN 978-3-8325-2900-0 38.00 €
30	Rodrigo Sigala	Investigating the neural representation of face categories ISBN 978-3-8325-2991-8 38.00 €
31	Elvira Fischer	Visual motion and self-motion processing in the human brain ISBN 978-3-8325-2994-9 40.00 €
32	Martin Breidt	Datenbasierte Gesichtsanimation ISBN 978-3-8325-3136-2 53.50 €
33	Frank M. Nieuwenhuizen	Changes in Pilot Control Behaviour across Stewart Platform Motion Systems ISBN 978-3-8325-3233-8 39.00 €
34	Florian Soyka	A Cybernetic Approach to Self-Motion Perception ISBN 978-3-8325-3335-9 35.50 €
35	Stephan Streuber	The Influence of Different Sources of Visual Information on Joint Action Performance ISBN 978-3-8325-3373-1 35.00 €
36	Samantha Alaimo	Novel Haptic Cues for UAV Tele-Operation ISBN 978-3-8325-3527-8 55.50 €
37	Hans-Joachim Bieg	On the Coordination of Saccades with Hand and Smooth Pursuit Eye Movements ISBN 978-3-8325-3648-0 38.00 €
38	Anette Giani	From multiple senses to perceptual awareness ISBN 978-3-8325-3656-5 38.00 €
39	Markus Leyrer	Understanding and Manipulating Eye Height to Change the User's Experience of Perceived Space in Virtual Reality ISBN 978-3-8325-3854-5 42.00 €
40	Katharina Dobs	Behavioral and Neural Mechanisms Underlying Dynamic Face Perception ISBN 978-3-8325-3910-8 37.00 €
41	Kathrin Kaulard	Visual Perception of Emotional and Conversational Facial Expressions ISBN 978-3-8325-3969-6 41.00 €
42	Ivelina Piryankova	The influence of a self-avatar on space and body perception in immersive virtual reality ISBN 978-3-8325-3978-8 41.50 €

43	Janina Esins	Face processing in congenital prosopagnosia ISBN 978-3-8325-3983-2 39.50 €
44	Alessandro Nesti	On the Perception of Self-motion: from Perceptual Laws to Car Driving Simulation ISBN 978-3-8325-4013-5 42.00 €
45	Joost Venrooij	Measuring, modeling and mitigating biodynamic feedthrough ISBN 978-3-8325-4105-7 53.50 €
46	Frank M. Drop	Control-Theoretic Models of Feedforward in Manual Control ISBN 978-3-8325-4354-9 45.00 €
47	Stefano Geluardi	Identification and augmentation of a civil light helicopter: transforming helicopters into Personal Aerial Vehicles ISBN 978-3-8325-4366-2 36.50 €
48	Laura Fademrecht	Action Recognition in the Visual Periphery ISBN 978-3-8325-4448-5 37.50 €
49	Aurelie Saulton	Understanding the nature of the body model underlying position sense ISBN 978-3-8325-4460-7 37.00 €
50	Burak Yüksel	Design, Modeling and Control of Aerial Robots for Physical Interaction and Manipulation ISBN 978-3-8325-4492-8 48.00 €
51	Christiane Glatz	Auditory cues for attention management ISBN 978-3-8325-4730-1 38.50 €
52	Sujit Rajappa	Towards Human-UAV Physical Interaction and Fully Actuated Aerial Vehicles ISBN 978-3-8325-4767-7 47.50 €
53	Johannes Lächele	Motion Feedback in the Teleoperation of Unmanned Aerial Vehicles ISBN 978-3-8325-4779-0 39.50 €
54	Albert H. van der Veer	Where are you? Self- and body part localization using virtual reality setups ISBN 978-3-8325-4987-7 42.50 €
55	Marianne Strickrodt	The impossible puzzle: No global embedding in environmental space memory ISBN 978-3-8325-4925-1 45.00 €
56	Anne Thaler	The Role of Visual Cues in Body Size Estimation ISBN 978-3-8325-5026-4 43.00 €

57	Diane Cleij	Measuring, modelling and minimizing perceived motion incongruence for vehicle motion simulation ISBN 978-3-8325-5044-8 49.00 €